Chemical Properties, Extraction Methods and Bioaccessibility of Bioactive Substances in Plant Foods

Chemical Properties, Extraction Methods and Bioaccessibility of Bioactive Substances in Plant Foods

Guest Editors

Cecilia E. Martínez-Sánchez
Efigenia Montalvo González

MDPI

Basel • Beijing • Wuhan • Barcelona • Belgrade • Novi Sad • Cluj • Manchester

Guest Editors

Cecilia E. Martínez-Sánchez
Departamento de Ingeniería
Química y Bioquímica
Tecnológico Nacional de
México/Campus Tuxtepec
Oaxaca
Mexico

Efigenia Montalvo González
Laboratorio Integral de
Investigación en Alimentos
Instituto Tecnológico de Tepic
Tepic
Mexico

Editorial Office
MDPI AG
Grosspeteranlage 5
4052 Basel, Switzerland

This is a reprint of the Special Issue, published open access by the journal *Foods* (ISSN 2304-8158), freely accessible at: https://www.mdpi.com/journal/foods/special_issues/3S41LB1GUF.

For citation purposes, cite each article independently as indicated on the article page online and as indicated below:

Lastname, A.A.; Lastname, B.B. Article Title. *Journal Name* **Year**, *Volume Number*, Page Range.

ISBN 978-3-7258-8367-7 (Hbk)
ISBN 978-3-7258-8368-4 (PDF)
https://doi.org/10.3390/books978-3-7258-8368-4

Contents

About the Editors

Cecilia E. Martínez-Sánchez

Cecilia E. Martínez-Sánchez is a Professor and Researcher at the Tecnológico Nacional de México, Campus Tuxtepec, with more than 35 years of experience in academic and scientific research in food science and technology. She has been a member of the Mexican National System of Researchers (SNI, by its Spanish acronym) since 2002 and currently holds Level II recognition for the 2026–2030 period. Her research focuses on food innovation, conservation technologies, bioactive compounds, functional foods, and sustainable food processing. Dr. Martínez Sánchez has participated in more than 35 funded research projects and has supervised numerous undergraduate, master's, and doctoral theses. She is the leader of the Academic Research Group "Food Development, Innovation and Preservation" (ITTUX-CA-2) and has authored more than 50 scientific publications in peer-reviewed journals. In addition, she has served as a reviewer and evaluator for national and international scientific programs, journals, and research projects. She also actively collaborates with international scientific organizations, including the Global Harmonization Initiative (GHI). Her expertise includes analytical methodologies, food chemistry and microbiology, development of innovative food products, scientific writing, and interdisciplinary collaboration. Through her academic leadership and research contributions, Dr. Martínez Sánchez has significantly advanced the field of food science, particularly in the development of sustainable technologies and functional ingredients for food and nutraceutical applications.

Efigenia Montalvo González

Efigenia Montalvo González holds a Bachelor of Science and a Master of Science in Biochemical Engineering from the Technological Institute of Tepic, and a Doctor of Science in Food Science from the Technological Institute of Veracruz. Since 2006, she has been a full-time professor and researcher at the Tecnológico Nacional de México, Campus Tepic. She received the 2013 National Award for Food Science and Technology. She served on the Board of Directors of the Mexican Society of Annonaceae from 2014 to 2016, and on the Board of Directors of the Mexican Association of Food Sciences (AMECA) from 2018 to 2021 and again from 2023 to 2025. She received an honorable mention within the 2021 National Food Science and Technology Award. She has led 32 funded research projects, 11 of which were conducted in collaboration with industry. She is the author of 165 publications and 19 book chapters on postharvest technology, food science and technology, functional foods and bioactive compounds. In addition, she has supervised nine doctoral, 44 master's, and 36 bachelor's thesis. Moreover, her research focuses on the formulation and preservation of fresh, functional, and/or nutraceutical foods, as well as the evaluation of their biological activity.

foods

MDPI

Article

Spinach (*Spinacia oleracea* L.) Flavonoids Are Hydrolyzed During Digestion and Their Bioaccessibility Is Under Stronger Genetic Control Than Raw Material Content

Michael P. Dzakovich [1,*], Alvin L. Tak [1], Elaine A. Le [1], Rachel P. Dang [2], Benjamin W. Redan [3] and Geoffrey A. Dubrow [4]

1 USDA-ARS Children's Nutrition Research Center, Department of Pediatrics, Baylor College of Medicine, 1100 Bates Ave., Houston, TX 77030, USA
2 Department of Biosciences, Rice University, 6100 Main St., Houston, TX 77005, USA
3 Thorne Research, 620 Omni Industrial Blvd., Summerville, SC 29483, USA
4 Independent Researcher, Columbus, OH 43210, USA
* Correspondence: michael.dzakovich@usda.gov

Abstract: Spinach (*Spinacia oleracea* L.) is a commonly consumed crop with a diverse array of unique flavonoids. These molecules likely contribute to the health benefits associated with spinach consumption. However, little is known about the genetic diversity of these molecules, their bioaccessibility, and the heritability of these traits. We assembled a diversity panel of 30 F_1 and open-pollinated spinach accessions and cultivated them under controlled conditions over two periods. Quantification of 39 flavonoids revealed that their concentration is largely influenced by environmental factors, and at least two divergent branches in the spinach flavonoid biosynthesis pathway may exist. Despite generally similar trends in the amounts of major flavonoids, open-pollinated and F_1 varieties of spinach could be distinguished based on the concentrations of minor flavonoid species. Broad-sense heritability estimates for absolute bioaccessibility accounted for more genetic variation than raw material content, suggesting that this trait is preferable for breeders seeking to alter the phytochemical profile of spinach. Lastly, we found that several spinach flavonoids are unstable under digestive conditions, which was made evident by the proportion of aglycones rising from 0.1% to approximately 15% of total flavonoids after digestion. Together, these data suggest that spinach flavonoid biosynthesis and bioaccessibility are complex and contextualize how these molecules may behave in vivo.

Keywords: flavonoids; spinach; bioaccessibility; bioavailability; breeding; mass spectrometry; LC-MS; polyphenols; health

1. Introduction

Flavonoids are among an array of phytochemicals associated with positive health benefits from fruit- and vegetable-rich diets. They have been hypothesized to have antioxidative, anti-inflammatory, and anticarcinogenic properties in vivo [1–3]. Spinach (*Spinacia oleracea* L.) is a dietary source of flavonoids with significant concentrations reported up to and beyond 224 mg/100 g fresh weight [4–7]. While concentrations are high, the profile of flavonoids in spinach departs from what is typically found in fruits and vegetables. Spinach flavonoids feature a patuletin, spinacetin, or jaceidin backbone that is further modified by glycosylation, glucuronidation, and conjugation with phenolic acids [6,8]. Although little is known about the specific roles of spinach flavonoids on human health, they remain a trait of interest due to their structural similarity to more common dietary flavonoids [9,10].

Plant breeding programs tend to focus on increasing nutrients and/or phytochemicals in crops. However, increased amounts of a given substance in a plant do not guarantee greater bioefficacy. Food matrix factors such as protein, starch, and fiber can affect the release of a compound from food (*bioaccessibility*) and the absorption into the body (*bioavailability*) [11,12]. Bioaccessibility and bioavailability are becoming key considerations in breeding programs focusing on mineral elements in rice [13], wheat [14], common bean [15], maize [16], and sorghum [17]; carotenoids in maize [18], banana [19], and spinach [20]; and polyphenols in blueberries [21]. While randomized clinical trials in humans are considered the gold standard to assess the impact of either foods or molecules on health outcomes, they are costly and logistically challenging to operate. Bioaccessibility estimates from in vitro models of digestion associate well with human clinical data and are frequently used as a cost-effective surrogate to estimate bioavailability [22,23]. While bioaccessibility does not necessarily guarantee uptake into the body, release from the food matrix is a required step prior to absorption. These models are crucial for efficiently phenotyping populations of plants for plant breeders to make selections that might ultimately be used in clinical applications.

Only limited reports in the context of post-harvest processing exist for spinach flavonoid bioaccessibility. Freezing spinach preserves flavonoid profiles while modestly enhancing bioaccessibility, ostensibly through changes in the food matrix [24]. Processing techniques, such as spray drying and mixing with a protein carrier, have also been shown to modulate spinach flavonoid bioaccessibility in a processing-dependent manner [25]. There are currently no data on the role of genetic background on flavonoid bioaccessibility in spinach, and whether this trait is heritable in this crop. We hypothesized that spinach flavonoid bioaccessibility is heritable and a distinct trait from raw material flavonoid content.

The primary goals of this study were to define the range and concentration of flavonoids in a diverse population of spinach using a comprehensive ultra-high performance liquid chromatography mass spectrometry (UHPLC-MS/MS) method, determine the bioaccessibility of spinach flavonoids using a three-stage static in vitro digestion model coupled with UHPLC-MS/MS, and to determine the broad-sense heritability of spinach bioaccessibility. These data would serve as a resource to plant breeders seeking to optimize the potential health benefits of spinach and to allow for the development of unique germplasm that can be tested in clinical settings.

2. Materials and Methods

2.1. Chemicals and Reagents

Reagents purchased from Sigma Aldrich (Sigma Chemical Co., St. Louis, MO, USA) included quercetin-3-glucoside (99%), naringin (99%), naringenin (99%), mucin, alpha-amylase (lot: SLCM1439), pepsin (lot: SLCG6556), lipase (lot: SLCJ9303), pancreatin (lot: SLCK9806), and bile (lot: SLCJ7934). Urea, uric acid, potassium chloride, sodium sulfate, sodium phosphate, sodium chloride, sodium bicarbonate, LC-MS grade water, acetonitrile, methanol, and formic acid were purchased from Fisher Scientific (Fisher Scientific, Waltham, MA, USA). Patuletin (99%) was purchased from Extrasynthese (Extrasynthese, Genay, France), and jaceidin (99%) and spinacetin (99%) were purchased from Key Organics (Key Organics, Camelford, United Kingdom). Taxifolin (99.99%) was purchased from Selleck Chemicals (Selleck Chemical LLC, Houston, TX, USA). StrataX polymeric reversed-phase solid-phase extraction (SPE) cartridges (30 mg, 33 μm pore size, 1 mL volume) were purchased from Phenomenex (Torrance, CA, USA).

2.2. Germplasm Selection and Growth Conditions

Thirty spinach accessions were selected to maximize genetic variation in our population based on previous sequencing and phenotyping efforts [20,26]. Accessions previously genotyped by Qin and colleagues were selected from unique sub-populations separated by sequence diversity as determined by STRUCTURE2 and MEGA6 [27,28]. Germplasm previously characterized by Hayes and colleagues was selected for this study based on variation in carotenoid bioaccessibility. Additional metadata for each accession can be found in Supplemental Table S1.

Spinach plants were cultivated according to Qin and others, with modifications described by Dzakovich and colleagues [6,26]. For each cultivar, seedlings were thinned to six individual plants per one-gallon pot. Pots were randomly distributed throughout a PGW36 walk-in growth chamber (Conviron; Winnipeg, Canada) maintained at 300 µmols/m^2/s photosynthetically active radiation for 12 h per day, 50 ± 10% relative humidity, and 20 ± 0.5 °C/15 ± 0.5 °C day and night temperatures, respectively. Plant positions were re-randomized each day to mitigate positional differences in environmental factors such as light intensity. Plants were considered mature at the six-to-eight leaf stage (five to six weeks after sowing), when edible tissues were harvested and stored at −80 °C until analysis. This population was grown and harvested as described above twice.

2.3. Sample Processing

Frozen tissue was homogenized 1:1 with MilliQ water for approximately one minute using a VWR 250 Homogenizer, 10032-766 (Radnor, PA, USA). Homogenate was immediately aliquoted for raw material analysis, in vitro digestion, and dry matter content. Homogenates were stored at -80 °C until use. All subsequent analyses were conducted in triplicate.

2.4. In Vitro Digestion

Spinach samples were subject to a three-phase static in vitro digestion method described previously for food products, including spinach and sorghum [20,29,30]. In the oral phase, spinach homogenate (0.4 g) was mixed with 1.2 mL of simulated saliva containing 31.8 mg/mL alpha-amylase, incubated at 37 °C for 10 min at 120 oscillations per minute (OPM). During the gastric phase, 0.4 mL of 10 mg/mL pepsin solution in 0.1 M HCl was added, and samples were further diluted with 2.7 mL of saline (0.9% NaCl). Sample pH was adjusted to 2.5 using 1.0 M HCl. Saline was added to dilute samples to 8 mL, and the sample headspace was replaced with nitrogen gas prior to incubation at 37 °C for 1 h at 120 OPM. Sample pH was then adjusted to 5.0 using 1.0 M $NaHCO_3$ to initiate the intestinal phase. Pancreatin-lipase solution (0.4 mL; 20 mg/mL) in 0.1 M $NaHCO_3$ and porcine bile extract (0.6 mL; 30 mg/mL) were added, and the sample pH was adjusted to 7.0. Sample volumes were adjusted to 10 mL using saline prior to an additional blanketing with nitrogen gas. Samples were then incubated at 37 °C for 2 h at 120 OPM. At the end of the intestinal phase, digesta was aliquoted, and 4 mL were stored at −80 °C for later analysis. The remaining digesta were blanketed with nitrogen gas and centrifuged at 3428× *g* for 75 min at 4 °C (Eppendorf 5920R). Approximately 4 mL of each aqueous fraction was filtered through 0.22 µm cellulose acetate filters and reserved to quantify bioaccessible flavonoids.

2.5. Extraction of Flavonoids from Fresh Spinach and Aqueous Fractions

Flavonoids were extracted in duplicate using the traditional method previously described by Dzakovich and others [6]. Dried spinach flavonoid extracts were redissolved and diluted to an equivalent volume of 25 mL of 1:1 methanol:water + 0.1% formic acid containing 4% 20 µM taxifolin as an internal standard to correct for variation in mass spectrometer response prior to filtration through 0.2 μm PTFE filters. Flavonoids from aqueous fractions were semi-purified using SPE. Aqueous fractions (500 μL) were pipetted into 3 mL Strata-X 33 μm, 30 mg polymeric reversed phase SPE cartridges (Phenomenex), and diluted with 2500 μL of 0.1% formic acid in water. Samples were loaded under vacuum and rinsed with one volume of 0.1% formic acid in water, followed by one volume of 1.0% formic acid in water. Cartridges were dried under vacuum for approximately 10 min and rinsed with one volume of hexane to remove nonpolar and semi-polar impurities (e.g., carotenoids). Cartridges were dried for 2 min under vacuum and analytes were eluted with 2 mL of methanol with 0.1% formic acid. Vials of eluent were dried under a stream of nitrogen gas and resolubilized in 500 μL of 1:1 methanol:water with 0.1% formic acid and 4% 20 μM taxifolin as an internal standard. Samples were passed through 0.2 μm PTFE filters and into LC vials prior to analysis.

2.6. Recovery Estimates from SPE

Separate spike addition experiments were conducted to optimize the recovery of flavonoids from aqueous fractions. Spinach-free digesta was created using the in vitro digestion protocol detailed above and spiked with jaceidin, patuletin, spinacetin, naringenin, and naringin. Various washing and elution conditions were tested using the methods of Mengist and colleagues as a starting point [21]. Methanol with 0.1% formic acid produced the most consistent and high-yielding results with estimated recoveries of 87%, 91%, 111%, 98%, and 99% for jaceidin, patuletin, spinacetin, naringenin, and naringin, respectively.

2.7. Analysis of Spinach Flavonoids

Flavonoids from both spinach fresh tissue and aqueous extracts were analyzed using a UHPLC-MS/MS method developed by Dzakovich and colleagues [6]. Briefly, flavonoids were separated on a Waters Acquity BEH C_{18} 2.1 × 100 mm (1.7 μm particle size) column using a gradient of 0.1% formic acid in water (A) and 0.1% formic acid in acetonitrile (B). Column eluent flowed into a TSQ Altis tandem mass spectrometer (ThermoFisher), and spinach flavonoids were quantified against the external standard quercetin-3-O-glucoside when an authentic standard was not available. Optimized multiple reaction monitoring (MRM) experiments for 39 spinach flavonoids, including isomers, to maximize the signal-to-noise of each analyte. Dwell times were automatically adjusted to maintain 12–15 points per peak, and the analyte signal was corrected for using taxifolin. Spinacetin was excluded from analyses due to being below the limit of quantification for all samples and sample types.

2.8. Statistical Analysis

Plants were grown in a randomized complete block design with two replications in time and analyzed with the following model parameters:

$$\hat{Y}_{ij} = \mu + G_i + B_j + G_i : B_j + \varepsilon_{ij}$$

where $\hat{Y}_{ij}$ represents an analyte estimate; μ is the population mean for an analyte; G_i represents genetic factors; B_j represents the contribution from block; G_i:B_j represents the interaction between genotype and block; and ε_{ij} is the residual error.

Relative bioaccessibility was defined as follows:

$$RB\ (\%) = \frac{Concentration\ of\ analyte\ in\ aqueous\ fraction}{Concentration\ of\ analyte\ in\ raw\ material} \times 100$$

Absolute bioaccessibility (mg/100 g FW) was defined as follows:

$$AB = RB \times Concentration\ of\ analyte\ in\ raw\ material$$

All statistical analyses and data visualization were performed using R statistical software v. 4.5.2 [31]. Linear modeling, including the calculation of best linear unbiased estimates (BLUEs) and random effects modeling, was conducted using the package 'lme4' [32]. The restricted maximum likelihood (REML) was used to estimate variance components to calculate broad-sense heritability [32]. Correlations were calculated and visualized using the 'Corrplot' package [33], while principal components were calculated using 'FactoMineR' and visualized with 'ggplot2' and 'plotly' [34,35]. Hierarchical clustering to determine chemical similarity using Tanimoto coefficients was conducted with the package 'ChemmineR' [36].

3. Results and Discussion

3.1. Spinach Flavonoid Profiles Are Diverse, but Dominated by Five Molecules

To determine the range and diversity of spinach flavonoid raw material content and bioaccessibility, we grew 30 unique accessions comprising both F_1 varieties and open-pollinated accessions over two periods of time. Approximately 56.5% of spinach flavonoids in raw material were patuletin-derived, while the remaining 31.6%, 6.7%, and 5.3% were derived from methylenedioxyflavones, spinacetin, and jaceidin. Concentrations in raw material ranged from 64.3 to 453.7 mg/100 g fresh weight (FW), with an average concentration of 185.8 mg/100 g FW (Supplemental Table S2). On average, approximately 67% of the total flavonoid profile was represented by five analytes, including 5,3′,4′-Trihydroxy-3-methoxy-6:7-methylendioxyflavone-4-β-D-glucuronide (519.08 m/z), Spinatoside (521.09 m/z), Patuletin-3-O-β-D-2-ρ-coumaroyl glucopyranosyl-(1→6)-β-D-glucopyranoside (801.21 m/z), 5,4 -Dihydroxy-3,3-dimethoxy-6:7-methylendioxyflavone-4-β-D-glucuronide (533.09 m/z), and Patuletin-3-O-β-D-(2-feruloyl glucopyranosyl)-(1→6)-[β-D-apiofuranosyl-(1→2)]-β-D-glucopyranoside (963.24 m/z) (Supplemental Table S2). The profile and concentration ranges of spinach flavonoids in our population were consistent with previous reports [6,7,37,38]. Ranges of aggregated flavonoid classes presented on a $\log_{10}$ transformed y-axis can be found in Figure 1.

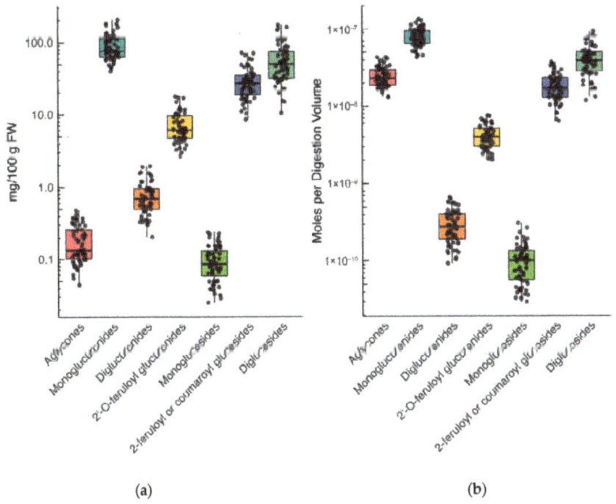

Figure 1. Box and whisker plots of aggregated spinach flavonoids in raw material (**a**) and aqueous fractions (**b**). Y-axes are log10 transformed to accommodate the wide range in concentrations calculated for these flavonoids. Each dot within a box and whisker plot represents an individual variety grown in a block. Of note is the increase in relative abundance of aglycones after digestion (**b**).

3.2. Minor Flavonoids Drive Differences Between F_1 and Open-Pollinated Spinach Lines

Multivariate analysis using PCA revealed a distinction between open-pollinated and F_1 spinach varieties (Figure 2; Supplemental Figure S1). Although PC1 explained a large proportion of total variance (53.4%, Figure 2), F_1 varieties and open-pollinated lines separated primarily on PC2 (17.95% total variation). The most influential variables explaining separation in PC2 are shown in Figure 2b and colored based on whether the analyte was higher in F_1 or open-pollinated spinach varieties. An interactive loadings plot of all variables can be found in Supplemental Figure S1. Individual loadings for the input variables indicated that several minor flavonoids, such as 5,4′-Dihydroxy-3,3′-dimethoxy-6:7-methylen-dioxyflavone-4′-β-D-2′-O-feurloyl-glucuronide (709.14 m/z), contributed to the differentiation between F_1 and open-pollinated lines (Figure 2b). Although low in concentration relative to other flavonoids (0.53–6.62 mg/100 g FW), 709.14 m/z was on average 53.7% higher in F_1 spinach varieties compared to open-pollinated lines with similar trends for other variables ($p < 0.001$; Supplemental Table S3). Similar trends were true for other minor flavonoids that were higher in F_1 spinach varieties (Figure 2b). These flavonoids may have been inadvertently selected for commercial germplasm. Flavonoids and other polyphenols have been reported to be influenced by human selection in other crops such as lettuce [39] and apple [40]. Alternatively, differences in minor flavonoids may be due to an overall increase in all flavonoids. Although statistically nonsignificant in our population ($p > 0.05$), our data shows that F_1 spinach varieties were, on average, lower in total flavonoids compared to open-pollinated lines (176.65 vs. 192.93 mg/100 g FW; Supplemental Tables S2 and S3). Increases in minor flavonoids seen in F_1 spinach varieties may be a consequence of lower competition for precursors and/or differences in enzymatic gene expression or enzyme efficiency. This phenomenon has been observed for carotenoids in crops such as maize [18,41], soybean [20,42], and tomato [43], as well as polyphenols in apple [44], maize [45], and rice [46–48]. Additional follow-up experiments coupling next-generation sequencing with an expanded population of wild and cultivated spinach are necessary to define the mechanism of this phenomenon in spinach.

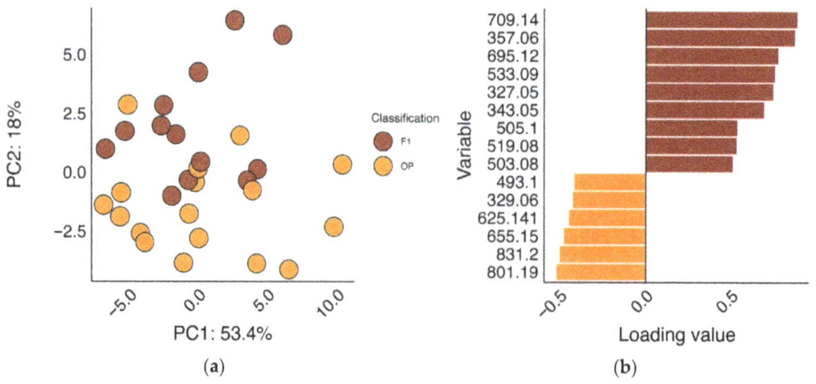

Figure 2. Principal components analysis scores plot (**a**) and stacked bar plots of the top 15 most influential variables from principal component 2 (**b**) of raw material content. BLUEs for each analyte within a cultivar were utilized as input data to more accurately reflect genetic effects and minimize environmental distortion. "F1" indicates F_1 varieties and "OP" indicates open-pollinated lines. The coloration of bars in (**b**) indicates if an analyte was higher in concentration in either F_1 varieties (red) or open-pollinated lines (orange).

3.3. Raw Material Content and Absolute Bioaccessibility Are Generally Associated, but Several Negative Associations Exist

To better understand the relationships between spinach flavonoids as well as raw content and absolute bioaccessibility, we performed correlation analyses (Figure 3). Matrices of Pearson's correlation coefficients and *p*-values can be found in Supplemental Table S4. Generally, individual flavonoids were positively associated with each other, indicating that concentrations rose and fell in concert. However, several analytes were negatively correlated with most other spinach flavonoids, including 5,4′-Dihydroxy-3,3′-dimethoxy-6,7-methylenedioxyflavone (357.06 m/z), 5,4′-Dihydrox-3-methoxy-6,7-methylenedioxyflavone (327.05 m/z), 709.14 m/z, 695.12 m/z, 5,3,4 Trihydroxy-3-methoxy-6:7-methylendioxyflavone (343.05 m/z), and, to a limited degree, 5,4-Dihydroxy-3,3-dimethoxy-6:7-methylendioxyflavone-4-β-D-glucuronide (533.09 m/z), as in Figure 3. Interestingly, these are the same analytes that appeared to drive differences between F_1 and open-pollinated germplasm (Figure 2). Absolute bioaccessibility correlated well with raw material content (Figure 3), which is consistent with other reports in the literature [21,25,49–51]. A few analytes, such as Spinacetin-3-O-β-D-glucopyranosyl-(1→6)-[β-D-apiofuranosyl-(1→2)]-β-D-glucopyranoside (801.19 m/z), displayed negative associations with the analytes mentioned above (e.g., 709.14 m/z), both in raw content and absolute bioaccessibility. Raw data for relative and absolute bioaccessibility can be found in Supplemental Tables S5 and S6, respectively.

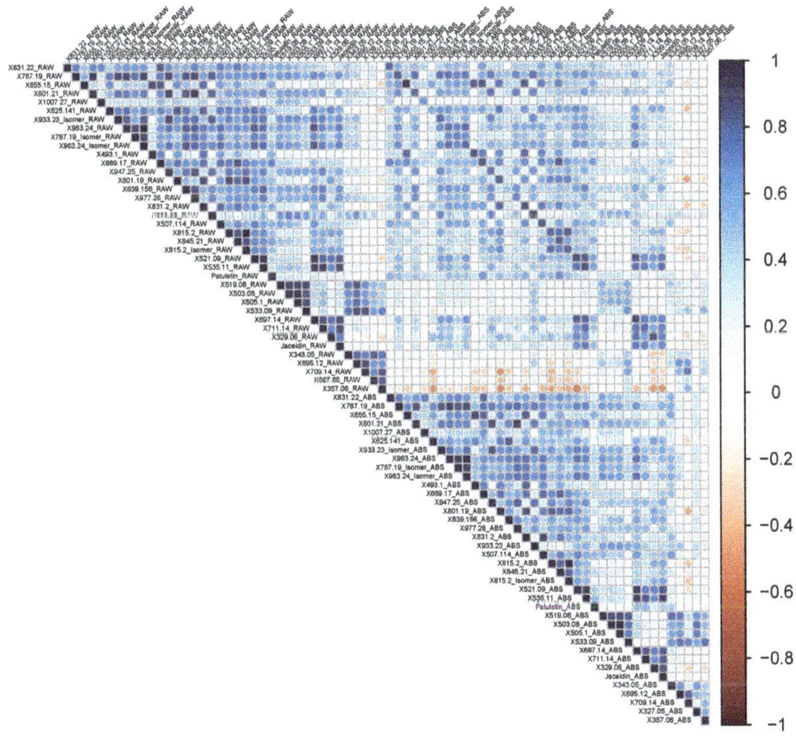

Figure 3. Correlation plot of raw material content and absolute bioaccessibility of the analytes measured in this study (n = 90). Dark blue to dark red indicates a Pearson correlation coefficient spanning from 1 to −1, respectively.

Given that the analytes with negative correlation coefficients are methylenedioxyflavones, we hypothesize that the trends we observed are due to a split in the spinach flavonoid biosynthesis pathway, whereby select methylenedioxyflavones compete for precursor molecules that feed into multiple segments of the pathway. While there have been strides to characterize the flavonoid profiles of spinach [6–8], the structural and regulatory aspects of this pathway remain largely unknown. Using Tanimoto coefficients, we sorted analytes profiled by our method using hierarchical clustering (Figure 4). Structures and SMILES abbreviations of each analyte in our UHPLC-MS/MS method can be found in Supplemental Table S7. These data were used in determining groups of aggregated flavonoids presented here (Figure 1; Supplemental Table S2). Clustering organized analytes by structural motifs and may suggest biosynthetic relationships among different spinach flavonoids. Similarity based on structural motifs may indicate common enzymes that act on multiple substrates. Additional studies leveraging genomic tools and techniques, such as virus-induced gene silencing, are necessary to definitively map the spinach flavonoid biosynthesis pathway.

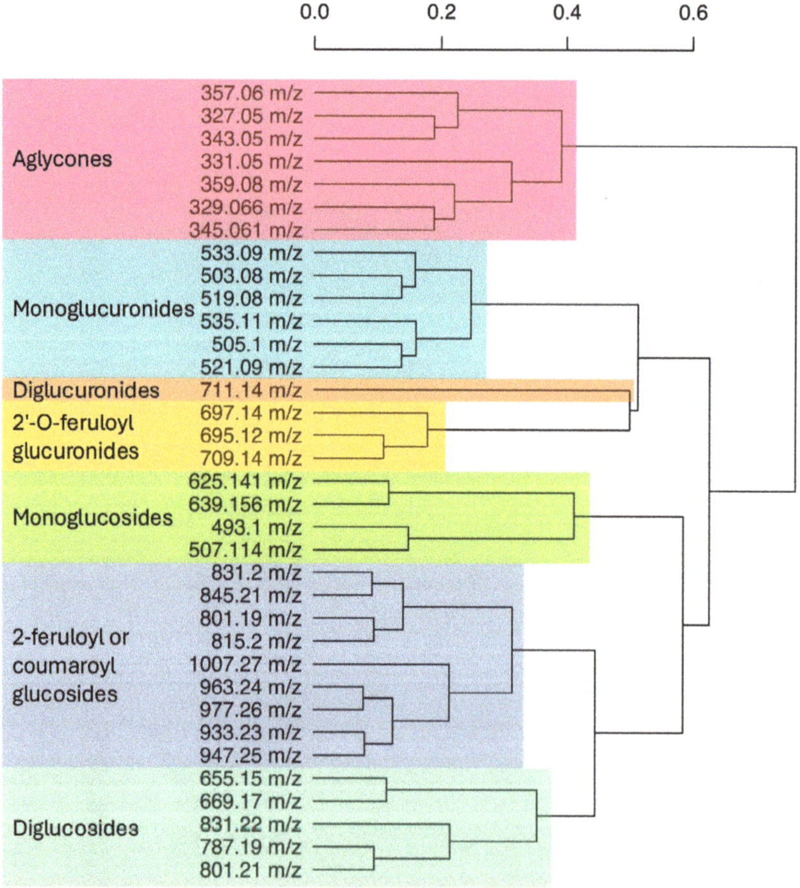

Figure 4. Dendrogram grouping spinach flavonoids based on Tanimoto coefficients. The scale bar above indicates Euclidean distance.

3.4. Spinach Flavonoid Absolute Bioaccessibility Is More Heritable than Flavonoid Content in Raw Material

To define the contribution of genetics relative to environmental factors, we calculated broad-sense heritability on a line-mean basis for raw material content, relative bioaccessibility, and absolute bioaccessibility [52]. A graphical illustration of variance partitioning for model components cultivar, block, block:cultivar interaction, and residual can be seen in Figure 5. Variance for each model component for every analyte can be found in Supplemental Table S8. Broad-sense heritability for raw material content ranged from 0 to 63%, with an average of 22% (Supplemental Table S8). Contrary to our expectations, relative bioaccessibility exhibited extremely low estimates of broad-sense heritability, 0–31%, with an average of only 2%. For virtually every analyte in relative bioaccessibility, most of the variation is partitioned into either block or block:cultivar interaction. This interaction was particularly dominant because relative bioaccessibility is a ratio between the concentration of an analyte in the aqueous fraction relative to the starting content in the raw material. Ratios can introduce distortions, as slight changes in the denominator (raw material concentration) can greatly amplify environmental effects. Given how prominent environmental effects are for raw material flavonoid concentration (Figure 5a), this phenomenon explains why the block:cultivar interaction dominated variance estimates contributing to relative bioaccessibility in our data, as well as other published studies. Reports of broad-sense heritability estimates for flavonoids in spinach, or other leafy greens, are scant. A similar trend of moderately high estimates of broad-sense heritability for relative content and absolute bioaccessibility, but extremely low for relative bioaccessibility, has been reported in blueberries [21]. Broad-sense heritability estimates for quercetin-3-arabidonside, quercetin-3-glucoside, and total flavonoids in blueberries were all 10% or less for relative bioaccessibility, but substantially higher for raw material and absolute bioaccessibility [21]. These findings echo our own and highlight a disconnect between raw material content and absolute bioaccessibility.

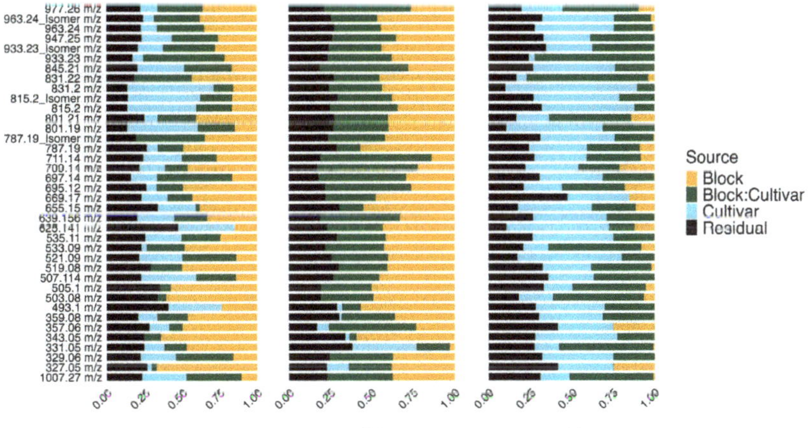

(a) (b) (c)

Figure 5. Variance partitioning estimates visualized for each flavonoid for raw material content (**a**), relative bioaccessibility (**b**), and absolute bioaccessibility (**c**).

Interestingly, absolute bioaccessibility, which is a product of raw material content multiplied by relative bioaccessibility, showed the highest estimates of broad-sense heritability (Supplemental Table S8). Broad-sense heritability for absolute bioaccessibility ranged from

4 to 76%, with an average of 34%. Four analytes were estimated to be greater than 50%, which included Spinacetin-3-O-β-D-(2-feruloyl glucopyranosyl)-(1→6)-[β-glucopyranosyl (1→2)]-β-D-glucopyranoside (625.141 m/z), (801.19 m/z), Patuletin-3-O-β-D-2-feruloyl glucopyranosyl-(1→6)-β-D-glucopyranoside (831.2 m/z), and 521.09 m/z (Supplemental Table S8). Only six analytes had broad-sense heritability estimates that were equal to or less than raw material content, supporting the assertion that absolute bioaccessibility is under stronger genetic control than raw material content. This outcome is intriguing given that relative bioaccessibility was estimated near 0% for 30 out of 38 analytes measured.

Absolute bioaccessibility is a product of raw material content and relative bioaccessibility [53]. Our findings that absolute bioaccessibility heritability exceeded raw material heritability suggest that traits related to digestion efficiency (e.g., cell wall composition) are under stronger genetic control and are more stable than flavonoid concentration. Similar findings have been reported for iron in maize [42] and carotenoids in spinach [20]. Heritability of absolute bioaccessibility may appear higher because absolute bioaccessibility combines the genetic differences in relative content with environmental variation in relative bioaccessibility, creating larger total differences in absolute bioaccessibility between genotypes. Even with low relative bioaccessibility, as is the case in our dataset, plants with slightly higher relative content tend to have higher absolute bioaccessibility on average, making absolute bioaccessibility more genetically predictable than relative content alone. Regardless, environmental factors tended to be a major influence on the phenotypes we measured.

While our data suggest that breeding for flavonoid content and profiles in spinach would be possible, but challenging, significant opportunities exist in controlled environment agriculture (CEA) to directly manipulate these traits by controlling plant growth conditions. Flavonoid biosynthesis is strongly controlled by light quantity and quality [54,55], temperature [56,57], and abiotic stress, such as salinity [58]. Advances such as light-emitting diodes (LEDs) allow for precise control of the light environment, and other environmental parameters can be manipulated in modern CEA systems [59–61]. Additional research is necessary to determine how environmental conditions can be leveraged to fine-tune the flavonoid profiles of spinach.

Although absolute bioaccessibility exhibited the highest broad-sense heritability of nearly all phenotypes measured in this study, estimates tended to be relatively low compared to other crops [21,62–64]. One possibility is due to the nature of spinach itself, which is a highly heterozygous, wind-pollinated outcrossing species [65–67]. Of the 30 accessions profiled in our study, 17 were open-pollinated (Supplemental Table S1). Open-pollinated accessions can breed true when cultivated under controlled conditions, but there is no guarantee of genetic uniformity seen in cultivated F_1 varieties. As such, it is distinctly possible that some of our open-pollinated or heirloom accessions, many have been segregating for any number of traits. Our population structure may have had unaccounted genetic variation that manifested as environmental variation. Future experiments can be conducted with larger populations comprising cultivated F_1 varieties to better understand the consequences of open pollination practices on flavonoid heritability. Regardless, heritability estimates for absolute bioaccessibility outperformed raw material content, and this trait may be preferable for plant breeders to select against when considering potential health implications of their spinach germplasm.

3.5. Some Spinach Flavonoids Are Hydrolyzed During Digestion

Our estimates of relative bioaccessibility align with the limited reports in the literature that only monitored parts of the spinach flavonoid biosynthetic pathway [24,25]. Estimates of relative bioaccessibility theoretically range from 0 to 100%. However, several analytes in our study exceeded 100%, such as 343.05 m/z, with relative bioaccessibility estimates ranging from 569.5 to 9489.1% (Supplemental Table S5). Over 99.9% of spinach flavonoids are O-glycosylated in raw spinach [6], yet this proportion shrinks to approximately 85% after digestion with a proportional rise in aglycones (Figure 1; Supplemental Table S2). Compared to C-linked glycosylated flavonoids, O-linked glycosylated flavonoids are considerably weaker and more prone to chemical hydrolysis [2]. In vitro digestions, such as the one conducted in this study, feature a 1 h gastric phase at pH 2.0 at 37 °C. While there is debate as to whether gastric conditions are sufficient to induce chemical hydrolysis [68,69], there is evidence to suggest that hydrolysis may occur with some polyphenols [21,70]. For example, Mengist and colleagues reported up to 4900% relative bioaccessibility for caffeic acid and over 100% relative bioaccessibility for catechin, epicatechin, flavon-3-ols, and total flavonoids. Given that some of these molecules have been found to be condensed with other analytes, like anthocyanins [45], it is feasible that analytes reported at >100% relative bioaccessibility may be degradation products formed during digestion from more complex phytochemicals. Grace and colleagues did not report relative bioaccessibility over 100% for any of the 18 spinach flavonoids they tracked, but reported 533.09 m/z above 90% in some of their treatments [25]. In animals and humans, we would anticipate this conversion to occur at an even greater rate due to enzymatic activity at the brush border [71,72] and interactions with the gut microbiome [1,73].

In vitro digestions similar to the one conducted in this study provide a cost-effective way to screen germplasm for bioaccessibility, a measure that generally associates well with bioavailability for many molecules [11]. Analytes we observed to increase after digestion increased at the expense of other analytes. For example, 695.12 m/z had an average relative bioaccessibility of 22.6%, the structurally related 519.08 m/z at 20.6%, and the aglycone 343.05 m/z at 2953.6% (Supplemental Table S5; Figure 6A). We hypothesize that the elevated appearance of 343.05 m/z was due to the successive loss of ferulic acid from 695.12 m/z and the loss of glucuronic acid from 519.08 m/z. Similar trends were noticed for other analytes, and hypothetical degradation reactions are proposed (Figure 6B,C). Additional proposed reactions not shown in Figure 6 include 709.14 m/z to 357.06 m/z and Spinatoside-4′-β-D-(2′-O-feruloyl-glucuronide) (697.14 m/z) to Patuletin (331.05 m/z). While the relative bioaccessibility estimates reported here are quite large for some aglycones, these analytes are among the least abundant flavonoids in raw spinach and are found in nanomolar concentrations. Small increases in these molecules result in relatively large percent increases. While these data suggest a simultaneous over- and slight under-estimation of some analytes, in vitro digestions still represent an invaluable tool to contextualize the potential health benefits of a food or food product. Studies leveraging isolated molecules and potentially intrinsic labeling would be necessary to elucidate their true absorption, distribution, metabolism, and excretion. Regardless, estimates provided here serve as the most comprehensive profiling of spinach flavonoids pre- and post-in vitro digestion to date and will help breeders and nutrition scientists make data-driven decisions about this crop.

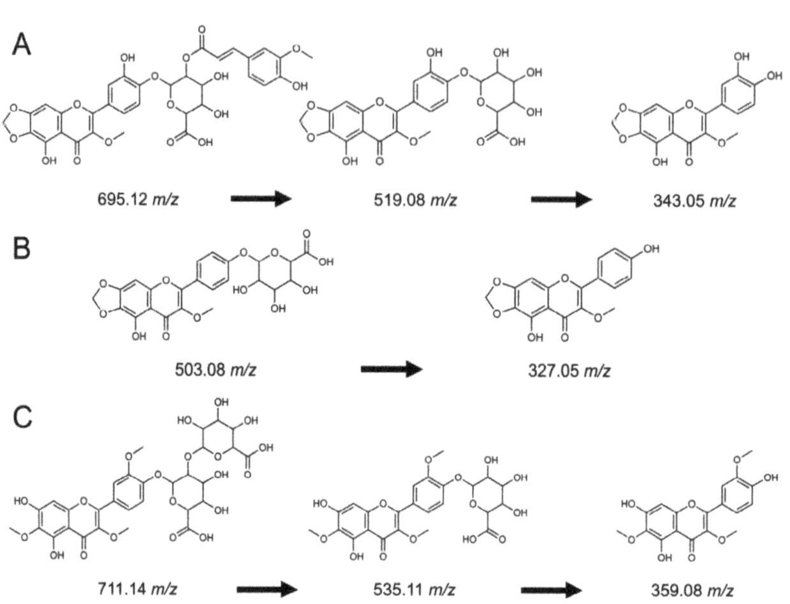

Figure 6. Proposed reactions that occurred during the in vitro digestion, justifying > 100% relative bioaccessibility for multiple spinach flavonoid aglycones not normally seen in high abundance in raw material. These reactions include 695.12 > 343.05 (**A**), 503.08 > 327.05 (**B**), and 711.14 > 359.08 (**C**).

4. Conclusions

Here, we reported the range and concentrations of flavonoids in a diverse population of spinach, estimated relative and absolute bioaccessibility of these molecules, and calculated the broad-sense heritability for these traits. These data were generated using a novel mass spectrometry methodology that allows for a comprehensive view of the spinach flavonoid pathway. While follow-up studies are necessary to determine bioavailability, these data address critical gaps in knowledge about the profiles and bioaccessibility of these molecules. Our data suggest that there may be at least one split in the spinach flavonoid biosynthetic pathway, whereby several methylenedioxyflavone species compete for common precursors required for the biosynthesis of other molecules.

We found that absolute bioaccessibility is more heritable than raw material content in the context of spinach flavonoids and will be of interest to plant breeders seeking to improve the phytochemical content of spinach. Indications that environmental factors play a major role in both flavonoid biosynthesis and bioaccessibility provide numerous opportunities for the CEA community to develop environmental manipulation strategies that could alter the phytochemical profile and potential health benefits of spinach. Additionally, we discovered that several spinach flavonoids appear to be susceptible to chemical transformation during digestion, which would be enhanced in vivo due to brush border enzymes and microbiome activity. Many opportunities exist to utilize genomic and metabolomic tools to fully characterize this pathway, define genetic and environmental cues that regulate this pathway, define the bioavailability of these molecules, and determine their impact on human health.

Supplementary Materials: The following supporting information can be downloaded at https://www.mdpi.com/article/10.3390/foods14244314/s1.

Author Contributions: Conceptualization, M.P.D.; methodology, M.P.D., A.L.T., E.A.L., B.W.R., and G.A.D.; software, M.P.D. and R.P.D.; validation, M.P.D.; formal analysis, M.P.D.; investigation, M.P.D.,

R.P.D., A.L.T., and E.A.L.; resources, M.P.D.; data curation, M.P.D.; writing—original draft preparation, M.P.D.; writing—review and editing, M.P.D., A.L.T., E.A.L., R.P.D., B.W.R., and G.A.D.; visualization, M.P.D. and R.P.D.; supervision, M.P.D.; project administration, M.P.D.; funding acquisition, M.P.D. All authors have read and agreed to the published version of the manuscript.

Funding: This research was funded by USDA-ARS CRIS project 3092-10700-066-001S, as well as by the Texas Children's Hospital Pediatric Pilot Award 69513-I.

Institutional Review Board Statement: Not applicable.

Informed Consent Statement: Not applicable.

Data Availability Statement: The original contributions presented in this study are included in the article/Supplementary Materials. Further inquiries can be directed to the corresponding author.

Acknowledgments: We greatly appreciate the advice from David Brenner and the USDA-ARS GRIN, as well as Massimo Iorizzo at North Carolina State University, for assistance with developing the spinach population used in this study. We thank Brian Chlouber for helping culture and harvest the plant material.

Conflicts of Interest: The findings and conclusions in this publication are those of the authors and should not be construed to represent any official USDA or U.S. Government determination or policy. Mention of trade names or commercial products in this publication is solely for the purpose of providing specific information and does not imply recommendation or endorsement by the U.S. Department of Agriculture. The USDA is an equal opportunity provider and employer. The author Dr. Benjamin W. Redan was employed by the company Thorne Research. He participated in method development as well as writing and reviewing the manuscript. Thorne Research played no role in the study's funding, execution, data analysis, or anyother aspects of the study. The remaining authors declare that the research wasconducted in the absence of any commercial or financial relationships that could beconstrued as a potential conflict of interest.

Abbreviations

The following abbreviations are used in this manuscript:

BLUE	Best linear unbiased estimate
CEA	Controlled environment agriculture
FW	Fresh weight
LED	Light-emitting diode
UHPLC-MS/MS	Ultra-high performance liquid chromatography mass spectrometry
REML	Restricted maximum likelihood
SPE	Solid-phase extraction

References

1. Williamson, G.; Kay, C.D.; Crozier, A. The Bioavailability, Transport, and Bioactivity of Dietary Flavonoids: A Review from a Historical Perspective. *Compr. Rev. Food Sci. Food Saf.* **2018**, *17*, 1054–1112. [CrossRef] [PubMed]
2. Xie, L.; Deng, Z.; Zhang, J.; Dong, H.; Wang, W.; Xing, B.; Liu, X. Comparison of Flavonoid O-Glycoside, C-Glycoside and Their Aglycones on Antioxidant Capacity and Metabolism during In Vitro Digestion and In Vivo. *Foods* **2022**, *11*, 882. [CrossRef]
3. Williamson, G. The Role of Polyphenols in Modern Nutrition. *Nutr. Bull.* **2017**, *42*, 226–235. [CrossRef]
4. Pandjaitan, N.; Howard, L.R.; Morelock, T.; Gil, M.I. Antioxidant Capacity and Phenolic Content of Spinach As Affected by Genetics and Maturation. *J. Agric. Food Chem.* **2005**, *53*, 8618–8623. [CrossRef]
5. Gil, M.I.; Ferreres, F.; Tomás-Barberán, F.A. Effect of Postharvest Storage and Processing on the Antioxidant Constituents (Flavonoids and Vitamin C) of Fresh-Cut Spinach. *J. Agric. Food Chem.* **1999**, *47*, 2213–2217. [CrossRef] [PubMed]
6. Dzakovich, M.P.; Le, E.A.; Tak, A.L.; Chacko, S.K. A Comprehensive UHPLC-MS/MS and Extraction Method Reveals Flavonoid Profile and Concentration Are Diverse in Spinach (*Spinacia oleracea* L.). *Front. Plant Sci.* **2025**, *16*, 1496200. [CrossRef]
7. Cho, M.J.; Howard, L.R.; Prior, R.L.; Morelock, T. Flavonoid Content and Antioxidant Capacity of Spinach Genotypes Determined by High-Performance Liquid Chromatography/Mass Spectrometry. *J. Sci. Food Agric.* **2008**, *88*, 1099–1106. [CrossRef]

8. Singh, A.; Singh, P.; Kumar, B.; Kumar, S.; Dev, K.; Maurya, R. Detection of Flavonoids from Spinacia Oleracea Leaves Using HPLC-ESI-QTOF-MS/MS and UPLC-QqQLIT-MS/MS Techniques. *Nat. Prod. Res.* **2019**, *33*, 2253–2256. [CrossRef]
9. Singh, J.; Jayaprakasha, G.K.; Patil, B.S. Extraction, Identification, and Potential Health Benefits of Spinach Flavonoids: A Review. In *Advances in Plant Phenolics: From Chemistry to Human Health*; American Chemical Society: Washington, DC, USA, 2018; pp. 107–136, ISBN 978-0-8412-3296-9.
10. Pedro, P.; Goncalo, J. Structural Analysis of Flavonoids and Related Compounds—A Review of Spectroscopic Applications. In *Phytochemicals—A Global Perspective of Their Role in Nutrition and Health*; Rao, V., Ed.; InTech: Berlin, Germany, 2012; ISBN 978-953-51-0296-0.
11. Ferruzzi, M.G.; Kruger, J.; Mohamedshah, Z.; Debelo, H.; Taylor, J.R.N. Insights from In Vitro Exploration of Factors Influencing Iron, Zinc and Provitamin A Carotenoid Bioaccessibility and Intestinal Absorption from Cereals. *J. Cereal Sci.* **2020**, *96*, 103126. [CrossRef]
12. Schweiggert, R.M.M.; Carle, R. Carotenoid Deposition in Plant and Animal Foods and Its Impact on Bioavailability. *Crit. Rev. Food Sci. Nutr.* **2017**, *57*, 1807–1830. [CrossRef]
13. Senguttuvel, P.; Padmavathi, G.; Jasmine, C.; Sanjeeva Rao, D.R.; Neeraja, C.; Jaldhani, V.; Beulah, P.; Gobinath, R.; Aravind Kumar, J.; Sai Prasad, S.; et al. Rice Biofortification: Breeding and Genomic Approaches for Genetic Enhancement of Grain Zinc and Iron Contents. *Front. Plant Sci.* **2023**, *14*, 1138408. [CrossRef]
14. Hakeem, S.; Ali, Z.; Saddique, M.A.B.; Habib-Ur-Rahman, M.; Wiehle, M. Harnessing Genetic Diversity in Wheat to Enhance Grain Nutrition and Yield for Biofortification Breeding. *Biol. Res.* **2025**, *58*, 33. [CrossRef]
15. Blair, M.W. Mineral Biofortification Strategies for Food Staples: The Example of Common Bean. *J. Agric. Food Chem.* **2013**, *61*, 8287–8294. [CrossRef]
16. Šimić, D.; Sudar, R.; Ledenčan, T.; Jambrović, A.; Zdunić, Z.; Brkić, I.; Kovačević, V. Genetic Variation of Bioavailable Iron and Zinc in Grain of a Maize Population. *J. Cereal Sci.* **2009**, *50*, 392–397. [CrossRef]
17. Kruger, J.; Oelofse, A.; Taylor, J.R.N. Effects of Aqueous Soaking on the Phytate and Mineral Contents and Phytate: Mineral Ratios of Wholegrain Normal Sorghum and Maize and Low Phytate Sorghum. *Int. J. Food Sci. Nutr.* **2014**, *65*, 539–546. [CrossRef]
18. Kandianis, C.B.; Stevens, R.; Liu, W.; Palacios, N.; Montgomery, K.; Pixley, K.; White, W.S.; Rocheford, T. Genetic Architecture Controlling Variation in Grain Carotenoid Composition and Concentrations in Two Maize Populations. *Theor. Appl. Genet.* **2013**, *126*, 2879–2895. [CrossRef] [PubMed]
19. Munoz, A.J.; Ard, J.D.; Heimburger, D.C. The Role of Preventive Nutrition in Clinical Practice. In *Preventive Nutrition*; Humana Press: Totowa, NJ, USA, 2010; pp. 793–821.
20. Hayes, M.; Pottorff, M.; Kay, C.; Van Deynze, A.; Osorio-Marin, J.; Lila, M.A.; Iorrizo, M.; Ferruzzi, M.G. In Vitro Bioaccessibility of Carotenoids and Chlorophylls in a Diverse Collection of Spinach Accessions and Commercial Cultivars. *J. Agric. Food Chem.* **2020**, *68*, 3495–3505. [CrossRef] [PubMed]
21. Mengist, M.F.; Burtch, H.; Debelo, H.; Pottorff, M.; Bostan, H.; Nunn, C.; Corbin, S.; Kay, C.D.; Bassil, N.; Hummer, K.; et al. Development of a Genetic Framework to Improve the Efficiency of Bioactive Delivery from Blueberry. *Sci. Rep.* **2020**, *10*, 17311. [CrossRef]
22. Neilson, A.P.; George, J.C.; Janle, E.M.; Mattes, R.D.; Rudolph, R.; Matusheski, N.V.; Ferruzzi, M.G. Influence of Chocolate Matrix Composition on Cocoa Flavan-3-Ol Bioaccessibility In Vitro and Bioavailability in Humans. *J. Agric. Food Chem.* **2009**, *57*, 9418–9426. [CrossRef] [PubMed]
23. Reboul, E.; Richelle, M.; Perrot, E.; Desmoulins-Malezet, C.; Pirisi, V.; Borel, P. Bioaccessibility of Carotenoids and Vitamin E from Their Main Dietary Sources. *J. Agric. Food Chem.* **2006**, *54*, 8749–8755. [CrossRef]
24. Kamiloglu, S. Industrial Freezing Effects on the Content and Bioaccessibility of Spinach (*Spinacia oleracea* L.) Polyphenols. *J. Sci. Food Agric.* **2020**, *100*, 4190–4198. [CrossRef]
25. Grace, M.H.; Hoskin, R.T.; Hayes, M.; Iorizzo, M.; Kay, C.; Ferruzzi, M.G.; Lila, M.A. Spray-Dried and Freeze-Dried Protein-Spinach Particles; Effect of Drying Technique and Protein Type on the Bioaccessibility of Carotenoids, Chlorophylls, and Phenolics. *Food Chemistry* **2022**, *388*, 133017. [CrossRef]
26. Qin, J.; Shi, A.; Mou, B.; Grusak, M.A.; Weng, Y.; Ravelombola, W.; Bhattarai, G.; Dong, L.; Yang, W. Genetic Diversity and Association Mapping of Mineral Element Concentrations in Spinach Leaves. *BMC Genom.* **2017**, *18*, 941. [CrossRef] [PubMed]
27. Pritchard, J.K.; Stephens, M.; Donnelly, P. Inference of Population Structure Using Multilocus Genotype Data. *Genetics* **2000**, *155*, 945–959. [CrossRef]
28. Tamura, K.; Stecher, G.; Peterson, D.; Filipski, A.; Kumar, S. MEGA6: Molecular Evolutionary Genetics Analysis Version 6.0. *Mol. Biol. Evol.* **2013**, *30*, 2725–2729. [CrossRef]

29. Dzakovich, M.P.; Debelo, H.; Albertsen, M.C.; Che, P.; Jones, T.J.; Simon, M.K.; Zhao, Z.-Y.; Glassman, K.; Ferruzzi, M.G. Trait Stacking Simultaneously Enhances Provitamin A Carotenoid and Mineral Bioaccessibility in Biofortified Sorghum Bicolor. *Food Funct.* **2023**, *14*, 7053–7065. [CrossRef]
30. Lipkie, T.E.; De Moura, F.F.; Zhao, Z.-Y.; Albertsen, M.C.; Che, P.; Glassman, K.; Ferruzzi, M.G. Bioaccessibility of Carotenoids from Transgenic Provitamin A Biofortified Sorghum. *J. Agric. Food Chem.* **2013**, *61*, 5764–5771. [CrossRef]
31. R Development Core Team. *R: A Language and Environment for Statistical Computing*; R Development Core Team: Vienna, Austria, 2018.
32. Bates, D.; Mächler, M.; Bolker, B.; Walker, S. Fitting Linear Mixed-Effects Models Using lme4. *J. Stat. Softw.* **2015**, *67*, 1–48. [CrossRef]
33. Wei, T.; Simko, V. R Package "Corrplot": Visualization of a Correlation Matrix 2017.
34. Lê, S.; Josse, J.; Husson, F. FactoMineR: An R Package for Multivariate Analysis. *J. Stat. Softw.* **2008**, *25*, 1–18. [CrossRef]
35. Wickham, H. *Ggplot2 Elegant Graphics for Data Analysis*; Springer: New York, NY, USA, 2016; ISBN 978-0-387-98140-6.
36. Cao, Y.; Charisi, A.; Cheng, L.-C.; Jiang, T.; Girke, T. ChemmineR: A Compound Mining Framework for R. *Bioinformatics* **2008**, *24*, 1733–1734. [CrossRef]
37. Bergquist, S.Å.M.; Gertsson, U.E.; Knuthsen, P.; Olsson, M.E. Flavonoids in Baby Spinach (*Spinacia oleracea* L.): Changes During Plant Growth and Storage. *J. Agric. Food Chem.* **2005**, *53*, 9459–9464. [CrossRef]
38. Koh, E.; Charoenprasert, S.; Mitchell, A.E. Effect of Organic and Conventional Cropping Systems on Ascorbic Acid, Vitamin C, Flavonoids, Nitrate, and Oxalate in 27 Varieties of Spinach (*Spinacia oleracea* L.). *J. Agric. Food Chem.* **2012**, *60*, 3144–3150. [CrossRef] [PubMed]
39. van Treuren, R.; van Eekelen, H.D.L.M.; Wehrens, R.; de Vos, R.C.H. Metabolite Variation in the Lettuce Gene Pool: Towards Healthier Crop Varieties and Food. *Metabolomics* **2018**, *14*, 146. [CrossRef] [PubMed]
40. Kschonsek, J.; Wolfram, T.; Stöckl, A.; Böhm, V. Polyphenolic Compounds Analysis of Old and New Apple Cultivars and Contribution of Polyphenolic Profile to the In Vitro Antioxidant Capacity. *Antioxidants* **2018**, *7*, 20. [CrossRef] [PubMed]
41. Suwarno, W.B.; Pixley, K.V.; Palacios-Rojas, N.; Kaeppler, S.M.; Babu, R. Genome-Wide Association Analysis Reveals New Targets for Carotenoid Biofortification in Maize. *Theor. Appl. Genet.* **2015**, *128*, 851–864. [CrossRef]
42. Katuuramu, D.N.; Wiesinger, J.A.; Luyima, G.B.; Nkalubo, S.T.; Glahn, R.P.; Cichy, K.A. Investigation of Genotype by Environment Interactions for Seed Zinc and Iron Concentration and Iron Bioavailability in Common Bean. *Front. Plant Sci.* **2021**, *12*, 670965. [CrossRef]
43. Enfissi, E.M.A.; Nogueira, M.; Bramley, P.M.; Fraser, P.D. The Regulation of Carotenoid Formation in Tomato Fruit. *Plant J.* **2017**, *89*, 774–788. [CrossRef]
44. Khan, S.A.; Chibon, P.-Y.; de Vos, R.C.H.; Schipper, B.A.; Walraven, E.; Beekwilder, J.; van Dijk, T.; Finkers, R.; Visser, R.G.F.; van de Weg, E.W.; et al. Genetic Analysis of Metabolites in Apple Fruits Indicates an mQTL Hotspot for Phenolic Compounds on Linkage Group 16. *J. Exp. Bot.* **2012**, *63*, 2895–2908. [CrossRef]
45. Chatham, L.A.; Juvik, J.A. Linking Anthocyanin Diversity, Hue, and Genetics in Purple Corn. *G3 Genes | Genomes | Genetics* **2021**, *11*, jkaa062. [CrossRef]
46. Furukawa, T.; Maekawa, M.; Oki, T.; Suda, I.; Iida, S.; Shimada, H.; Takamure, I.; Kadowaki, K. The Rc and Rd Genes Are Involved in Proanthocyanidin Synthesis in Rice Pericarp. *Plant J.* **2007**, *49*, 91–102. [CrossRef] [PubMed]
47. Yang, W.; Chen, L.; Zhao, J.; Wang, J.; Li, W.; Yang, T.; Dong, J.; Ma, Y.; Zhou, L.; Chen, J.; et al. Genome-Wide Association Study of Pericarp Color in Rice Using Different Germplasm and Phenotyping Methods Reveals Different Genetic Architectures. *Front. Plant Sci.* **2022**, *13*, 841191. [CrossRef] [PubMed]
48. Sweeney, M.T.; Thomson, M.J.; Pfeil, B.E.; McCouch, S. Caught Red-Handed: Rc Encodes a Basic Helix-Loop-Helix Protein Conditioning Red Pericarp in Rice. *Plant Cell* **2006**, *18*, 283–294. [CrossRef]
49. Bouayed, J.; Hoffmann, L.; Bohn, T. Total Phenolics, Flavonoids, Anthocyanins and Antioxidant Activity Following Simulated Gastro-Intestinal Digestion and Dialysis of Apple Varieties: Bioaccessibility and Potential Uptake. *Food Chem.* **2011**, *128*, 14–21. [CrossRef]
50. García-Pérez, P.; Tomas, M.; Rivera-Pérez, A.; Patrone, V.; Giuberti, G.; Capanoglu, E.; Lucini, L. Exploring the Bioaccessibility of Polyphenols and Glucosinolates from Brassicaceae Microgreens by Combining Metabolomics Profiling and Computational Chemometrics. *Food Chem.* **2024**, *452*, 139565. [CrossRef]
51. Tomas, M.; Zhang, L.; Zengin, G.; Rocchetti, G.; Capanoglu, E.; Lucini, L. Metabolomic Insight into the Profile, In Vitro Bioaccessibility and Bioactive Properties of Polyphenols and Glucosinolates from Four Brassicaceae Microgreens. *Food Res. Int.* **2021**, *140*, 110039. [CrossRef]
52. Cotterill, P. Short Note: On Estimating Heritability According to Practical Applications. *Silvae Genet.* **1987**, *36*, 46–48.
53. Garrett, D.A.; Failla, M.L.; Sarama, R.J. Development of an In Vitro Digestion Method To Assess Carotenoid Bioavailability from Meals. *J. Agric. Food Chem.* **1999**, *47*, 4301–4309. [CrossRef]

54. Zoratti, L.; Karppinen, K.; Luengo Escobar, A.; Häggman, H.; Jaakola, L. Light-Controlled Flavonoid Biosynthesis in Fruits. *Front. Plant Sci.* **2014**, *5*, 534. [CrossRef]
55. Falcone Ferreyra, M.L.; Rius, S.P.; Casati, P. Flavonoids: Biosynthesis, Biological Functions, and Biotechnological Applications. *Front. Plant Sci.* **2012**, *3*, 222. [CrossRef]
56. Sytar, O.; Zivcak, M.; Bruckova, K.; Brestic, M.; Hemmerich, I.; Rauh, C.; Simko, I. Shift in Accumulation of Flavonoids and Phenolic Acids in Lettuce Attributable to Changes in Ultraviolet Radiation and Temperature. *Sci. Hortic.* **2018**, *239*, 193–204. [CrossRef]
57. Neugart, S.; Kläring, H.-P.; Zietz, M.; Schreiner, M.; Rohn, S.; Kroh, L.W.; Krumbein, A. The Effect of Temperature and Radiation on Flavonol Aglycones and Flavonol Glycosides of Kale (*Brassica oleracea* var. *sabellica*). *Food Chem.* **2012**, *133*, 1456–1465. [CrossRef]
58. Xu, C.; Mou, B. Responses of Spinach to Salinity and Nutrient Deficiency in Growth, Physiology, and Nutritional Value. *J. Am. Soc. Hortic. Sci.* **2016**, *141*, 12–21. [CrossRef]
59. Kaiser, E.; Kusuma, P.; Vialet-Chabrand, S.; Folta, K.; Liu, Y.; Poorter, H.; Woning, N.; Shrestha, S.; Ciarreta, A.; van Brenk, J.; et al. Vertical Farming Goes Dynamic: Optimizing Resource Use Efficiency, Product Quality, and Energy Costs. *Front. Sci.* **2024**, *2*. [CrossRef]
60. van Delden, S.H.; SharathKumar, M.; Butturini, M.; Graamans, L.J.A.; Heuvelink, E.; Kacira, M.; Kaiser, E.; Klamer, R.S.; Klerkx, L.; Kootstra, G.; et al. Current Status and Future Challenges in Implementing and Upscaling Vertical Farming Systems. *Nat. Food* **2021**, *2*, 944–956. [CrossRef]
61. Mitchell, C.A.; Dzakovich, M.P.; Gomez, C.; Lopez, R.; Burr, J.F.; Hernández, R.; Kubota, C.; Currey, C.J.; Meng, Q.; Runkle, E.S.; et al. *Light-Emitting Diodes in Horticulture*; Wiley: Hoboken, NJ, USA, 2015; Volume 43. [CrossRef]
62. Simko, I. Spatio-Temporal Dynamics of Lettuce Metabolome: A Framework for Targeted Nutritional Quality Improvement. *Plants* **2024**, *13*, 3316. [CrossRef]
63. Cartea, M.E.; Francisco, M.; Soengas, P.; Velasco, P. Phenolic Compounds in Brassica Vegetables. *Molecules* **2010**, *16*, 251–280. [CrossRef]
64. Cho, E.; Gurdon, C.; Zhao, R.; Peng, H.; Poulev, A.; Raskin, I.; Simko, I. Phytochemical and Agronomic Characterization of High-Flavonoid Lettuce Lines Grown under Field Conditions. *Plants* **2023**, *12*, 3467. [CrossRef]
65. Berlingeri, J.; Fuentes, A.; Ranario, E.; Yun, H.; Rim, E.Y.; Garrett, O.; Howard, A.; LaPorte, M.-F.; Lo, S.; Pauli, D.; et al. Integration of Crop Modeling and Sensing into Molecular Breeding for Nutritional Quality and Stress Tolerance. *Theor. Appl. Genet.* **2025**, *138*, 205. [CrossRef] [PubMed]
66. Ma, X.; Yu, L.; Fatima, M.; Wadlington, W.H.; Hulse-Kemp, A.M.; Zhang, X.; Zhang, S.; Xu, X.; Wang, J.; Huang, H.; et al. The Spinach YY Genome Reveals Sex Chromosome Evolution, Domestication, and Introgression History of the Species. *Genome Biol.* **2022**, *23*, 75. [CrossRef]
67. Hulse-Kemp, A.M.; Bostan, H.; Chen, S.; Ashrafi, H.; Stoffel, K.; Sanseverino, W.; Li, L.; Cheng, S.; Schatz, M.C.; Garvin, T.; et al. An Anchored Chromosome-Scale Genome Assembly of Spinach Improves Annotation and Reveals Extensive Gene Rearrangements in Euasterids. *Plant Genome* **2021**, *14*, e20101. [CrossRef]
68. Wojtunik-Kulesza, K.; Oniszczuk, A.; Oniszczuk, T.; Combrzyński, M.; Nowakowska, D.; Matwijczuk, A. Influence of In Vitro Digestion on Composition, Bioaccessibility and Antioxidant Activity of Food Polyphenols—A Non-Systematic Review. *Nutrients* **2020**, *12*, 1401. [CrossRef]
69. Hao, Y.; Yang, J.; Cui, J.; Fan, Y.; Li, N.; Wang, C.; Liu, Y.; Dong, Y. Stability and Mechanism of Phenolic Compounds from Raspberry Extract under In Vitro Gastrointestinal Digestion. *LWT* **2021**, *139*, 110552. [CrossRef]
70. Bouayed, J.; Deußer, H.; Hoffmann, L.; Bohn, T. Bioaccessible and Dialysable Polyphenols in Selected Apple Varieties Following In Vitro Digestion vs. Their Native Patterns. *Food Chem.* **2012**, *131*, 1466–1472. [CrossRef]
71. Redan, B.W.; Albaugh, G.P.; Charron, C.S.; Novotny, J.A.; Ferruzzi, M.G. Adaptation in Caco-2 Human Intestinal Cell Differentiation and Phenolic Transport with Chronic Exposure to Blackberry (*Rubus* sp.) Extract. *J. Agric. Food Chem.* **2017**, *65*, 2694–2701. [CrossRef] [PubMed]
72. Redan, B.W.; Chegeni, M.; Ferruzzi, M.G. Differentiated Caco-2 Cell Monolayers Exhibit Adaptation in the Transport and Metabolism of Flavan-3-Ols with Chronic Exposure to Both Isolated Flavan-3-Ols and Enriched Extracts. *Food Funct.* **2017**, *8*, 111–121. [CrossRef] [PubMed]
73. Goodrich, K.M.; Dorenkott, M.R.; Ye, L.; O'Keefe, S.F.; Hulver, M.W.; Neilson, A.P. Dietary Supplementation with Cocoa Flavanols Does Not Alter Colon Tissue Profiles of Native Flavanols and Their Microbial Metabolites Established During Habitual Dietary Exposure in C57BL/6J Mice. *J. Agric. Food Chem.* **2014**, *62*, 11190–11199. [CrossRef]

foods

MDPI

Article

In Vitro Gastrointestinal Bioaccessibility of the Phenolic Fraction from *Agave inaequidens* Flower

Imelda N. Monroy-García [1], Laura Lucely González-Galván [2], Catalina Leos-Rivas [2], Mayra Z. Treviño-Garza [2], Eduardo Sánchez-García [2] and Ezequiel Viveros-Valdez [2,*]

1 Departamento de Ingeniería Química y Bioquímica, Tecnológico Nacional de Mexico, Instituto Tecnológico de Los Mochis, Juan de Dios Bátiz y 20 de Noviembre, Los Mochis 81259, Sinaloa, Mexico; imelda.mg@mochis.tecnm.mx

2 Facultad de Ciencias Biológicas, Universidad Autónoma de Nuevo León, Av. Pedro de Alba S/N, San Nicolás de los Garza 66450, Nuevo León, Mexico; lucely.gonzalezg@uanl.edu.mx (L.L.G.-G.); catalina.leosrs@uanl.edu.mx (C.L.-R.); mayra.trevinogrz@uanl.edu.mx (M.Z.T.-G.); eduardo.sanchezgrc@uanl.edu.mx (E.S.-G.)

* Correspondence: jose.viverosvld@uanl.edu.mx

Abstract: Edible flowers are gaining recognition as rich sources of nutrients and phytochemicals. In Mexico, the flower of *Agave inaequidens* has been traditionally consumed since pre-Hispanic times. This study investigated its nutritional profile and the in vitro gastrointestinal bioaccessibility of its phenolic fraction. During in vitro digestion (oral, gastric, and intestinal), the total phenolic content of *A. inaequidens* significantly decreased from 138 to 21 mg GAE/100 g DW (15.22% bioaccessibility), while total flavonoid content dropped from 8 to 4.6 mg CE/100 g DW (57.5% bioaccessibility). Consequently, antioxidant activity, assessed by ABTS, DPPH, and hemolysis inhibition assays, also declined post-digestion. Interestingly, the digestive process modulated the flower's inhibitory activity against digestive enzymes before and after in vitro digestion: α-amylase inhibition slightly decreased (IC_{50} 1.8 to 2.1 mg/mL), but α-glucosidase (IC_{50} 2.7 to 1.6 mg/mL) and lipase (IC_{50} > 3 to 1.4 mg/mL) inhibition increased. The *A. inaequidens* flower is a good source of fiber and low in fat. These findings underscore its potential as a functional food ingredient, offering bioaccessible phenolic compounds with antioxidant and enzyme inhibitory properties.

Keywords: edible flowers; biological activity; enzyme inhibition; in vitro digestion

1. Introduction

Bioprospecting, defined as the systematic exploration of natural resources for valuable compounds, is crucial for discovering novel applications of plant-based materials across various industries, including food, cosmetics, pharmaceuticals, and agriculture [1]. Within the food sector, bioprospecting aids in identifying bioactive ingredients and developing functional foods that offer health benefits beyond basic nutrition through biologically active compounds, potentially reducing the risk of chronic diseases [2]. In relation to the above, edible flowers are increasingly recognized as integral components of such foods [3].

The global interest in edible flowers, or floriphagy, is expanding, notably in Mexico, due to their decorative, nutritional, and functional attributes [4]. Scientific research underscores their health-promoting potential, attributing it to essential vitamins, minerals, proteins, amino acids, and a wide range of bioactive compounds [5]. Among these, phenolic compounds are especially noted for their strong antioxidant activity [6].

The agave flower, known by regional names like flor de maguey, bayusas, gualumbos, and cacayas, is a traditional Mexican delicacy, deeply embedded in the culinary practices of several states [7]. Edible agave species include *A. salmiana*, *A. mapisaga*, *A. marmorata*, *A. durangensis*, *A. inaequidens*, *A. americana*, *A. applanata*, *A. lechuguilla*, and *A. striata*, among others [7]. Despite its traditional use, comprehensive research on the composition and biological activity of these flowers remains limited. Previous studies have analyzed the nutritional content of *A. salmiana*, revealing high levels of protein, fiber, and minerals [8]. Furthermore, *A. durangensis* and *A. americana* flowers have been identified as excellent sources of antioxidants [9,10]. However, detailed information regarding the specific functional properties of the agave flower is largely unexplored, necessitating further in-depth investigation.

It is well-known that flowers contain high amounts of phenols and flavonoids, two key groups of plant secondary metabolites known for their significant biological activities. In edible flowers, these compounds exhibit anti-inflammatory, anticancer, and antidiabetic properties [11]. The Agave genus is particularly rich in phenolic compounds, including flavonoids, homoisoflavonoids, and phenolic acids, which are associated with diverse biological activities such as antioxidant, antibacterial, antifungal, antinematode, and immunomodulatory effects [12]. Agave phenols also serve as valuable chemotaxonomic markers for species classification [12].

Furthermore, certain flavonoid-rich flowers have demonstrated the ability to inhibit key digestive enzymes, such as α-amylase and α-glucosidase, which is significant for glycemic control and potential antidiabetic effects [13]. Inhibition of lipase, another crucial digestive enzyme modulated by phenolic compounds, is linked to a reduced risk of cardiovascular diseases, hypertension, and obesity [13]. Therefore, it is hypothesized that agave flower extract will also exhibit inhibitory activity against lipase.

Since the efficacy of bioactive compounds depends on their bioaccessibility (their release from the food matrix during digestion and subsequent absorption) [14], this study evaluates the biological effects of enriched phenolic fraction from the agave flower before and after simulated digestion. This approach enables a more accurate assessment of its health potential. The research aims to uncover new insights into the antioxidant and enzyme-inhibitory properties of *Agave inaequidens* flowers, supporting their valorization as a functional food ingredient.

2. Materials and Methods

2.1. Plant Material Collection and Preparation

Agave flowers were collected in the locality of San Dimas (Durango, Mexico) during their flowering period (April–May). Flowers were harvested from three different agave plants of the same species at an intermediate stage of maturity, when the flowers were still closed and exhibited a greenish-yellow coloration. The taxonomic identification of the plant material was performed by Dr. Martha González Elizondo, Head of the CIIDIR Durango Herbarium, Durango, Mexico, (Voucher Number: 55902). Subsequently, a cleaning process was performed, which involved removing only the immature inflorescence (buds) and discarding the receptacle, peduncle, and branching. The cleaned flowers were maintained under frozen conditions until analysis. Given that agave flowers are traditionally consumed cooked, they were pre-cooked in water at 70 °C for 30 min under atmospheric conditions, prior to analysis.

2.2. Proximate Analysis

A proximate analysis was performed to identify the following parameters: moisture (NOM-116-SSA-1994), crude protein (NMX-F-608-NORMEX-2011), ether extract (NOM-086-SSA1-1994), fiber (NMX-F-613-NORMEX-2017), ash (NMX-F-607-NORMEX-2020), and nitrogen-free extract. The moisture and total solids content of the flower samples was determined by drying at 110 °C in an oven until a constant weight was reached (Thermo Scientific Heratherm, Waltham, MA, USA). Crude protein was quantified using the macro-Kjeldahl method, applying a nitrogen-to-protein conversion factor of 6.25. Total fat was extracted via a Soxhlet procedure. Ash content was measured following incineration of the samples at 525 °C for 24 h in a muffle furnace (Nabertherm L 3/11, Lilienthal, Germany). Total carbohydrates were calculated by subtracting the percentages of moisture, crude protein, total fat, and ash from 100%. Finally, the total energy content was estimated using Atwater conversion factors. Results were expressed as a percentage on a dry basis (DB) and wet basis (WB).

2.3. Extraction and Fractionation

The extraction method was based on the methodology described by Jimenez-Aspee et al., 2015 [15] with minor modifications. Briefly, pre-cooked flowers were comminuted to reduce particle size. This was followed by three sequential extractions using acidified methanol (MeOH) 0.01% at a 1:3 (*w*/*v*) ratio of plant material to solvent. Each extraction step included sonication for 15 min. The three extracts were then combined, and the solvent was removed under reduced pressure (rotary evaporation) at 37 °C to yield the first concentrated MeOH extract. A portion of this concentrated MeOH extract was subsequently adsorbed onto pre-conditioned Amberlite XAD-7-HP resin. The concentrated extract was dissolved in distilled water and thoroughly mixed with the Amberlite resin at a 1:5 (extract to Amberlite) ratio, with constant agitation for 40 min. The extract-resin mixture was then filtered and washed with water to eliminate undesired components, such as proteins, sugars, and lipids. Finally, MeOH was added to desorb the phenolic compounds from the resin. After recovering the phenolic compounds with MeOH, the solvent was evaporated under reduced pressure at 37 °C to obtain the second phenolic extract (PE). Finally, a fractionation step with ethyl acetate was performed. A portion of the PE was dissolved in distilled water, followed by liquid-liquid extraction using ethyl acetate at a 1:2 (*v*/*v*) ratio. The separated ethyl acetate fraction was dried over sodium sulfate and filtered, and the solvent was evaporated by simple distillation under reduced pressure at 37 °C to yield the third enriched phenolic extract (EPE).

The extraction yield for all four obtained extracts was calculated using the following formula:

$$\text{Yield (\%)} = (\text{Weight of extract}/\text{Weight of sample}) \times 100$$

2.4. In Vitro Gastrointestinal Digestion

The bioaccessibility of phenolic compounds during in vitro gastrointestinal digestion is hypothesized to be influenced by a complex interplay of factors, including food matrix interactions, chemical structure of phenolics (e.g., glycosylation, polymerization, lipophilicity), and digestive conditions (changes in pH, enzymes, bile salts). These factors significantly impact the stability, solubility, and potential degradation or transformation of phenolic compounds [16].

A portion of the EPE extract was subjected to in vitro gastrointestinal digestion, employing a three-phase sequential digestion system: oral, gastric, and intestinal. This methodology was based on the approach described by Morais et al., 2020 [16] with some modifications. A fraction of the EPE extract was vigorously mixed and dissolved in distilled water. For the oral phase, 25 mL of the sample were mixed with 5 mL of simulated oral fluid (2.38 g of Na_2HPO_4, 0.19 g of KH_2PO_4, 8 g of NaCl, and 200 U/L α-amylase dissolved in one liter of distilled water). This mixture was incubated under light protection with constant agitation at 90 rpm for 10 min at 37 ± 2 °C and pH = 7. The gastric phase was conducted by adding 1 mL of simulated gastric fluid (0.108 g of pepsin dissolved in 10 mL of 0.1 M HCl), and the pH was adjusted to 2.0 with 6 M HCl. The mixture was then re-incubated under the same conditions for 2 h. Finally, for the intestinal phase, 2.5 mL of the pancreatin solution (0.08 g of pancreatin dissolved in 10 mL of 0.5 M $NaHCO_3$) and 2.5 mL of the bile salts solution (0.5 g of bile salts were dissolved in 10 mL of 0.5 M $NaHCO_3$) were added. The pH was adjusted to 7 with 6 M NaOH. The mixture was then re-incubated under the same conditions for 2 h.

Upon completion of the digestion, a liquid-liquid extraction with ethyl acetate (100 mL × 3 times) was performed on the digested material. The phenolic compounds present in the ethyl acetate phase were concentrated by simple distillation under reduced pressure at 37 °C to obtain the fourth extract, the digested enriched phenolic extract (DEPE).

To determine the bioaccessibility of the bioactive compounds from the *A. inaequidens* flower, the following equation was used:

$$\text{Bioaccessibility (\%)} = (\text{Digested Compounds}/\text{Undigested Compounds}) \times 100$$

2.5. Total Phenolic Content

Total phenolic content was determined using the Folin–Ciocalteu (FC) method. This method relies on the transfer of electrons from phenolic groups to phosphomolybdic and phosphotungstic acids in an alkaline environment. Extracts (MeOH, PE, EPE, DEPE) were initially dissolved in distilled water (10 mg/mL) [17]. In triplicate, 100 μL of each extract were combined with 250 μL of 1N FC reagent, 1250 μL of 20% (*w*/*v*) Na_2CO_3, and 400 μL of distilled water. The mixtures were incubated for 2 h at room temperature in darkness. Subsequently, 200 μL of each reaction mixture were transferred to a 96-well microplate, and absorbance was measured at 760 nm using a microplate spectrophotometer (Agilent BioTek Epoch Microplate Spectrophotometer, Winooski, VT, USA). Gallic acid (10–1000 μg/mL) was used as a standard ($y = 10.329x + 0.1454$; $R^2 = 0.9952$), and results were expressed as mg of gallic acid equivalents (GAE) per 100 g of fresh weight.

2.6. Total Flavonoid Content

Total flavonoid content was quantified using the colorimetric aluminum chloride ($AlCl_3$) method, which involves the formation of stable aluminum-flavonoid complexes in a basic medium. The procedure was performed based on Monroy-García et al., 2024 [17]. Extracts (MeOH, PE, EPE, DEPE) were dissolved in MeOH (10 mg/mL). Subsequently, 200 μL of each extract were mixed with 1250 μL of distilled water and 75 μL of 5% (*w*/*v*) $NaNO_2$. The mixture was incubated for 5 min at room temperature, protected from light. Then, 150 μL of 10% (*w*/*v*) $AlCl_3$ were added, followed by a one min incubation. Finally, 500 μL of 1M NaOH and 325 μL of distilled water were added. Two hundred μL of the final solution were transferred to a 96-well microplate, and absorbance was measured at 510 nm using a microplate spectrophotometer (Agilent BioTek Epoch Microplate Spectrophotometer). Catechin (50–500 μg/mL) served as the standard ($y = 0.0124x + 0.0173$; $R^2 = 0.9995$), and results were expressed as mg of catechin equivalents (CE) per 100 g of fresh weight.

2.7. Antioxidant Activity

2.7.1. DPPH• Radical Scavenging Activity Assay

The 2,2-diphenyl-1-picrylhydrazyl (DPPH•) assay is a colorimetric method used to evaluate the free radical scavenging potential of antioxidant molecules. The procedure was conducted following the method proposed by Monroy-García et al., 2024 [17]. Extracts (MeOH, PE, EPE, DEPE) were initially dissolved in MeOH (1 mg/mL). Then, 100 µL of each serially diluted extract were added to wells of a 96-well microplate, followed by the addition of 100 µL of 150 µM DPPH• radical solution (80% MeOH). The mixtures were incubated for 30 min at room temperature, protected from light. Finally, absorbance was measured at 517 nm using a microplate spectrophotometer (Agilent BioTek Epoch Microplate Spectrophotometer). The absorbance values were converted to percentage of anti-radical activity (AA%) using the following formula:

$$AA\% = 100 - [(Am - Ac)/Ac] \times 100$$

where Ac and Am represent the absorbance of the control and the sample, respectively. Trolox was used as a positive control, and MeOH served as a negative control. The assay was performed in triplicate, and radical scavenging activity was expressed as half the maximal effective concentration (EC_{50}) in mg/mL.

2.7.2. ABTS•+ Radical Scavenging Ability Assay

The ABTS•+ radical scavenging ability assay is a spectrophotometric technique that assesses antioxidant capacity based on the inhibition of the 2,2′-azino-bis-3-ethylbenzothiazoline-6-sulfonic acid radical cation (ABTS•+). The procedure followed modified methods by Monroy-García et al., 2021 [17]. The ABTS•+ radical was prepared by mixing 7 mM ABTS in an aqueous solution with 2.45 mM potassium persulfate ($K_2S_2O_8$), followed by a 16 h incubation at room temperature in the dark. Once formed, the ABTS•+ radical mixture was diluted with distilled water until an absorbance of 0.7 ± 0.02 was achieved at 734 nm using a microplate spectrophotometer. Subsequently, the reactive mixture was prepared in a 96-well microplate by combining 20 µL of each extract (MeOH, PE, EPE, DEPE), dissolved in MeOH (1 mg/mL), with 200 µL of the ABTS•+ radical solution. The mixtures were incubated for 20 min in the dark, and, finally, absorbance was measured at 734 nm (Agilent BioTek Epoch Microplate Spectrophotometer). Trolox was used as a standard, and results were expressed as µM Trolox equivalents per 100 g of fresh weight.

2.7.3. Protective Hemolysis Erythrocyte Induced by the Radical 2,2′-Azobis(2-amidinopropane) Dihydrochloride (AAPH)

This method evaluates the protective effect of compounds against hemolysis induced by peroxyl radicals generated by AAPH. AAPH-induced oxidation of erythrocyte membrane lipids and proteins leads to hemolysis.

To obtain the blood, the following procedure was followed. The research protocol was reviewed and approved by the Department of Chemistry of the FCB/UANL, protocol code ATL-03-2023, approved in May 2023. Special consideration was given to participant privacy and data security. The participants received a comprehensive explanation of the study objectives and procedures and provided voluntary, written informed consent.

Initially, a 10 mL blood sample was collected from a healthy individual via venipuncture into EDTA-containing tubes. The blood sample was centrifuged at 1500 rpm for 10 min to separate red blood cells, white blood cells, and plasma. The red blood cells (erythrocytes) were then separated from the plasma and washed three times with 5 volumes of saline buffer (PBS; 120 mM NaCl, 22.21 mM Na_2HPO_4, 5.59 mM KH_2PO_4, and 5.5 mM glucose *w*/*v*) at pH 7.4. Each wash was followed by centrifugation at 1500 rpm for 2 min. Subsequently, an erythrocyte suspension was prepared in 4 volumes of PBS solution to achieve a concentration of 8×10^9 cells/mL [18].

The anti-hemolytic activity of the extracts (MeOH, PE, EPE, DEPE) was determined following a methodology adapted from Yang et al., 2017 [19]. Briefly, 250 μL of erythrocyte suspension were mixed with 250 μL of extracts diluted in PBS at various concentrations (250, 500, 1000, and 2000 μg/mL). Finally, 250 μL of 300 mM AAPH in PBS were added, and the mixture was gently shaken for 3 h at 37 °C. After incubation, the mixture was diluted to 8 volumes in PBS and centrifuged at 4000 rpm for 5 min. The absorbance of the supernatant was measured at 540 nm using a microplate spectrophotometer (Agilent BioTek Epoch Microplate Spectrophotometer). The percentage of inhibition was calculated using the following equation:

$$\% \text{ of inhibition} = [\text{AAAPH} - \text{Am}]/[\text{AAAPH}] \times 100$$

where AAAPH is the absorbance of AAPH at 540 nm, and Am is the absorbance of the extracts at 540 nm. The half maximal inhibitory concentration (IC_{50}) of the extract was calculated from a dose-response curve by plotting the percentage of hemolysis inhibition against the extract concentration. Ascorbic acid was used as a positive control, and plain water served as a negative control. Results were expressed as the IC_{50}.

2.8. *Inhibitory Activity of Digestive Enzymes*

2.8.1. α-Amylase Inhibitory Activity

The α-amylase inhibitory assay was conducted to assess the capacity of the extracts (MeOH, PE, EPE, DEPE) to inhibit starch hydrolysis by the α-amylase enzyme. The method was adapted from Burgos et al., 2017 [20]. Briefly, 50 μL of serially diluted extracts (5 mg/mL in PBS, pH 6.8) were added to a 96-well microplate, followed by 50 μL of α-amylase solution (1 U/mL in PBS). The mixture was incubated at 37 °C for 15 min. Subsequently, 50 μL of 0.5% (*w*/*v*) starch in PBS were added, and incubation continued for another 20 min at 37 °C. The reaction was terminated by adding 20 μL of 1M HCl, and, finally, 50 μL of iodine reagent (3 mM I_2 and 30 mM KI) were added. Absorbance was measured at 750 nm using a microplate spectrophotometer (Agilent BioTek Epoch Microplate Spectrophotometer). The percentage of inhibition was calculated using the following formula:

$$\% \text{ of activity inhibition} = [(\text{Ac} - \text{Am})/\text{Ac}] \times 100$$

where Ac and Am represent the absorbance of the control and the sample, respectively. Acarbose was used as a positive control, and PBS served as a negative control. Results were expressed as the half maximal inhibitory concentration (IC_{50}), calculated using logistic regression analysis.

2.8.2. α-Glucosidase Inhibitory Activity

The α-glucosidase inhibitory assay evaluated the ability of the extracts (MeOH, PE, EPE, DEPE) to inhibit the enzymatic hydrolysis of a synthetic substrate, *p*-nitrophenyl-α-D-glucopyranoside (*p*-NPG), by α-glucosidase. Following procedures by Monroy-García et al., 50 μL of serially diluted extracts (5 mg/mL in PBS, pH 6.8) were placed in a 96-well microplate [18]. Then, 50 μL of α-glucosidase solution (0.8 U/mL in PBS) were added, and the mixture was incubated at 37 °C for 15 min. Subsequently, 50 μL of 625 mM *p*-NPG substrate were added, followed by another 15 min incubation at 37 °C. The reaction was stopped by adding 100 µL of 0.2 M Na_2CO_3. Absorbance was measured at 405 nm using a microplate spectrophotometer (Agilent BioTek Epoch Microplate Spectrophotometer). The percentage of inhibition was calculated using the following formula:

$$\% \text{ of activity inhibition} = [(Ac - Am)/Ac] \times 100$$

where Ac and Am represent the absorbance of the control and the sample, respectively. Acarbose was used as a positive control, and PBS served as a negative control. Results were expressed as the half maximal inhibitory concentration (IC_{50}), determined via logistic regression analysis.

2.8.3. Pancreatic Lipase Inhibitory Activity

The inhibitory activity against pancreatic lipase was assessed using *p*-nitrophenol palmitate (*p*-NPP) as a substrate. For this analysis, pancreatic lipase was dissolved (20 mg/mL) in 60 mM PBS (pH 8) and centrifuged at 3500 rpm for 10 min at 4 °C to collect the supernatant for the assay. Following the modified method by Monroy-García et al., 100 µL of serially diluted extracts (5 mg/mL in PBS) were mixed with 30 µL of the lipase solution in a 96-well microplate and incubated for 15 min at 37 °C [18]. After incubation, 10 µL of 10 mM *p*-NPP (in DMSO) were added, and the mixture was re-incubated at 37 °C for 30 min. Absorbance was measured at 405 nm using a microplate spectrophotometer (Agilent BioTek Epoch Microplate Spectrophotometer). The percentage of inhibition was calculated using the following formula:

$$\% \text{ of activity inhibition} = [(Ac - Am)/Ac] \times 100$$

where Ac and Am represent the absorbance of the control and the sample, respectively. Orlistat was used as a positive control, and PBS served as a negative control. Results were expressed as the half maximal inhibitory concentration (IC_{50}), calculated through logistic regression analysis.

2.9. Statistical Analysis

All values were expressed as the mean ± standard deviation (n = 3). Differences between means were determined using factorial analysis of variance (ANOVA), followed by mean comparison using Student's t-test. Statistical analysis was performed using OpenStat statistical software (version 11.9.08). Statistical significance was set at $p < 0.05$.

3. Results and Discussion

The macronutrient content of cooked *A. inaequidens* flowers was determined, as boiling is a common practice for their preparation and consumption [7]. As shown in Table 1, the primary component was water (88.70%), followed by nitrogen-free extract (NFE) at 81.24%, fiber at 7.52%, ash at 6.11%, protein at 3.19%, and ether extract at 1.95%. Compared to raw *Agave salmiana* flowers, which reported 88.10% moisture, 9.65% fiber, 5.65% ash, 11.58% protein, 1.58% ether extract, and 71.58% NFE, the cooked *A. inaequidens* flowers showed no significant nutritional loss despite the cooking process [21]. The most notable differences were observed in protein (11.58% vs. 3.19%) and fiber (9.65% vs. 7.52%).

Table 1. Results of the proximate analysis of the agave flower.

Nutritional Information	Wet Basis (%)	Dry Basis (%)
Moisture	88.70 ± 12.6	-
Crude Fiber	0.85 ± 0.09	7.52 ± 0.98
Ash	0.69 ± 0.05	6.11 ± 0.52
Crude Protein	0.36 ± 0.02	3.19 ± 0.26
Crude Fat	0.22 ± 0.03	1.95 ± 0.16
Nitrogen-Free Extract	10.03 ± 1.2	81.24 ± 10.8
Energy content (Kcal/100g)	43.54 ± 2.8	

Values represent the samples mean ± SD (standard deviation, n = 3).

While cooking can alter nutrient concentration and bioavailability, leading to potential reductions in water-soluble components like proteins and minerals, the nutritional profile of boiled *A. inaequidens* flowers remained comparable to other raw edible flowers, such as *Tropaeolum majus*, *Tagetes erecta*, and *Rosa grandiflora*. These comparisons revealed similar values for ash (0.63–0.80% FW), ether extract (0.23–0.33% FW), and carbohydrates (7.14–14.15% FW), although protein (1.32–1.99% FW) and fiber (3.20–9.20% FW) content varied more significantly [22,23]. Furthermore, the macronutrient composition of boiled *A. inaequidens* flowers was comparable to that of commonly cooked vegetables like cabbage, green pepper, and carrots, which typically contain 0.77–1.27% protein, 0.31–0.59% ash, 0.06–0.20% ether extract, and 5.51–6.70% NFE on a WB [24].

The yield of Agave flower extracts and fractions is presented in Table 2. A decreasing trend in yield was observed with successive fractionation steps from the initial MeOH extract. The solid-liquid extraction with acidified MeOH (0.01%) showed the highest yield at 3.68%. These extracts are known to contain various phytochemicals, including phenolic compounds, carotenoids, coumarins, tannins, alkaloids, terpenoids, steroids, and carbohydrates [25,26]. Subsequent fractionation using Amberlite polymeric resin for phenolic compound adsorption yielded the phenolic extract (PE) with a lower yield of 0.45%. Further partitioning of the PE with ethyl acetate (EtOAc) to isolate more polar phenolic compounds resulted in a significantly reduced yield of 0.15%. Finally, the yield of the ethyl acetate extract after an in vitro digestion (DEPE) process was 0.03%.

Table 2. Percentage yield of Agave flower extracts and fractions.

Sample	Yield
Methanolic extract (MeOH)	3.68 ± 0.2
Enriched Phenolic Extract (PE)	0.45 ± 0.08
EE fraction in ethyl acetate (EPE)	0.15 ± 0.01
Digested fraction of EPE (DEPE)	0.03 ± 0.005

Values represent the samples mean ± SD (standard deviation, n = 3).

Phenolic compounds from plant sources are widely recognized for their diverse health-beneficial effects, including antioxidant, anti-inflammatory, and anticarcinogenic properties. However, the ultimate impact of these compounds on human health is heavily dependent on their bioaccessibility and bioavailability, which are significantly influenced by the complex processes occurring during gastrointestinal digestion [27,28].

During digestion, phenolic compounds undergo various transformations due to changes in pH, enzymatic activity, and interactions with other dietary components. These processes can lead to structural modifications, including hydrolysis, glycosylation, and polymerization. Such alterations can profoundly impact the biological activity of the compounds. For instance, some phenolic compounds may experience a reduction in their biological activity due to degradation or conversion into less active forms [29]. Conversely, digestive processes can sometimes enhance the activity of certain phenolics by releasing them from their complex matrices or transforming them into more bioavailable and potent metabolites [30].

The total phenolic content (TPC) and total flavonoid content (TFC) of *A. inaequidens* flower extracts, both before and after in vitro digestion, are presented in Table 3. Significant variations in TPC and TFC were observed across the extracts. TPC ranged from 16.4 to 46.1 mg GAE/g extract, while TFC ranged from 8.4 to 25.3 mg CE/g extract. Solvent type notably impacted extraction efficiency: the EPE exhibited the highest concentrations of both phenolics (46.1 mg GAE/g) and flavonoids (25.3 mg CE/g), while MeOH extract had the lowest (16.4 mg GAE/g and 8.4 mg CE/g, respectively). Despite the PE's lower soluble fraction compared to MeOH, its phenolic content was nearly threefold higher.

Evaluation of the PE after in vitro digestion (DPE) revealed a significant decrease in TPC from 46.1 to 7.6 mg GAE/g extract. TFC also decreased from 25.3 to 14.6 mg CE/g extract. Gastrointestinal conditions are known to decrease or transform phenolic compounds, though some may increase [16]. The observed bioaccessibility values for *A. inaequidens* flowers align with reported ranges for other edible flowers, where catechin bioaccessibility varies from 50–172%, quercetin from 1.7–70%, caffeic acid from 4–110%, and syringic acid from 6.6–30% [16]. Similarly, reported bioaccessibility for phenolics and flavonoids in fruit juices is approximately 12.8% and 20.1%, respectively, further supporting our findings for *A. inaequidens* [31].

Table 3. Total phenolic and flavonoid content and bioaccessibility percentage of phenolic compounds from undigested and digested Agave flower extracts.

Sample	Phenols [mg Eq de GA/g de Sample]	Flavonoids [mg Eq de Cat/g de Sample]
MeOH	c 16.4 ± 2.0	c 8.4 ± 0.8
PE	b 21.8 ± 3.0	b 11.2 ± 2.0
EPE	a 46.1 ± 7.0	a 25.3 ± 1.0
DEPE	d 7.6 ± 0.5	b 14.6 ±1.0
* FW	138.0 mg ± 21.0/100 g	8.0 mg ± 1.2/100 g
* DFW	21.0 mg ± 0.4/100 g	4.6 mg ± 0.3/100 g
Bioaccessibility (%)	15.22 ± 1.9	57.5 ± 2.5

Letters (a–d) represent significant differences ($p < 0.5$) between treatments. * Results expressed in fresh weight. MeOH: methanolic extract; PE: phenolic extract; EPE: enriched phenolic extract fractionated with ethyl acetate; DEPE: enriched digested phenolic extract, FW: fresh weight; DFW: digested fresh weight.

The TPC of *A. inaequidens* flowers (FW) was 138 mg GAE/100 g, decreasing significantly to 21 mg GAE/100 g after digestion. This fresh weight TPC is comparable to values reported for other edible flowers like sunflower (*Helianthus annuus*), morning glory (*Ipomoea cairica*), iris (*Iris japonica*), lily (*Lilium brownii*), bean flower (*Phaseolus vulgaris*), and sage (*Salvia splendens*), which range from 50 to 156 mg GAE/100 g FW [32]. However, some edible flowers, such as begonia (*Begonia × tuberhybrida*), nasturtium (*Tropaeolum majus*), marigold (*Calendula officinalis*), french marigold (*Tagetes patula*), jasmine (*Jasminum officinale*), and petunia (*Ruellia simplex*), exhibit considerably higher TPC (323–758 mg GAE/100 g FW) [25,33].

The TFC of *A. inaequidens* flowers was 8 mg CE/100 g FW, decreasing to 4.6 mg CE/100 g FW post-digestion. This value is comparable to basil flower (*Ocimum basilicum*) (3.99–21.72 mg CE/100 g FW, depending on region) [34]. In contrast, flowers like begonia (*Begonia boliviensis*), cornflower (*Centaurea cyanus*), rose (*Rosa odorata*), French marigold (*Tagetes patula*), nasturtium (*Tropaeolum majus*), and pansy (*Viola x wittrockiana*) report higher TFCs (135–204 mg rutin equivalents (RE)/100 g FW) [35].

While *A. inaequidens* flowers exhibit similar phenolic concentrations to some edible flowers, significant differences exist compared to others. Notably, the *A. inaequidens* flowers underwent two aggressive processing steps (boiling and in vitro digestion), which can lead to the loss of compounds, particularly water-soluble and thermolabile phenolics. This likely explains the observed lower phenolic content compared to some raw edible flowers and suggests that the *A. inaequidens* results more accurately reflect actual dietary intake. Furthermore, the phenolic content of *A. inaequidens* flowers was found to be higher than that of common cruciferous vegetables like Chinese cabbage (*Brassica rapa* var. chinensis), cabbage (*B. oleracea* var. capitata), cauliflower (*B. oleracea* var. botrytis), broccoli (*B. oleracea* var. italica), and red radish (*Raphanus sativus*), which reported TPC (uncooked) between 35 and 106 mg GAE/100g FW [36].

Antioxidants are crucial for delaying or preventing cellular oxidation. Diets rich in phenolic compounds exhibit significant antioxidant activity, primarily through free radical scavenging, interruption of radical chain reactions, and metal chelation [37]. To assess the antioxidant/anti-radical capacity of edible *A. inaequidens* flowers, three distinct methods were employed: ABTS•+ scavenging ability assay, DPPH• radical scavenging assay, and AAPH-induced hemolysis inhibition.

The results, presented in Table 4, indicate a clear correlation between the concentration of phenolics and flavonoids and antioxidant activity. Higher concentrations of these compounds, achieved through specific extraction methods, consistently led to increased antioxidant efficacy. In the TEAC assay, activity ranged from 19 to 42 μM/mg extract. The phenolic-enriched extract (EPE), with the highest phenolic concentration, demonstrated the strongest activity (42 μM/mg). After in vitro digestion, this activity significantly decreased to 8 μM/mg. Similarly, for the DPPH• radical assay, IC_{50} values ranged from 0.43 to 0.93 mg/mL extract. EPE showed the lowest IC_{50} (0.43 mg/mL), signifying superior activity. Post-digestion, the IC_{50} for EPE increased to 4.2 mg/mL, clearly indicating a reduction in activity. The AAPH-induced hemolysis assay was performed only on EPE and its digested counterpart (DEPE). EPE exhibited an IC_{50} of 70.43 μg/mL. Following in vitro gastrointestinal digestion, the IC_{50} increased to >100 μg/mL, confirming a notable decrease in antioxidant activity, consistent with the other assays.

Table 4. Antioxidant/anti-radical activity of undigested and digested agave flower extracts by three different methods.

Sample	ABTS [μM/mgExt]	DPPH (IC_{50}) [mg/mL]	AAPH (IC_{50}) [μg/mL]
MeOH	^{c} 19.0 ± 3.0	^{b} 0.93 ± 0.08	-
PE	^{b} 27.0 ± 2.0	^{c} 0.61 ± 0.07	-
EPE	^{a} 42.0 ± 7.0	^{d} 0.43 ± 0.05	^{b} 70.43 ± 2.8
DEPE	^{d} 8.0 ± 0.6	^{a} 4.2 ±0.1	>100
Trolox	-	^{e} 0.012 ± 0.01	-
Ascorbic acid	-	-	^{a} 19.37 ± 0.5
* FW	5.7 μM/100 g	-	-
* DFW	1.9 μM/100 g	-	-

Letters (a–e) represent significant differences ($p < 0.5$). * Results expressed in fresh weight. Trolox: positive control (polar analog of Vitamin E). Ascorbic acid: positive control. MeOH: methanolic extract; PE: phenolic extract; EPE: phenolic extract enriched with ethyl acetate; DEPE: enriched digested phenolic extract; FW: fresh weight; DFW: digested fresh weight.

These findings suggest that the digestion process significantly reduces both the concentration and antioxidant activity of phenolic compounds, aligning with previous research [27,38]. While antioxidant activity may increase during the acidic gastric phase, a considerable decrease often occurs during alkaline intestinal digestion. This is likely due to pH-induced structural changes, including racemization and deprotonation of hydroxyl groups on aromatic rings, which can alter biological responses [39]. Li et al. evaluated the structural stability and antioxidant activity of 27 phenolic compounds (phenolic acids, flavonols, flavonoids, and flavanones) during the in vitro simulated digestion. The results showed lower values of most phenolic compounds after in vitro digestion. Nevertheless, the DPPH, ABTS, and ferric reducing antioxidant power (FRAP) values of phenolic acids and flavonols decreased after in vitro simulated digestion ($p < 0.05$), while the values of DPPH, ABTS, and FRAP of most flavonoids ($p < 0.05$) increased [40]. Similar results were observed on phenolic extracts from olive mill wastewaters. After the digestion process, a massive loss of phenols was found (50–89.3%), coupled with a decrement in antioxidant activity (ABTS•+ and ORAC) [41].

The radical ABTS•+ scavenging ability assay of *A. inaequidens* (5.7 μM/100 g FW) is comparable to that of common vegetables like bitter gourd (7 μM/100 FW) and green cucumber (5 μM/100 FW) [42]. In the DPPH assay, Agave flower showed higher antioxidant activity than three out of four edible flower species analyzed by Pires et al., 2023 (rose: 0.18 mg/mL; dahlia: 0.63 mg/mL; centaurea: 0.83 mg/mL; calendula: 1.37 mg/mL) [43]. For the AAPH-induced hemolysis assay, *A. inaequidens* offered greater protection (IC_{50} of 70.43 μg/mL) before digestion compared to chamomile and pennyroyal (IC_{50} of 127.48 to 129.52 μg/mL) [44]. However, after digestion, the Agave flower's IC_{50} > 100 μg/mL became similar to these other edible flowers.

Metabolic syndrome, characterized by risk factors like obesity, dyslipidemia, and hyperglycemia, is often managed by inhibiting lipolytic enzymes (lipase) and carbohydrate-metabolizing enzymes (α-amylase and α-glucosidase) [45,46]. While commercial inhibitors like acarbose, voglibose, and orlistat are available, their use is associated with side effects [47]. Consequently, natural inhibitors, particularly phenolic compounds, are being investigated as safer alternatives due to their ability to non-specifically bind and inactivate these enzymes [48]. The inhibitory activity of agave flower extracts, both before and after in vitro digestion, are presented in Table 5. As can be observed, the extraction method influences the phenolic metabolites with enzyme-inhibitory activity. MeOH and PE extracts showed lower activity against the enzymes, while the EPE exhibited notable inhibitory effects.

Table 5. Half maximal inhibitory concentration (IC_{50}) of undigested and digested agave flower extracts on three digestive enzymes.

Sample	α-Amilase (IC_{50}) [mg/mL]	α-Glucosidase (IC_{50}) [mg/mL]	Lipase (IC_{50}) [mg/mL]
MeOH	>3.0	>3.0	>3.0
PE	>3.0	^{c} 1.2 ± 0.1	>3.0
EPE	^{b} 1.8 ± 0.1	^{a} 2.7 ± 0.6	>3.0
DEPE	^{a} 2.1 ± 0.4	^{b} 1.6 ± 0.2	^{a} 1.4 ± 0.2
* Control	^{c} 0.9 ± 0.1	^{d} 0.2 ± 0.05	^{b} 0.25 ± 0.07

Letters (a–d) represent significant differences ($p < 0.5$) between treatments. * Acarbose was used as a control for α-amylase and α-glucosidase, and orlistat was used for lipase. MeOH: methanolic extract; PE: phenolic extract; EPE: enriched phenolic extract fractionated with ethyl acetate; DEPE: enriched phenolic extract digested.

For α-amylase, EPE showed the highest inhibition (IC_{50} of 1.8 mg/mL), consistent with its higher phenolic content. However, after in vitro gastrointestinal digestion, the IC_{50} increased to 2.1 mg/mL, indicating reduced inhibitory activity. In contrast, α-glucosidase and lipase activities showed an opposite trend post-digestion, with an increase in inhibitory activity. EPE exhibited an α-glucosidase IC_{50} of 2.7 mg/mL, which decreased to 1.6 mg/mL after digestion. Similarly, lipase inhibition by EPE, initially with an IC_{50} > 3 mg/mL, improved to 1.4 mg/mL post-digestion.

This increase in activity is likely attributable to the transformation of phenolic glycosides, prevalent in plant foods, into their corresponding aglycones during digestion [49]. Aglycones, often more bioavailable and potent than their glycoside forms, exhibit stronger bioactivities, including antidiabetic effects [50]. In this regard, the flavonoid profile of the flowers of *A. durangensis* demonstrated the presence of five quercetin glycosides and three kaempferol glycosides; however, several in vitro studies have shown that quercetin and kaempferol aglycones are more potent enzyme inhibitors than their corresponding glycosides [9,51]. Aglycones are generally more lipophilic than their glycosides due to the absence of the polar sugar moiety. This increased lipophilicity facilitates their interaction with the hydrophobic active sites of enzymes, which can lead to stronger binding and, consequently, greater in vitro inhibitory activity [52]. In addition, the presence or absence of sugars at specific positions can influence the availability of hydroxyl groups or other structures crucial for interaction with amino acids in the enzyme's active site. In many cases, the addition of a glycosyl group can "block" or alter the molecule's conformation, reducing its ability to fit into the enzyme active site [53]. Thus, in vitro digestion appears to convert phenolic compounds in *A. inaequidens* flowers into more active aglycones, enhancing their α-glucosidase and lipase inhibitory capabilities.

Compared to other edible flowers and legumes, *A. inaequidens* demonstrates competitive or superior enzyme inhibitory activity. For instance, α-amylase inhibition by *A. inaequidens* (EPE IC_{50} = 1.8 mg/mL) is comparable to *Passiflora tripartita* (1.72 mg/mL), *Phaseolus vulgaris* (2.69 mg/mL), and *Glycine max* (2.25 mg/mL) [54,55]. For α-glucosidase, *A. inaequidens* (DEPE IC_{50} = 1.6 mg/mL) outperformed *Hylocereus megalanthus* (8.69 mg/mL), *Dregea volubilis* (3.78 mg/mL), *Cosmos sulphureus* (5.62 mg/mL), and *Bougainvillea glabra* (5.21 mg/mL) [55–57]. Similarly, *A. inaequidens* also demonstrated higher lipase inhibition (DEPE IC_{50} = 1.4 mg/mL) than *C. sulphureus* (4.6 mg/mL) and *B. glabra* (5.14 mg/mL) [57]. These findings suggest that *A. inaequidens* flowers possess significant potential as natural inhibitors of key digestive enzymes, especially after simulated digestion.

4. Conclusions

Our findings highlight the *A. inaequidens* flower as a highly promising candidate for development as a natural functional food. Its nutritional profile marked by high fiber content and low levels of calories and fat makes it well suited for inclusion in health-conscious diets. Although the digestive process significantly reduced the levels of phenolic compounds, flavonoids, and antioxidant activity, a substantial degree of bioaccessibility was retained, particularly for flavonoids. Interestingly, digestion had a differential effect on the flower's enzyme inhibitory properties: while its ability to inhibit amylase decreased, its inhibitory effects on glucosidase and lipase increased. These results suggest potential benefits for glycemic control and lipid metabolism. Overall, the evidence strongly supports the incorporation of *A. inaequidens* as a functional ingredient with valuable bioactive properties that may contribute to improved health outcomes.

Author Contributions: Bioassays, L.L.G.-G.; conceptualization, I.N.M.-G. and E.V.-V.; proximal methodology and supervision, C.L.-R. and M.Z.T.-G.; antioxidant assays and supervision, I.N.M.-G. and E.S.-G.; enzymatic inhibition assays and supervision, E.V.-V.; project administration, E.V.-V.; I.N.M.-G. and E.V.-V., writing—review and editing. All authors have read and agreed to the published version of the manuscript.

Funding: This research study was funded by ProACTI-UANL: 8-CAAFE-2024 and Proyectos de Investigación Científica, Desarrollo Tecnológico e Innovación TECNM: 23214.25-P.

Institutional Review Board Statement: The study was carried out in accordance with the Declaration of Helsinki and was approved by the Institutional Review Board of the Department of Chemistry of the FCB/UANL, protocol code: ATL-03-2023, approved in May 2023.

Informed Consent Statement: Informed consent was obtained from all subjects involved in the study.

Data Availability Statement: The original contributions generated for this study are included in the article; the data presented in this study are available on request from the corresponding author.

Acknowledgments: Author Laura Lucely Gonzalez-Galvan thanks CONAHCYT (Mexico) for her master scholarship, MEXICO.

Conflicts of Interest: The authors declare no conflicts of interest.

References

1. Singh, K.; Kumar, A.; Kumar, S.; Gairola, S. Bioprospecting of Plants for Phytochemicals: Important for Drugs. In *Phytochemical Genomics*; Swamy, M.K., Kumar, A., Eds.; Springer: Singapore, 2022; pp. 63–83. [CrossRef]
2. Sun, W.; Shahrajabian, M.H.; Lin, M. Research progress of fermented functional foods and protein factory-microbial fermentation technology. *Fermentation* **2022**, *8*, 688. [CrossRef]
3. Pensamiento-Niño, C.A.; Castañeda-Ovando, A.; Añorve-Morga, J.; Hernández-Fuentes, A.D.; Aguilar-Arteaga, K.; Ojeda-Ramírez, D. Edible flowers and their relationship with human health: Biological activities. *Food Rev. Int.* **2024**, *40*, 620–639. [CrossRef]
4. Wilczyńska, A.; Kukułowicz, A.; Lewandowska, A. Effect of packaging on microbial quality of edible flowers during refrigerated storage. *Pol. J. Food Nutr. Sci.* **2023**, *73*, 32–38. [CrossRef]
5. Monroy-García, I.N.; González-Galván, L.L.; Viveros-Valdez, E. Edible flowers as a source of bioactive compounds. In *Improving Health and Nutrition through Bioactive Compounds*; Segura Campos, M.R., Ed.; Woodhead Publishing: Cambridge, UK, 2025; pp. 169–190. [CrossRef]
6. Breda, C.; Nascimento, A.; Meghwar, P.; Lisboa, H.; Aires, A.; Rosa, E.; Ferreira, L.; Barros, A.N. Phenolic composition and antioxidant activity of edible flowers: Insights from synergistic effects and multivariate analysis. *Antioxidants* **2025**, *14*, 282. [CrossRef]
7. Figueredo, C.; Álvarez, G.; Cortés, L. Edible flowers commercialized in local markets of Pachuca de Soto, Hidalgo, Mexico. *Bot. Sci.* **2022**, *100*, 120–138. [CrossRef]
8. Sotelo, A.; López, S.; Basurto, F. Content of nutrient and antinutrient in edible flowers of wild plants in Mexico. *Plant Foods Hum. Nutr.* **2007**, *62*, 133–138. [CrossRef]
9. Barriada, L.; Almaraz, N.; Delgado, E.; Gallardo, T.; Ávila, J. Flavonoid composition and antioxidant capacity of the edible flowers of *Agave durangensis* (Agavaceae). *CyTA-J. Food* **2013**, *12*, 105–114. [CrossRef]
10. Enríquez Estrella, M.Á.; Enríquez Ujukam, T.J.; Serrano Torres, G.J.; Torres Rodriguez, S.H.; Cuadrado Ayala, D.G. Comportamiento antioxidante y polifenólico de una conserva de flor de cabuya negra (*Agave americana*). *Perfiles* **2024**, *1*, 54–60. [CrossRef]
11. Wen, H.; Li, S.; Wei, Y.; Dong, J.; Liang, Z.; Guo, L.; Zhang, Y. Yunnan edible flowers and their potential in future foods: Focus on ethnological applications, chemical and pharmacological research. *J. Future Foods* **2025**, *2*, 119–133. [CrossRef]
12. Almaraz-Abarca, N.; Delgado-Alvarado, E.A.; Ávila-Reyes, J.A.; Uribe-Soto, J.N.; González-Valdez, L.S. The phenols of the genus Agave (Agavaceae). *J. Biomater. Nanobiotechnol.* **2013**, *4*, 9–16. [CrossRef]
13. Zheng, J.; Lu, B.; Xu, B. An update on the health benefits promoted by edible flowers and involved mechanisms. *Food Chem.* **2021**, *340*, 127940. [CrossRef]
14. Barros, H.D.D.F.Q.; Junior, M.R.M. Phenolic compound bioavailability using in vitro and in vivo models. In *Bioactive Compounds*; Segura Campos, M.R., Ed.; Woodhead Publishing: Cambridge, UK, 2019; pp. 113–126. [CrossRef]
15. Jimenez-Aspee, F.; Thomas, S.; Schulz, A.; Ladio, A.; Theoduloz, C.; Schmeda, G. Antioxidant activity and phenolic profiles of the wild currant *Ribes magellanicum* from Chilean and Argentinean Patagonia. *Food Sci. Nutr.* **2015**, *4*, 595–610. [CrossRef]
16. Morais, J.; Sant'Ana, A.; Dantas, A.; Silva, B.; Lima, M. Antioxidant activity and bioaccessibility of phenolic compounds in white, red, blue, purple, yellow and orange edible flowers through a simulated intestinal barrier. *Food Res. Int.* **2020**, *131*, 109046. [CrossRef]

17. Monroy-García, I.N.; Carranza-Rosales, P.; Carranza-Torres, I.E.; Castro-Ochoa, L.D.; González-Villasana, V.; Islas-Rubio, A.R.; Viveros-Valdez, E. Antioxidant and biological activity of mexican madroño Fruit (*Arbutus arizonica*). *Foods* **2024**, *13*, 2982. [CrossRef]
18. Monroy-García, I.N.; Carranza-Torres, I.E.; Carranza-Rosales, M.P.; Oyón, M.; García-Estevez, I.; Viveros-Valdez, E. Phenolic profiles and biological activities of extracts from edible wild fruits *Ehretia tinifolia* and *Sideroxylon lanuginosum*. *Foods* **2021**, *10*, 2710. [CrossRef]
19. Yang, H.L.; Korivi, M.; Lin, M.K.; Chang, H.C.W.; Wu, C.R.; Lee, M.S.; Hseu, Y.C. Antihemolytic and antioxidant properties of pearl powder against 2, 2′-azobis (2-amidinopropane) dihydrochloride-induced hemolysis and oxidative damage to erythrocyte membrane lipids and proteins. *J. Food Drug Anal.* **2017**, *25*, 898–907. [CrossRef]
20. Burgos, A.; Jimenez, F.; Thomas, S.; Schmeda, G.; Theoduloz, C. Qualitative and quantitative changes in polyphenol composition and bioactivity of *Ribes magellanicum* and *R. punctatum* after in vitro gastrointestinal digestion. *Food Chem.* **2017**, *237*, 1073–1082. [CrossRef]
21. Pinedo-Espinoza, J.; Gutierrez-Tlahque, J.; Santiago-Saenz, Y.O.; Aguirre-Mancilla, C.L.; Reyes-Fuentes, M.; Lopez-Palestina, C.U. Nutritional composition, bioactive compounds and antioxidant activity of wild edible flowers consumed in semiarid regions of Mexico. *Plant Food Hum Nutr.* **2020**, *75*, 413–419. [CrossRef]
22. De Lima, F.; Rodrigues, M.; Fernando, H.; Farias, J.; Martins, L. Composición química de pétalos de flores de rosa, girasol y caléndula para su uso en la alimentación humana. *Cienc. Tecnol. Agropecu.* **2019**, *20*, 149–158. [CrossRef]
23. Navarro-González, I.; González-Barrio, R.; García-Valverde, V.; Bautista-Ortín, A.B.; Periago, M.J. Nutritional composition and antioxidant capacity in edible flowers: Characterization of phenolic compounds by HPLC-DAD-ESI/MSn. *Int. J. Mol. Sci.* **2015**, *16*, 805–822. [CrossRef]
24. Arroyo Uriarte, P.; Mazquiaran Bergera, L.; Rodríguez Alonso, P.; Valero Gaspar, T.; Ruiz Moreno, E.; Ávila Torres, J.M.; Varela Moreiras, G. *Informe de Estado de Situación Sobre "Frutas y Hortalizas: Nutrición y Salud en la España del S. XXI"*; Fundación Española de Nutrición: Madrid, Spain, 2018.
25. Kritsi, E.; Tsiaka, T.; Ioannou, A.; Mantanika, V.; Strati, I. In Vitro and In Silico Studies to Assess Edible Flowers' Antioxidant Activities. *J. Appl. Sci.* **2022**, *12*, 7331. [CrossRef]
26. Lu, B.; Li, M.; Yin, R. Phytochemical content, health benefits, and toxicology of common edible flowers: A Review (2000–2015). *Crit. Rev. Food Sci. Nutr.* **2015**, *56*, S130–S148. [CrossRef]
27. Grgić, J.; Šelo, G.; Planinić, M.; Tišma, M.; Bucić-Kojić, A. Role of the encapsulation in bioavailability of phenolic compounds. *Antioxidants* **2020**, *9*, 923. [CrossRef]
28. Wojtunik-Kulesza, K.; Oniszczuk, A.; Oniszczuk, T.; Combrzyński, M.; Nowakowska, D.; Matwijczuk, A. Influence of in vitro digestion on composition, bioaccessibility and antioxidant activity of food polyphenols—A non-systematic review. *Nutrients* **2020**, *12*, 1401. [CrossRef]
29. Karaś, M.; Jakubczyk, A.; Szymanowska, U.; Złotek, U.; Zielińska, E. Digestion and bioavailability of bioactive phytochemicals. *Int. J. Food Sci. Technol.* **2017**, *52*, 291–305. [CrossRef]
30. Lewandowska, U.; Szewczyk, K.; Hrabec, E.; Janecka, A.; Gorlach, S. Overview of metabolism and bioavailability enhancement of polyphenols. *J. Agric. Food Chem.* **2013**, *61*, 12183–12199. [CrossRef]
31. Rodríguez-Roque, M.J.; Rojas-Graü, M.A.; Elez-Martínez, P.; Martín-Belloso, O. In vitro bioaccessibility of health-related compounds from a blended fruit juice–soymilk beverage: Influence of the food matrix. *J. Funct. Foods* **2014**, *7*, 161–169. [CrossRef]
32. Li, A.N.; Li, S.; Li, H.B.; Xu, D.P.; Xu, X.R.; Chen, F. Total phenolic contents and antioxidant capacities of 51 edible and wild flowers. *J. Funct. Foods* **2013**, *6*, 319–330. [CrossRef]
33. Mlcek, J.; Plaskova, A.; Jurikova, T.; Sochor, J.; Baron, M.; Ercisli, S. Chemical, nutritional and sensory characteristics of six ornamental edible flowers species. *Foods* **2021**, *10*, 2053. [CrossRef]
34. Rashid, A.; Anwar, F.; Qadir, R.; Sattar, R.; Akhtar, M.T.; Nisar, B. Characterization and biological activities of essential oil from flowers of sweet basil (*Ocimum basilicum* L.) Selected from different regions of Pakistan. *J. Essent. Oil-Bear. Plants* **2023**, *26*, 95–107. [CrossRef]
35. Rop, O.; Mlcek, J.; Jurikova, T.; Neugebauerova, J.; Vabkova, J. Edible flowers: A new promising source of mineral elements in human nutrition. *Molecules* **2012**, *17*, 6672–6683. [CrossRef]
36. Li, Z.; Lee, H.W.; Liang, X.; Liang, D.; Wang, Q. Profiling of phenolic compounds and antioxidant activity of 12 cruciferous vegetables. *Molecules* **2018**, *5*, 1139. [CrossRef]
37. Moreno Muñoz, M.T. Nuevas Técnicas Electroquímicas Para la Determinación de la Capacidad Antioxidante en Extractos Alimentarios Basadas en el Método CUPRAC. Ph.D. Thesis, Universidad D Córdoba, Córdoba, Spain, 15 July 2021. Available online: https://helvia.uco.es/handle/10396/21901 (accessed on 10 May 2025).

38. Rodríguez-Roque, M.J.; Rojas-Graü, M.A.; Elez-Martínez, P.; Martín-Belloso, O. Soymilk phenolic compounds, isoflavones and antioxidant activity as affected by in vitro gastrointestinal digestion. *Food Chem.* **2013**, *136*, 206–212. [CrossRef]
39. Chen, G.L.; Chen, S.G.; Xie, Y.Q.; Chen, F.; Zhao, Y.Y. Total phenolic, flavonoid and antioxidant activity of 23 edible flowers subjected to in vitro digestion. *J. Funct. Foods* **2015**, *17*, 243–259. [CrossRef]
40. Li, C.X.; Wang, F.R.; Zhang, B.; Deng, Z.Y.; Li, H.Y. Stability and antioxidant activity of phenolic compounds during in vitro digestion. *J. Food Sci.* **2023**, *88*, 696–716. [CrossRef]
41. Mercatante, D.; Ansorena, D.; Taticchi, A.; Astiasarán, I.; Servili, M.; Rodriguez-Estrada, M.T. Effects of in vitro digestion on the antioxidant activity of three phenolic extracts from olive mill wastewaters. *Antioxidants* **2022**, *12*, 22. [CrossRef]
42. Deng, G.F.; Lin, X.; Xu, X.R.; Gao, L.L.; Xie, J.F.; Li, H.B. Antioxidant capacities and total phenolic contents of 56 vegetables. *J. Funct. Foods* **2013**, *5*, 260–266. [CrossRef]
43. Pires, T.C.S.P.; Dias, M.I.; Barros, L.; Calhelha, R.C.; Alves, M.J. Edible flowers as sources of phenolic compounds with bioactive potential. *Food Res. Int.* **2017**, *105*, 580–588. [CrossRef]
44. Mustapha Tacherfiout, M.; Kherbachi, S.; Kheniche, M.; Mattonai, M.; Degano, I. HPLC-DAD and HPLC-ESI-MS-MS profiles of hydroalcoholic extracts of *Chamaemelum nobile* and *Mentha pulegium*, and study of their antihemolytic activity against AAPH-induced hemolysis. *S. Afr. J. Bot.* **2022**, *150*, 678–690. [CrossRef]
45. Subramaniyan, V.; Hanim, Y.U. Role of pancreatic lipase inhibition in obesity treatment: Mechanisms and challenges towards current insights and future directions. *Int. J. Obes.* **2025**, *49*, 492–506. [CrossRef]
46. Proença, C.; Ribeiro, D.; Freitas, M.; Fernandes, E. Flavonoids as potential agents in the management of type 2 diabetes through the modulation of α-amylase and α-glucosidase activity: A review. *Crit. Rev. Food Sci. Nutr.* **2022**, *62*, 3137–3207. [CrossRef]
47. Choudhary, N.; Kalra, S.; Unnikrishnan, A.G.; Ajish, T.P. Preventive pharmacotherapy in type 2 diabetes mellitus. *Indian J. Endocrinol. Metab.* **2012**, *16*, 33–43. [CrossRef]
48. Rincon, N.; Tincon, J.; Acosta, J. Inhibición de la α-glucosidasa mediante flavonoides de origen natural como vía de control en el desarrollo de diabetes mellitus. *Biociencias* **2019**, *14*, 162–181. [CrossRef]
49. Barrios Silva, I.; Bravo Muñoz, J. Biodisponibilidad y Bioaccesibilidad de Polifenoles y Flavonoides. Bachelor's Thesis, TALCA Universidad, Talca, Chile, 2020. Available online: http://dspace.utalca.cl/handle/1950/12420 (accessed on 14 May 2025).
50. Guo, N.; Zhu, Y.W.; Jiang, Y.W.; Li, H.K.; Liu, Z.M. Improvement of flavonoid aglycone and biological activity of mulberry leaves by solid-state fermentation. *Ind. Crops Prod.* **2018**, *148*, 112287. [CrossRef]
51. Xiao, J. Dietary flavonoid aglycones and their glycosides: Which show better biological significance? *Crit. Rev. Food Sci. Nutr.* **2017**, *57*, 1874–1905. [CrossRef]
52. Křen, V. Glycoside vs. Aglycon: The Role of Glycosidic Residue in Biological Activity. In *Glycoscience*; Fraser-Reid, B.O., Tatsuta, K., Thiem, J., Eds.; Springer: Berlin/Heidelberg, Germany, 2008; pp. 2589–2644. [CrossRef]
53. He, M.; Zhou, X.; Wang, X. Glycosylation: Mechanisms, biological functions and clinical implications. *Signal Transduct. Target. Ther.* **2024**, *9*, 194. [CrossRef]
54. Tan, Y.; Chang, S.; Zhang, Y. Comparison of alpha-amylase, alpha-glucosidase and lipase inhibitory activity of the phenolic substances in two black legumes of different genera. *Food Chem.* **2017**, *214*, 259–268. [CrossRef]
55. Coral, E.; Calixto, M.; Soberón, M. Actividad inhibitoria in vitro de los extractos acuosos de los frutos de *Hylocereus megalanthus* y *Passiflora tripartita* var. mollisima sobre las enzimas α-amilasa y α-glucosidasa. *Rev. Soc. Quím. Perú* **2020**, *86*, 93–104. [CrossRef]
56. Das, B.; De, A.; Das, M.; Das, S.; Samanta, A. A new exploration of *Dregea volubilis* flowers: Focusing on antioxidant and antidiabetic properties. *S. Afr. J. Bot.* **2017**, *109*, 16–24. [CrossRef]
57. Kaisoon, O.; Konczak, I.; Siriamornpun, S. Potential health enhancing properties of edible flowers from Thailand. *Food Res. Int.* **2012**, *46*, 563–571. [CrossRef]

foods

Article

Bioaccessibility and Functional Food Potential of *Equisetum telmateia* Ehrh. Against Diabetes-Induced Kidney Disorders

Timur Hakan Barak [1,*], İnci Kurt-Celep [2] and Engin Celep [1]

[1] Department of Pharmacognosy, Faculty of Pharmacy, Acibadem Mehmet Ali Aydınlar University, Ataşehir, İstanbul 34752, Türkiye; engin.celep@acibadem.edu.tr

[2] Department of Biotechnology, Faculty of Pharmacy, İstanbul Okan University, Tuzla, İstanbul 34940, Türkiye; inci.celep@okan.edu.tr

* Correspondence: timur.barak@acibadem.edu.tr

Abstract: Various species from the genus *Equisetum* are recorded as food and folk medicine against both kidney complications and diabetes. *Equisetum telmateia* Ehrh. is documented as a folk remedy in Türkiye against several kidney disorders. This study was designed to evaluate the possible protective mechanisms of *E. telmateia* EtOH extract (ETE) against kidney disorders and diabetes through different routes, such as the prevention of ROS formation, inhibitory potential against various DM-related enzymes, and a reduction in the amount of the mediators leading to disorders in both systems at the cellular level. The objective was to achieve advanced precision for in vitro results while considering the effect of GIS on oral consumption. Both phytochemical and bioactivity studies were conducted before and after simulated digestion. The results showed that ETE is a rich source of flavonoids and phenolic acids. In addition, it has significant antioxidant and enzyme inhibitory potential. Treatment also yielded promising results at the cellular level for both antioxidative and inhibitor proteins, which may play a role in the pathogenesis of kidney disorders and diabetes. Following the in vitro digestion procedure, both the number of phytochemical ingredients and bioactivity parameters showed a considerable decreasing trend; however, the results are still significant enough to justify the traditional utilization of the genus *Equisetum*. This investigation demonstrated that ETE has noteworthy potential as a functional food for protection against diabetic kidney disease.

Keywords: *Equisetum telmateia*; LC-MS/MS; in vitro digestion; anti-diabetic; diuretic

1. Introduction

The single genus in the family Equisetaceae, also known as horsetail in English and as atkuyrugu or kirkkilit in Turkish, is *Equisetum* L. There are seven species in the Turkish flora, compared to about 32 identified species worldwide. They are typically perennial plants with silica-rich cell walls that thrive in humid and damp environments. *Equisetum telmateia* Ehrh. is widely distributed in continents such as Europe, West Asia, Northwest Africa, and North America [1]. Various studies have explored the potential of the *Equisetum* genus as a promising source of functional foods due to its diverse bioactive compounds [2]. Furthermore, extracts prepared from different *Equisetum* species have been proposed as valuable ingredients for nutraceutical formulations owing to their potential health benefits [3]. However, certain species, including *E. telmateia*, still require further investigation to assess their full potential as functional food sources. In addition, all species of *Equisetum* are important biological remedies in Turkish folk medicine against various diseases. Among these species, *E. telmateia* has an important place in Turkish folk medicine. It is used in Anatolia for various diseases such as cough, hemorrhoids, and asthma; to treat stomach pain or peptic ulcers; in the case of eczema, acne, and rheumatism; against the pain of fractures; to strengthen hair, skin, and nails against oral infections; against fungal and parasitic infections; and against heart failure and atherosclerosis. Together with other utilizations, *E. telmateia* is prominent as an ailment against urinary disorders. Ethnobotanical records showed that it is used as a diuretic, thus being used to pass kidney stones or sand, and against pyelonephritis, prostatic hypertrophy, and cystitis [4]. Previous studies demonstrated that species from the genus *Equisetum* are significantly rich in phenolic acids and flavonoids. It is known that these compounds are responsible for the potent antioxidative effects of numerous plants. In addition, plants that possess high contents of these compounds are known for their anti-inflammatory and anti-diabetic properties. Hence, several species from the genus *Equisetum* are recorded as anti-diabetic ailments in traditional medical systems, and various scientific studies have confirmed such anti-diabetic uses [5]. However, *E. telmateia* has not been investigated in that manner, to our knowledge.

In vitro studies are reliable, low-cost, and reproducible; thus, they are widespread in scientific investigations and serve as valuable starting points for the systemic assessment of a variety of bioactivities of plant extracts. However, the direct application of plant extracts in in vitro assays neglects the pharmacokinetic effects induced by the gastrointestinal system and overlooks the bioavailability phenomenon. Previous studies showed that these effects may lead to significant alterations in bioactivity. This outcome takes place due to significant modifications of phytochemical ingredients after the application of the simulated digestion process [6].

Diabetes is the most prevalent metabolic disease, affecting hundreds of millions of patients worldwide [7]. In addition, kidney problems are among the most widespread manifestations of diabetes. Various studies have shown that hyperglycemia-induced advanced glycation end product (AGE) formation plays a central role in the pathogenesis of diabetes-related kidney disorders. AGE accumulation on receptors (RAGE) is known to generate oxidative stress with increased ROS formation and inflammation, resulting in deleterious effects on diabetic conditions [8]. In addition, previous studies reported that increased ROS in the target tissues lead to the activation of NF-kB and AP-1, which are both involved in the pathogenesis of diabetic nephropathy [9]. Moreover, diuretics are commonly used in kidney disorders, and studies have shown that they may be beneficial in DM-related kidney disorders [10]. Thus, investigating herbal medicines for preventing ROS formation, inhibiting enzymes related to diuresis, and acting as molecular triggers in the pathway for diabetic kidney damage may provide accurate knowledge to verify their ethnobotanical utilizations.

In light of this information, the anti-diabetic and diuretic properties of ethanolic extracts from the aerial parts of *E. telmateia* were investigated using several in vitro assays. To obtain better accuracy, an in vitro human digestion simulation procedure was employed on the extract prior to chromatographic analysis and bioactivity studies. Even though in vivo assays may yield more accurate outcomes, economic and ethical considerations have increased the importance of the alternative methods. It is known that the bioactivity of plant extracts is based on their phytochemical ingredients. Thus, extensive LC-MS/MS analysis was employed on the extract and fractions obtained via in vitro gastric simulation. For the evaluation of the in vitro antioxidant potential, FRAP, TOAC, and DPPH assays were conducted on all fractions. In addition, α-amylase, α-glucosidase, PTP1B, advanced glycation end product (AGE), and urease inhibition assays were conducted on all fractions to evaluate the anti-diabetic and diuretic properties and variations after digestion simulation. Moreover, cell culture studies were employed to determine *Equisetum telmateia* ethanolic extract (ETEBFR) and its bioavailable fraction (ETEIN) on HEPG2 cells in order to elucidate its possible mechanism of action at the cellular level. A WST-1 assay was performed to determine the non-toxic concentration, and a DCFDA assay was conducted to determine the antioxidant capacity of the fractions at a cellular level. Furthermore, Western blot analysis was employed to determine the levels of important proteins involved in the apoptosis pathway, including phosphorus-NfκB and RAGE in the HEPG2 cell line, which are all induced with diabetes and kidney disorders.

2. Materials and Methods

2.1. Plant Material

Aerial parts of *Equisetum telmateia* Ehrh. were collected from Demirköy/Kırklareli by Dr. Engin Celep in September 2022. Botanical identification was completed by Dr. Ahmet Doğan, and the voucher specimens were added to the Marmara University Faculty of Pharmacy Herbarium (MARE 23084). Then, the plant material was dried and powdered before extraction. Next, 700 mL of 96% EtOH was added to 50 g of the powder and macerated on a shaker for 24 h. After that, the remaining EtOH was filtered through filter paper and EtOH was added on the remains. This procedure was repeated three times. Afterwards, the filtered parts were combined, and the EtOH was evaporated with a rotary evaporator and then lyophilized. At the end of the extraction, the dry extract (2.6 g 5.2%, yield) was weighed. The dry extract was stored at −20 °C until further analysis. All chemicals, enzymes and references used in the experiment protocols were purchased from Sigma Chemical Co. (St. Louis, MO, USA).

2.2. In Vitro Digestion Simulation Procedure

As mentioned previously, a human GI digestion simulation model was used for this study [11]. After the sample solution was added to 20 mL of the gastric environment, the mixture was incubated for two hours in a shaking water bath to simulate peristaltic movement. To represent the post-gastric (PG) sample, the mixture was set aside. The content that diffused into the dialysis membrane (i.e., modeling the absorption process) was named IN after two hours of incubation in intestinal medium with cellulose dialysis tubing and a sufficient amount of powdered $NaHCO_3$.

2.3. Phytochemical Profiling and LC-MS/MS Analysis

Fifty-three phytochemicals containing various phenolic compounds were analyzed according to a previously validated method using LC-MS/MS for all fractions [12]. A Nexera model UHPLC (Shimadzu) combined with tandem MS equipment was used to perform the quantitative investigations. Shimadzu Lab Solutions software (https://www.shimadzu.com/an/products/software-informatics/labsolutions-series/index.html accessed on 24 October 2024) was used to process the LC-ESI-MS/MS results. The column temperature was set to 40 °C. The elution gradient was composed of eluent A (water, 5 mM ammonium formate, and 0.1% formic acid) and eluent B (methanol, 5 mM ammonium formate, and 0.1% formic acid). The gradient elution profiles used were 20–100% B (0–25 min), 100% B (25–35 min), and 20% B (35–45 min). In addition, the solvent flow rate and injection volume were determined to be 0.5 mL/min and 5 μL, respectively.

2.4. In Vitro Phenolic and Antioxidant Profiling

The total phenolic and flavonoid contents of the samples were evaluated according to a previously described method. Folin–Ciocalteu reagent (FCR) was used, and the absorbance at 765 nm was measured to determine the total phenol content of all the fractions. After the incubation period, the total flavonoid content was measured at 415 nm with a previously described method [13]. Antioxidant capacity assays were carried out as previously explained [14]. Butylated hydroxytoluene (BHT) was used as the reference compound for DPPH radical scavenging activity, and the results are given as BHT equivalents. To determine the cupric reducing antioxidant capacity (CUPRAC), ascorbic acid was used as a reference compound, and the absorbance was measured with a 450 nm spectrophotometer. Ascorbic acid was similarly used as the reference antioxidant compound in the total antioxidant capacity (TOAC) assay. Following a 90-min incubation period, absorbance was measured at 695 nm. The ferric reducing antioxidant power (FRAP) was determined via a redox-linked colorimetric method, and the absorbance of the mixture was read at 593 nm.

2.5. Enzyme Inhibition Assays

Urease: The optimized protocol in the literature was used to detect the inhibitory effects of 1 mg/mL of the ETBFR, ETEIN, and ETEPG fractions on the urease enzyme. The sample group was obtained by mixing 100 µL of BFR, IN, and PG fractions at a concentration of 1 mg/mL with 15 µL of urea enzyme and 850 µL of 0.1 M urine. Unlike the sample groups, the blanks contained 0.1 M phosphate buffer (at pH 7.4) instead of the same volume of enzyme, while the control groups did not contain plant samples. A reference concentration of thiourea ranging from 62.5 to 1000 µg/µL was used in the experiment. After all the experimental groups were established, the samples were incubated for 60 min at 37 °C for pre-incubation. Then, 500 µL of solution A (50 mL of phenol and 2.5 g of sodium nitroprusside, prepared in distilled water) and 500 µL of solution B (50 mL of 250 mg sodium hydroxide and 820 µL of 5% sodium chloride) were added and kept for another 30 min at 37 °C, followed by an absorbance measurement at 640 nm [15].

α-Amylase: The α-amylase inhibitory activity was measured according to the method previously described by Inan et al. with slight modifications [16]. Twenty microliters of BFR, IN, and PG samples, obtained from the ETE and subjected to the gastric simulation method, and 10 µL of α-amylase (2 units/mL) in buffer solution were added to each well and allowed to incubate for 45 min at 37 °C. Then, 20 µL of starch solution (1% in water (*w*/*v*)) was added, and the mixture was incubated again. At the end of the incubation period, 3,5-dinitrosalicylic acid (DNS) solution was added, and the mixture was boiled at 90 °C for 20 min. The absorbance at 540 nm was read using a 96-well microplate reader. Acarbose was used as a positive control. The negative control was performed without adding the sample. The α-amylase inhibitory activity is expressed as the percentage of inhibition.

α-Glucosidase: The α-glucosidase inhibitory activity was determined according to a previously described procedure with minor modifications [16]. In each well of a 96-well plate, 20 µL of α-glucosidase and 20 µL of BFR, IN, and PG samples were placed in phosphate buffer and incubated for 15 min at 37 °C. Then, 20 µL of substrate (2.5 mM 4-nitrophenyl α-D-glucopyranoside) was added and allowed to incubate again. Next, 0.2 M Na_2CO_3 was added to terminate the reaction and the absorbance was read at 405 nm using a Thermo Scientific™ Varioskan 96-well microplate reader. Quercetin was used as a positive control. The negative control was performed without adding ETE.

PTP1B: PTP1B has a negative regulatory effect on insulin signaling through the phosphorylation of thyroxine residues in the insulin receptor downstream pathway. For this reason, a literature-presented and optimized experimental protocol was followed to detect the inhibitory effect of ETE on PTP1B. Using ursolic acid as a positive control in the experiment, the effect of ETE on the inhibition of PTP1B by the BFR, IN, and PG fractions obtained after a gastric simulation was calculated [17].

2.6. Advanced Glycation End Products (AGEs)

The glycated bovine serum albumin (BSA) formation method (with a minor modification) was conducted as previously described. In brief, BSA (10 mg/mL) was incubated with 0.5 M glucose in 0.1 M phosphate-buffered saline (PBS) containing sodium azide in the dark at 37 °C for 40 h. Before incubation, the gastrointestinal digestion products (BFR, IN, and PG) of ETE were added to the mixtures. Quercetin was used as a positive control. Glycose-BSA formation, also known as AGE formation, was measured in a Thermo Scientific™ Varioskan™ Lux multimode microplate reader using fluorescence intensity (FI) at a 370 nm excitation wavelength and a 450 nm emission wavelength After measurement, the previously given equation was used to calculate the AGE inhibition activity at different concentrations [18].

2.7. Cell Culture

The hepatocellular cell line HepG2 was selected as the target cell line for the insulin resistance (IR) model for the investigation of molecular pathways involved in diabetes metabolism. HepG2 cells were cultured for increasing durations (6, 12, 18, 24, 32, and 48 h) in a low-glucose medium supplemented with increasing insulin concentrations (0.005, 0.05, 0.5, 5, and 50 µmol) at various concentrations determined in previous studies on the same cell line. A commercially available HepG2 hepatocellular carcinoma cell line was purchased from the American Tissue Culture Collection (ATTC). Then, an IR model was created by treating the cells with 50 µmol of insulin every 24 h for insulin resistance. After developing IR in HepG2 cells, 20 µg/mL metformin was used as a positive control [19,20].

2.8. WST-1 Assay

After the formation of insulin resistance in commercial HepG2 cells supplied by ATCC, as described in Section 2.7, the WST-1 test was used to determine the non-toxic concentrations of ETEBFR and ETEIN in these cells. WST is a technique in which a color change occurs depending on the vitality of the cells; in other words, it depends on metabolic activity and can be measured spectroscopically. The cells were plated in a 96-well plate at 10,000 cells/well. The next day, increasing doses of ETEBFR and ETEIN were applied to the HepG2-IR cells for 24, 48, and 72 h. The control group consisted of HepG2-IR cells not treated with ETEBFR or ETEIN, and the cell viability was calculated at 100% confluency at all time points compared to the control group [21].

2.9. Western Blot Analysis

ETEBR and ETEIN were administered at 125 µg/mL separately for 48 h to insulin-resistant (IR) HepG2-IR cells to investigate the molecular pathways involved in diabetes, followed by variable metabolic syntheses of the phospho-NFκB p65, RAGE, and AP-1 proteins via the Western blot protocol previously described in the literature. β-actin was used as a loading control to ensure that the molecular weight of the proteins transferred to the membrane of 15% nitrocellulose was loaded with the same volume of cell lysates. The membranes were incubated overnight with the target antibodies. Then, the membranes were incubated with a horseradish peroxidase-conjugated secondary antibody at room temperature for two hours. After the incubation, a commercially available enhanced chemiluminescence (ECL) solution, known as a Western blotting detection reagent, was diluted 1:10 and photographed using a ChemiDOC device after 1 min of treatment with the membrane. The band densities of each protein were compared to beta-actin, and quantitative analysis was carried out using the ImageJ program [22].

2.10. DCFDA Assay

Following the development of the IR model in HepG2 cells, reactive oxygen species (ROS), which cause oxidative stress in liver cells, reached high levels and caused mitochondrial dysfunction. For this reason, the DCFDA method described previously was applied to detect the increased amount of ROS in HepG2-IR cells and the changing ROS levels in the cells after the application of ETEBFR and ETEIN [22].

2.11. Statistical Analysis

The experiments were performed at least in triplicate. The results are presented as the means ± standard deviations. Statistical comparisons were made using one-way analysis of variance (ANOVA) for the enzyme inhibition experiments, followed by 2-way ANOVA with a multiple comparisons test for the WST-1, Western blot, and DCFDA assays. A statistically significant difference was defined as $p < 0.05$.

3. Results and Discussion

3.1. Phytochemical Evaluation

A validated LC-MS/MS method was used for detailed qualitative and quantitative analyses of all fractions of the ETE. In this method, 53 different compounds were analyzed, with 21 different compounds detected in at least one of the fractions and 6 of them quantified in significant amounts (Table 1 and Figure 1). Kaempferol and its glycosides were detected as the major flavonoids in the main extract before digestion; the quantified amounts were 7.02 μg/g for kaempferol-3-O-rutinoside, 3.11 μg/g for kaempferol-3-O-glucoside, and 1.29 μg/g for kaempferol in its aglycon form. These findings correspond to our previous findings. Milovanovic et al. quantified flavonoid glycosides in five different species of *Equisetum* from Serbia and reported that *E. telmateia* has one of the richest contents of kaempferol and its glycosides [23]. Moreover, there are other reports demonstrating high amounts of kaempferol and its glycosides in *E. telmateia* from different regions [24,25]. However, after simulated digestion, the bioavailability of these flavonoids was relatively low. Kaempferol slightly decreased in the IN fraction and the amounts of kaempferol-3-O-rutinoside and kaempferol-3-O-glucoside significantly decreased. Parallel outcomes were observed in previous studies. Feng et al. showed that the kaempferol-3-O-glucoside content of *Nelumbo nucifera* Gaertn. leaves remained unchanged after gastric digestion but was significantly reduced after simulated intestinal digestion [26]. In addition, Bortolini et al. reported that the kaempferol content of Nastrutium decreased following a digestion simulation with zero bioavailability [27]. Protocatechuic acid and protocatechuic aldehyde were observed in significant amounts from the BFR fraction of the ETE (1.17 and 0.68 μg/g, respectively). In previous studies, significant protocatechuic acid contents were detected in *E. telmateia* extracts. Nunes et al. quantified 2.85 ± 0.41 mg/100 g dry weight protocatechuic acid from samples collected in Portugal, and Yeganegi et al. detected significant amounts of this acid in samples from Iran [24,28]. In contrast to kaempferol derivates, these phenolic compounds have shown resistance to in vitro gastric simulation and were quantified in equivalent amounts in the PG and IN fractions. Previous studies showed intricate results on the effects of in vitro digestion on protocatechuic acid content and its bioavailability. Hilary et al. investigated the effects of in vitro digestion on phenolic composition and bioaccessibility in different preparations from *Phoenix dactylifera* seeds [29]. The results showed conflicting alterations in protocatechuic acid content, with a significant increment detected in the protocatechuic acid content in seed extracts, while a strong reduction was seen in seed bread. On the contrary, equivalent amounts were measured for seed powder. An interpretation of this can be that the bioaccessibility and durability of phenolic ingredients are highly dependent on complex interactions between the ingredients of the extracts and the unique characteristics of plant parts. Furthermore, epicatechin was found to be absent in the BFR fraction, while after the gastric digestion procedure, a significant amount was measured in the PG fraction; it was also detected in lower amounts in the bioaccessible phase (5.63 and 1.13 μg/g, respectively). It has been shown in the literature that *E. telmateia* is rich in proanthocyanidins [25], and these compounds consist of condensations of catechins, including epicatechin [30]. The abrupt occurrence of epicatechin in the PG phase together with its absence in the extract may result from the disruptive gastric environment with a low pH, high temperature, and enzymes leading to the disintegration of proanthocyanidins into their monomers, such as epicatechin. These rich phenolic profiles and their alterations after the simulated digestion are expected to immensely influence the bioactivities of the extract. The same alterations were also observed in the total phenol and flavonoid content assays. According to the results of the in vitro evaluation, the total phenolic and flavonoid amounts were decreased significantly after the simulated digestion and showed lower bioavailability (Table 2).

Table 1. Phytochemicals of ETE and its fractions based on LC-MS/MS analysis.

No	Analytes	RT [a]	M.I. (m/z) [b]	F.I. (m/z) [c]	Ion. Mode	Amounts (μg/g)		
						BFR	*PG*	*IN*
1	Quinic acid	3.0	190.8	93.0	Neg	0.262	0.36	0.351
2	Fumaric aid	3.9	115.2	40.9	Neg	N.D.	N.D.	N.D.
3	Aconitic acid	4.0	172.8	129.0	Neg	0.232	0.377	0.224
4	Gallic acid	4.4	168.8	79.0	Neg	N.D.	N.D.	N.D.
5	Epigallocatechin	6.7	304.8	219.0	Neg	N.D.	N.D.	N.D.

Table 1. *Cont.*

No	Analytes	RT [a]	M.I. (m/z) [b]	F.I. (m/z) [c]	Ion. Mode	Amounts (μg/g)		
						BFR	*PG*	*IN*
6	Protocatechuic acid	6.8	152.8	108.0	Neg	1.17	1.48	1.18
7	Catechin	7.4	288.8	203.1	Neg	N.D.	N.D.	N.D.
8	Gentisic acid	8.3	152.8	109.0	Neg	N.D.	N.D.	N.D.
9	Chlorogenic acid	8.4	353.0	85.0	Neg	0.046	0.052	0.046
10	Protocatechuic aldehyde	8.5	137.2	92.0	Neg	0.676	1.06	0.90
11	Tannic acid	9.2	182.8	78.0	Neg	N.D.	N.D.	N.D.
12	Epigallocatechin gallate	9.4	457.0	305.1	Neg	N.D.	N.D.	N.D.
13	1,5-dicaffeoylquinic acid	9.8	515.0	191.0	Neg	N.D.	N.D.	N.D.
14	4-OH Benzoic acid	10.5	137.2	65.0	Neg	N.D.	N.D.	N.D.
15	Epicatechin	11.6	289.0	203.0	Neg	N.D.	5.63	1.13
16	Vanilic acid	11.8	166.8	108.0	Neg	N.D.	N.D.	N.D.
17	Caffeic acid	12.1	179.0	134.0	Neg	0.099	0.14	0.097
18	Syringic acid	12.6	196.8	166.9	Neg	N.D.	N.D.	N.D.
19	Vanillin	13.9	153.1	125.0	Poz	N.D.	N.D.	N.D.
20	Syringic aldehyde	14.6	181.0	151.1	Neg	N.D.	N.D.	N.D.
21	Daidzin	15.2	417.1	199.0	Poz	N.D.	N.D.	N.D.
22	Epicatechin gallate	15.5	441.0	289.0	Neg	N.D.	N.D.	N.D.
23	Piceid	17.2	391.0	135/106.9	Poz	N.D.	N.D.	N.D.
24	*p*-Coumaric acid	17.8	163.0	93.0	Neg	N.D.	N.D.	N.D.
25	Ferulic acid-D3-IS [h]	18.8	196.2	152.1	Neg	N.A.	N.A.	N.A.
26	Ferulic acid	18.8	192.8	149.0	Neg	N.D.	N.D.	N.D.
27	Sinapic acid	18.9	222.8	193.0	Neg	N.D.	N.D.	N.D.
28	Coumarin	20.9	146.9	103.1	Poz	0.064	0.083	0.052
29	Salicylic acid	21.8	137.2	65.0	Neg	N.D.	N.D.	N.D.
30	Luteolin 7-O-β-D-glucoside	23.7	447.0	284.0	Neg	0.187	0.167	0.053
31	Quercetin 3-glucuronide	24.1	477.0	150.9	Neg	N.D.	N.D.	N.D.
32	Rutin-D3-IS [h]	25.5	612.2	304.1	Neg	N.A.	N.A.	N.A.
33	Rutin	25.6	608.9	301.0	Neg	0.083	0.102	0.076
34	Isoquercitrin	25.6	463.0	271.0	Neg	0.062	0.077	0.048
35	Hesperidin	25.8	611.2	449.0	Poz	0.033	0.041	0.024
36	*o*-Coumaric acid	26.1	162.8	93.0	Neg	N.D.	N.D.	N.D.
37	Genistin	26.3	431.0	239.0	Neg	N.D.	N.D.	N.D.
38	Rosmarinic acid	26.6	359.0	197.0	Neg	N.D.	N.D.	N.D.
39	Ellagic acid	27.6	301.0	284.0	Neg	N.D.	N.D.	N.D.
40	Apigenin 7-glucoside	28.2	431.0	269.0	Neg	0.035	0.026	0.013
41	Quercitrin	29.8	447.0	301.0	Neg	N.D.	N.D.	N.D.
42	Kaempferol 3-O-glucoside	30.4	447.0	255.0	Neg	3.11	3.49	1.70
43	Kaempferol-3-O-rutinoside	30.6	592.9	255.0/284.0	Neg	7.02	8.41	5.34
44	Fisetin	30.6	285.0	163.0	Neg	N.D.	N.D.	N.D.
45	Daidzein	34.0	253.0	223.0	Neg	N.D.	N.D.	0.173
46	Quercetin-D3-IS [h]	35.6	304.0	275.9	Neg	N.A.	N.A.	N.A.
47	Quercetin	35.7	301.0	272.9	Neg	0.036	N.D.	N.D.
48	Naringenin	35.9	270.9	119.0	Neg	0.242	0.107	0.036
49	Hesperetin	36.7	301.0	136.0/286.0	Neg	N.D.	N.D.	N.D.
50	Luteolin	36.7	284.8	151.0/175.0	Neg	0.098	0.033	N.D.
51	Genistein	36.9	269.0	135.0	Neg	N.D.	N.D.	0.003
52	Kaempferol	37.9	285.0	239.0	Neg	1.285	0.258	0.02
53	Apigenin	38.2	268.8	151.0/149.0	Neg	0.049	0.008	N.D.
54	Amentoflavone	39.7	537.0	417.0	Neg	N.D.	N.D.	N.D.
55	Chrysin	40.5	252.8	145.0/119.0	Neg	N.D.	N.D.	N.D.
56	Acacetin	40.7	283.0	239.0	Neg	0.026	N.D.	N.D.

[a] N.D.: not detected; N.A.: not applicable; IS: internal standard; RT: retention time; [b] MI: multiple ion; [c] FI: fragment ion; [h]: internal standards

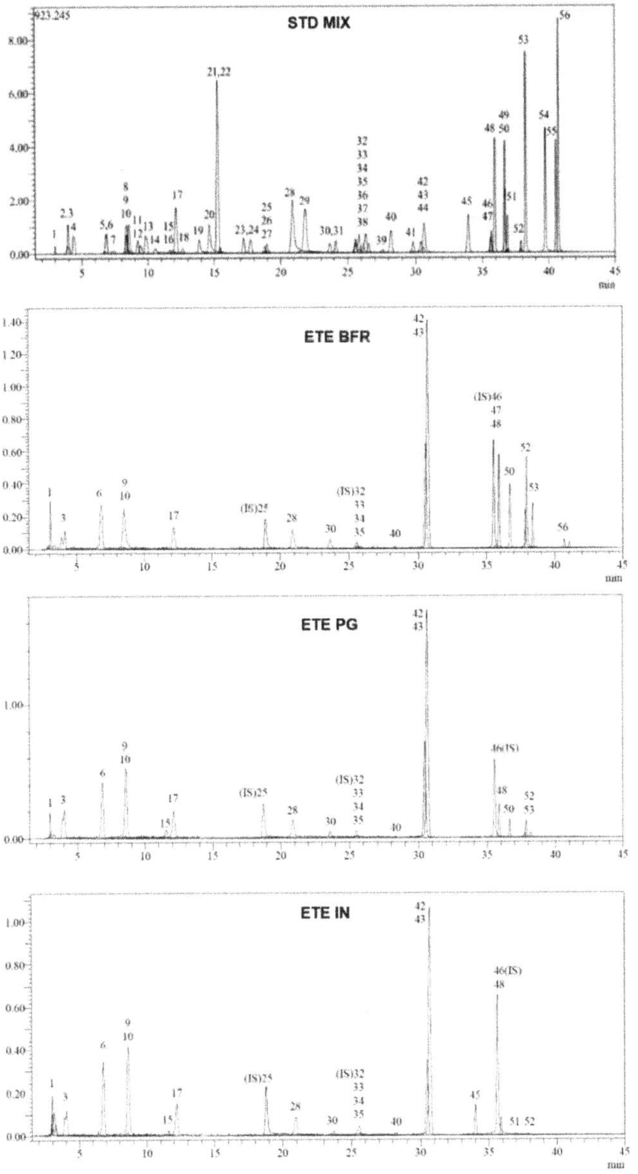

Figure 1. LC/MS/MS chromatogram of the samples and the mixture of standards.

Table 2. Results of in vitro evaluation for phytochemical profile.

	ETE BFR [A]	ETE PG	ETE IN
TPC [BC*]	300 ± 7 [a]	263 ± 11 [b]	190 ± 5 [c]
TFC [D]	93.7 ± 0.3 [a]	21.3 ± 0.1 [b]	7.7 ± 0.01 [c]

[A] Abbreviations = BFR: non-digested; PG: post-gastric; IN: bioavailable. [B] Abbreviations = TPC: total phenolic content; TFC: total flavonoid content. [C] Results are expressed as the mean of triplicates ± standard deviation (S.D.) and as mg gallic acid equivalents (GAE) in 1 g sample. [D] Results are expressed as the mean of triplicates ± standard deviation (S.D.) and as mg quercetin equivalents (QE) in 1 g sample. * Different lowercase letters in the same row indicate significance ($p < 0.05$).

3.2. Evaluation of Antioxidant Capacity

It is well established that elevated oxidation with oxidative stress leads to increased ROS formation, which plays an indisputable role in the pathogenesis of various conditions, including diabetes and kidney disorders [31]. Thus, revealing the antioxidative properties of herbal extracts may provide key information about their possible role in preventing chronic complications. For this reason, several in vitro and cell culture assays were conducted in this study to determine the antioxidant capacity of the ETE before and after simulated digestion. Similarly, numerous studies have been carried out on the antioxidant potential of *Equisetum* species, including *E. telmateia* [5]. Stajner et al. investigated the antioxidant potential of three different species of *Equisetum* from Serbia using several different assays, including FRAP and DPPH, and their results showed that *E. telmateia* extracts showed the greatest potential among all the studied extracts [32,33]. Likewise, other studies have shown the significant antioxidant potential of different in vitro methods in *E. telmateia* [24,28,34]. Even though the antioxidant potential of *E. telmateia* has been shown through various in vitro studies, none of these studies considered the effect of the gastrointestinal tract on the extracts or the bioavailability of the active ingredients. Hence, in the current study, in vitro assays such as DPPH, TOAC, FRAP, and CUPRAC were conducted on fractions of ETE considering the effects of the GI tract and bioaccessibility. The results given in Table 3 show that the ETE showed significant in vitro antioxidant potential prior to digestion; however, the bioactivity decreased significantly through the steps of the digestion simulation. DPPH radical scavenging activity was measured as 272 $\pm$ 1 mg BHTE/g dry extract, but it was significantly reduced after gastric digestion in the bioaccessible fraction. The same results were observed in the TOAC, FRAP, and CUPRAC assays: after gastric and intestinal digestion, TOAC decreased from 853 $\pm$ 15 to 329 $\pm$ 18, FRAP decreased from 1.83 $\pm$ 0.14 to 0.38 $\pm$ 0.03, and CUPRAC from 387 $\pm$ 12 to 86.6 $\pm$ 1.5 in the bioavailable fraction.

Table 3. In vitro preliminary tests for the antioxidant potential.

	ETE BFR ^{A}	ETE PG	ETE IN
DPPH ^{BC*}	272 $\pm$ 1 ^{a}	140 $\pm$ 5 ^{b}	72 $\pm$ 3 ^{c}
TOAC ^{D}	852 $\pm$ 15 ^{a}	398 $\pm$ 11 ^{b}	329 $\pm$ 18 ^{c}
FRAP ^{E}	1.82 $\pm$ 0.14 ^{a}	1.23 $\pm$ 0.04 ^{b}	0.37 $\pm$ 0.03 ^{c}
CUPRAC ^{D}	387 $\pm$ 12 ^{a}	253 $\pm$ 3 ^{b}	86 $\pm$ 2 ^{c}

^{A} Abbreviations = BFR: non-digested; PG: post-gastric; IN: bioavailable. ^{B} Abbreviations = DPPH: 2,2-diphenyl-1,1-picrylhydrazyl; TOAC: total antioxidant capacity; FRAP: ferric reducing antioxidant power; CUPRAC: cupric reducing antioxidant capacity. ^{C} Results are expressed as the mean of triplicates $\pm$ standard deviation (S.D.) and as mg BHT equivalents (BHTE) in 1 g sample. ^{D} Results are expressed as the mean of triplicates $\pm$ standard deviation (S.D.) and as mg ascorbic acid (AAE) in 1 g sample. ^{E} Results are expressed as the mean of triplicates $\pm$ standard deviation (S.D.) and as mM $FeSO_4$ equivalents in 1 g sample. * Different lowercase letters in the same row indicate significance ($p < 0.05$).

These results are analogous to the alterations in the phytochemical profile, where similar reductions were observed in terms of phenolics. As expected, phenolics are responsible for the antioxidant potential of ETE, and decreases in their amounts also led to a reduction in the bioavailability of the extract.

For better accuracy, the current study also investigated the antioxidant potential of ETE at the cellular level with the DCFDA assay, a frequently used method because of its specific ROS staining properties. HepG2 cells treated with H_2O_2 for extensive ROS formation were used for the assay, and the reduction in fluorescence in the cell culture after 24 and 48 h was measured in comparison to metformin, which is a standard anti-diabetic agent; the results are given in Figure 2. A high fluorescence intensity was measured in the non-treated cells at both time points. The standard substance, ETEBFR, and ETEIN significantly reduced the fluorescence, indicating antioxidant bioactivity at the cellular level. ETEBFR at a dose of 125 μg/mL had weaker bioactivity than did metformin at 20 μg/mL. However, the bioaccessible fraction showed a lower yet significant antioxidant capacity at both time points. These observations are consistent with the alterations in the phytochemical profile and in vitro antioxidant capacities of the fractions. The DCFDA assay has been used to determine the cellular antioxidant levels of various herbal medicines [22]. Although this is the first study to investigate the cellular antioxidant capacity of *E. telmateia* to our knowledge, several previous studies have investigated the differentiation of the cellular antioxidant capacities of herbal extracts before and after simulated digestion. Ge et al. investigated the antioxidant activity of an ethanol extract from Ligusticum after being subjected to digestion simulation [35]. A significant reduction in the cellular antioxidant level was reported after digestion simulation, which is consistent with our findings. In light of these results, it was established that ETE had significant antioxidant capacity, even though this potential decreased after digestion, and the bioaccessible fraction still showed significant antioxidant bioactivity in both in vitro studies and at the cellular level when compared with the results from the literature.

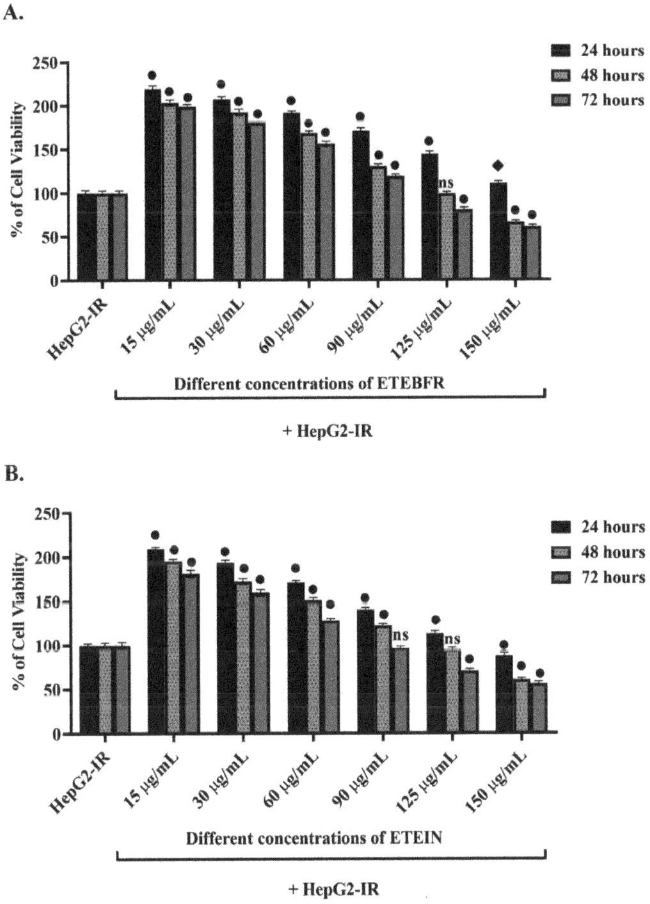

Figure 2. Determination of non-toxic concentrations of ETEBFR and ETEIN extracts in insulin-resistant HepG2 (HepG2-IR) cells by WST-1. (**A**) The % value of cell viability with increasing concentrations of ETEBFR applied to HepG2-IR cells for 24, 48, and 72 h. (**B**) The % value of cell viability with increasing concentrations of ETEIN applied to HepG2-IR cells for 24, 48, and 72 h. HepG2-IR cells were used as the control cell group and analyzed using 2-way ANOVA with multiple comparisons in the GraphPad Prism program. • Statistically significant value at $p \leq 0.0001$; ♦ statistically significant at $p \leq 0.001$. ns = statistically similar results (not significant).

3.3. In Vitro Enzyme and AGE Inhibition Studies

There are various ethnobotanical records of the effects of several species from the *Equisetum* genus on kidney disorders and their diuretic potential. In addition, *Equisetum telmateia* is used for treating kidney disorders such as kidney stones, nephritis, and urinary tract infections in folk medicine in various countries, including Türkiye, Spain, Italy, and Bosnia [5]. It is well known that diuretics may be beneficial for treating kidney disorders, especially kidney stones. Moreover, diuretic utilization of *E. telmateia* in traditional medicine in Türkiye has also been recorded [36]. Thus, an anti-urease inhibition assay was conducted on ETE to determine the potential diuretic properties of the extract. It was previously shown that different extracts of *E. telmateia* showed significant urease inhibition activity [37]. However, that study disregarded pharmacokinetic alterations of the extract after oral consumption. Hence, in this study, all the fractions of ETE were investigated along with in vitro digestion simulation for more accurate determination of the in vitro anti-urease inhibition potential, and the results are given in Table 4. The results show that ETE notably inhibited urease enzyme activity before digestion (92.5 ± 1.9%). However, a significant reduction was detected after the digestion simulation. Nonetheless, significant inhibition was still detected, which provides a meaningful foundation for traditional utilization.

Table 4. In vitro enzyme and AGE inhibition results.

	ETE BFR ^{AB}	ETE PG	ETE IN	STD ^{C}
Urease ^{C*}	92.5 ± 1.9 ^{a}	78.8 ± 1.6 ^{b}	87.3 ± 1.6 ^{c}	99.1 ± 0.2 ^{d}
α-Amylase	76.6 ± 2.2 ^{a}	46.4 ± 0.9 ^{b}	60.9 ± 0.9 ^{c}	94.6 ± 0.9 ^{d}
α-Glucosidase	82.6 ± 2.0 ^{a}	55.9 ± 0.3 ^{b}	71.6 ± 2.1 ^{c}	90.8 ± 1.0 ^{d}
PTP1B ^{D}	79.0 ± 1.4 ^{a}	61.8 ± 1.4 ^{b}	66.8 ± 1.9 ^{c}	96.7 ± 2.1 ^{d}
AGE	87.6 ± 3.6 ^{a}	62.2 ± 1.6 ^{b}	71.5 ± 0.9 ^{c}	95.3 ± 2.9 ^{d}

^{A} Abbreviations = BFR: non-digested; PG: post-gastric; IN: bioavailable. ^{B} All results are given as percent inhibition ± standard deviation. ^{C} All standard substances and samples were used at 1 mg/mL concentration. Standard substances: thiourea for urease assay, acarbose for α-amylase assay, quercetin for α-glucosidase and AGE assay, and ursolic acid for PTP1B assay. ^{D} PTP1B: Protein tyrosine phosphatase 1B; AGE: advanced glycation end product. * Different lowercase letters in the same row indicate significance ($p < 0.05$).

Diabetes is a widespread chronic disease that leads to various complications in the human body. Some of the most prevalent and malicious outcomes of diabetes are kidney disorders. Thus, diabetic control is essential for kidney homeostasis. The anti-diabetic potential of the genus *Equisetum* has been shown in various studies. In one study, a methanolic extract of *E. arvense* was studied on streptozotocin-induced diabetic male rats, and the results showed that blood glucose and urinary microalbumin and creatinine levels were reduced significantly [38]. Moreover, *E. giganteum* and *E. myriochaetum* showed significant anti-diabetic activity in vivo [39,40]. α-Amylase and α-glucosidase enzymes are important targets for diabetes control due to their significance in postprandial diabetes [41]. Likewise, PTP1B is an important target enzyme in the control of diabetes, which is a negative regulator of insulin pathway signaling and is deliberated as a likely potential therapeutic target [42]. For the above reasons, the inhibition potential of ETE against α-amylase, α-glucosidase, and PTP1B enzymes was investigated in the present study. Previous studies showed that another *Equisetum* species, namely *E. arvense*, showed significant potential for inhibiting α-amylase and α-glucosidase enzymes [43]. This is the first study investigating the α-amylase and α-glucosidase enzyme inhibition potential of *E. telmateia* and the PTP1B inhibitory activity of a species in the *Equisetum* genus to our knowledge. The results from Table 4 show that ETE had significant inhibitory activity against the three aforementioned enzymes before the simulated digestion. Although a significant decrease was observed after the in vitro simulation process, noteworthy inhibitory activity was still detected in both the PG and IN phases. These reductions are consistent with the alterations in the phenolic profile and antioxidant activity. It is well known that phenolic ingredients are generally considered as active secondary metabolites of plant extracts due to their antioxidant and enzyme inhibitory bioactivities [44]. Thus, slight reductions may be explained via these phytochemical alterations.

Advanced glycation end products (AGEs) are formed non-enzymatically between carbonyl groups of sugars and amino groups of some macromolecules. The most important harmful effect of AGEs is the generation of oxidative stress and aggregation of inflammation. AGEs are associated with various pathological conditions such as diabetes and its complications such as diabetic kidney disease [8]. Thus, inhibiting AGE formation is a valuable target to preventing the damage of hyperglycemia on kidney health. Accordingly, in this study, the AGE-inhibiting potential of the ETE before and after in vitro digestion was investigated. The results demonstrated that the ETE had a notable AGE inhibition ability, similar to the standard substance quercetin (Table 4). Corresponding to former results, this activity slightly diminished yet was still noteworthy following the digestion procedure. When the results of the in vitro inhibition studies are evaluated altogether, ETE appears to have valuable in vitro potential for utilization against diabetes and related kidney problems, which is in agreement with ethnopharmacological records.

3.4. Determination of Non-Toxic ETE Concentration

The WST-1 test (Figure 3) was performed to detect the non-toxic concentrations of ETEBFR and ETEIN sub-extracts obtained after gastric simulation via the ETE method was applied to HepG2-IR model cells. HepG2-IR cells were treated with ETEBFR and ETEIN sub-extracts for 24, 48, and 78 h. Metformin at 20 µg/mL, which is used to overcome insulin resistance and was previously optimized in HepG2-IR cells, was used as a positive control [20]. When the non-toxic concentrations of ETEBFR and ETEIN in Figure 3 were compared to the untreated HepG2-IR control cells, the viability of the cells appeared to decrease at an acceptable rate due to the increase in concentration at each time point. As shown in Figure 3A, ETEBFR at 150 µg/mL increased the cellular vitality by 10.35% at the end of the 24th hour, which then decreased to 66% after 48 h and to 61.04% after the 72nd hour. As shown in Figure 3B, these values were 87.5%, 60.43%, and 56%, respectively. A non-toxic concentration was determined for progressive experiments with a concentration-dependent decrease in the vitality rate of not less than 60% in the cells. In this case, 125 µg/mL was selected as a promising concentration for both ETEBFR and ETEIN. *Equisetum* species, which are commonly used in folk medicine, have been found to be free of cytotoxic effects [45]. The antimicrobial activity and toxicity of *Equisetum hyemale* extracts have been shown both in vitro and in vivo [5]. Several studies in the literature support our findings, as shown in Figure 3. In a study by Alavarce et al., no toxic effects were observed as a result of increased concentrations of extracts obtained from aerial parts of *E. giganteum*, both in human palatine epithelial cells and in human monocyte cells [46]. Similar results were obtained in another study involving a clinical trial of *Equisetum bogotense*, in which 0.75 g of herbal infusion was administered daily to 25 people and no side effects were revealed other than toxicity [47]. Our findings, which are consistent with these data in the literature, suggest that ETE is a promising extract for the treatment of insulin resistance without causing cytotoxicity at the end of gastric simulation and may be beneficial for kidney health.

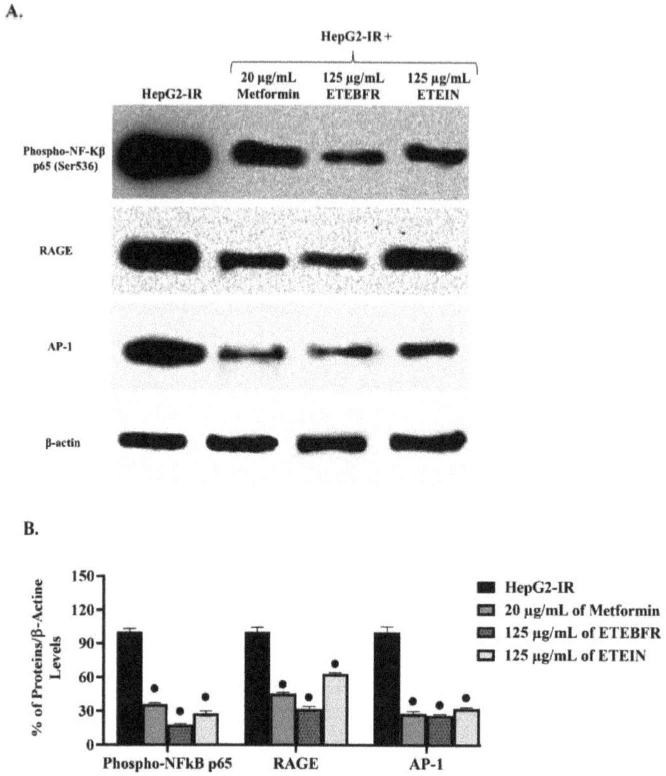

Figure 3. Detection of the changing protein levels in the molecular mechanism following the application of ETEBFR and ETEIN extracts at a non-toxic concentration of 125 µg/mL to HepG2-IR cells by Western blot. (**A**) Membrane image showing expression of phospho-NfκB, RAGE, and AP-1 proteins in the cellular pathway of HepG2-IR cells treated with 20 µg/mL of metformin (positive control), 125 µg/mL of ETEBFR, and 125 µg/mL of ETEIN extracts. (**B**) Statistical analysis of each protein compared to beta-actin, which has its own loading control. • Significant at $p \leq 0.0001$.

3.5. Determination of Alterations in Protein Synthesis via Western Blot

After the application of 125 µg/mL ETEBFR and ETEIN to HepG2-IR cells, Western blotting (Figure 3) was performed to determine the changes in phospho-NFκB, RAGE, and AP-1 protein synthesis in these cells. After the experiments were independently repeated three times, the proteins were quantitatively analyzed using the β-actin ratio, which is the loading control of each protein. Phospho-NFκB was reduced by 35.88% in metformin-treated cells compared to HepG2-IR control cells, and similar decreases were found for RAGE (45%) and AP-1 (28.06). On the other hand, after ETEBFR application, phospho-NFkB protein levels decreased by 18%, RAGE levels decreased by 32.06%, and AP-1 protein decreased by 27%. Similar results were obtained following the application of BFRIN to HepG2-IR cells: phospho-NfκB decreased by 28%, RAGE by 63.07%, and AP-1 by 32.55%. It is well known that creating insulin resistance in HepG2 cells causes instability and the accumulation of cellular reactive oxygen species (ROS) [48]. An increase in ROS leads to the development of many metabolic disorders, especially inflammation, in cells. In addition, ROS are an insidious factor underlying the pathogenic effects of insulin resistance, hyperglycemia, and type 2 diabetes. Phospho-NfκB, RAGE, and AP-1 proteins play crucial roles in regulating this molecular mechanism and restoring healthy insulin metabolism [49]. Studies have shown that NfκB is inactive in balanced metabolic pathways and under normal conditions, depending on the presence of IKB in the cytosol. However, NfκB is translocated to the nucleus and transformed into phospho-NfκB due to the phosphorylation of IKB in cases of inflammation in cells, increased ROS levels, the deterioration of insulin metabolism, type 2 diabetes, atherosclerosis, and many other metabolic disorders [50]. As a result of the accumulation of phospho-NF-κB, the active form of phospho-NFκB, in the nucleus, it has been found that it especially impairs the function of pancreatic β-cells and increases the production of lipotoxic products such as ceramide and diacylglycerol, which damage vascular reactivity and mitochondrial metabolism and cause obesity by inducing muscle mass [51]. In parallel with this information, the application of ETEBFR and ETEIN to HepG2-IR cells indicated that these compounds are candidates for natural phospho-NFκB inhibitors because they reduce the level of phospho-NFkB protein.

As presented in Table 1, due to the high phytochemical content of ETE and its antioxidant capacity and AGE inhibition, the ROS balance, which was disrupted by reducing the level of phospho-NFκB in the molecular pathway, was restored. It has been revealed that ROS-induced insulin resistance can be eliminated by decreasing RAGE and AP-1 protein levels. It has been demonstrated that the increase in RAGE, the receptor of AGE, which occurs as a result of the irreversible reaction between reducing sugars and amino groups of proteins, lipids, and nucleic acids, is involved in the development of type 2 diabetes, hyperglycemia, and insulin resistance. It is known that the stress pathway is triggered in the endoplasmic reticulum by the binding of AGEs to the cell receptor RAGE, after which the expression of pro-inflammatory molecules such as interleukin-6 (IL-6) and tumor necrosis factor alpha (TNF-α) increases and ROS-induced diabetes develops. The AGE/RAGE interaction results in insulin metabolism impairment as well as glomerular hypertrophy and nephropathy [52]. The fact that the sub-fractions of ETEBFR and ETEIN obtained as a result of the gastric digestion model applied to ETE have AGE inhibition activity as well as the fact that the RAGE level decreased after these extracts were applied to HepG2-IR cells are shown in our results in Figure 3. These data show that healthy insulin metabolism may be restored by preventing the formation of the AGE/RAGE complex of ETE. On the other hand, the increased ROS level as a result of the AGE/RAGE interaction causes the activation of some transcription factors, such as activator protein-1 (AP-1) and cytokines, in the diabetes and inflammation pathways [53]. AP-1 activation, which is controlled by the mitogen-activated protein kinase (MAPK) and NFκB pathways, may lead to various diseases such as diabetic retinopathy, cancer, asthma, psoriasis, renal dysfunction, and liver disorders [54]. In accordance with this information, HepG2 cells have increased levels of phospho-NfκB and AP-1 proteins, which are over-synthesized after IR development, as described before, and play a crucial role in the development of diabetes. Studies supporting the findings in this study with other extracts have shown that after application, the levels of nuclear NfκB and AP-1 proteins in cells decrease [55]. After the administration of ETEBFR and ETEIN to the HepG2-IR cell model, significantly lower levels of the phospho-NFκB, RAGE, and AP-1 proteins were detected in these cells than in control cells. These results indicate that the ETE can be used to treat insulin resistance and diabetic kidney disorders in a gastric digestion model, which is consistent with ethnobotanical claims. It should be noted that the limited number of phenolics studied here can be regarded as a limitation to our study. In addition, confirming the results with in vivo studies may be the subject of the upcoming research.

4. Conclusions

Our present study investigated the benefits of ETE on diabetes and kidney disorders through various mechanisms at the cellular level. First, to increase the accuracy of the bioactivity assays by considering the possible pharmacokinetic effects on oral consumption, a human digestion simulation procedure was conducted on the ETE. To elucidate the relationship between chemical compounds and bioactivity, an extensive LC-MS/MS procedure was used on the extracts and fractions, together with in vitro phytochemical assays, and major phenolic compounds and their alterations were determined through simulated digestion. In vitro antioxidant screening was accomplished via several methods, supported by the DCFDA assay to evaluate antioxidant potential at the cellular level. Various enzyme inhibition assays related to ethnobotanical utilization were performed, and alterations in protein synthesis were investigated. This study confirmed that even after gastric digestion, the bioavailable part of ETE has significant potential for inhibiting various enzymes, including urease, α-amylase, α-glucosidase, and PTP1B; reducing AGE formation; and reducing the synthesis of proteins such as AP-1, NFκB, and RAGE, which are related to the pathogenesis of diabetes and kidney disorders. In conclusion, the present study supports the ethnobotanical use of *Equisetum telmateia* Ehrh. in traditional medicine in Türkiye.

Author Contributions: Methodology, T.H.B.; investigation, T.H.B. and İ.K.-C.; resources, T.H.B. and E.C.; writing—original draft; T.H.B. and İ.K.-C.; writing—review and editing, E.C.; visualization, İ.K.-C.; supervision, E.C.; project administration, T.H.B.; funding acquisition, T.H.B. All authors have read and agreed to the published version of the manuscript.

Funding: This project was funded by Acıbadem Mehmet Ali Aydınlar University Commission of Scientific Projects (ABAPKO: E-2022/04-02).

Institutional Review Board Statement: Not applicable.

Informed Consent Statement: Not applicable.

Data Availability Statement: The original contributions presented in the study are included in the article, further inquiries can be directed to the corresponding author.

Acknowledgments: We would like to thank Mustafa Abdullah Yılmaz at the Research and Application of Science and Technology Center (DUBTAM), Dicle University, for the support on chromatographic analysis and Ahmet Doğan for the botanical identification of the plant material.

Conflicts of Interest: The authors declare no conflicts of interest.

References

1. Husby, C.E.; Walkowiak, R.J. An Introduction to the Genus Equisetum (Horsetail) and the Class Equisetopsida (Sphenopsida) as a whole. *Plant Sci. Bull.* **2012**, *1*, 2–10.
2. Salazar-Campos, J.; Salazar-Campos, O.; Gálvez-Ruiz, O.; Gavidia-Chávez, H.; Gavidia-Chávez, M.; Irigoin-Guevara, L.; Obregón-Domínguez, J. Functional Properties and Acceptability of Potentially Medicinal Tea Infusions Based on *Equisetum arvense, Desmodium molliculum,* and *Mentha piperita. Prev. Nutri. Food Sci.* **2023**, *28*, 444. [CrossRef] [PubMed]
3. Chaiyana, W.; Punyoyai, C.; Somwongin, S.; Leelapornpisid, P.; Ingkaninan, K.; Waranuch, N.; Mueller, M. Inhibition of 5α-reductase, IL-6 secretion, and oxidation process of *Equisetum debile* Roxb. ex vaucher extract as functional food and nutraceuticals ingredients. *Nutrients* **2017**, *9*, 1105. [CrossRef] [PubMed]
4. Gürbüz, İ.; Yeşilada, E. In vivo anti-ulcerogenic activity of *Equisetum telmateia* Ehrh. extracts used in Turkish folk medicine. *Turk. J. Bio.* **2008**, *32*, 259–263.
5. Sureshkumar, J.; Jenipher, C.; Sriramavaratharajan, V.; Gurav, S.S.; Gandhi, G.R.; Ravichandran, K.; Ayyanar, M. Genus Equisetum L: Taxonomy, Toxicology, Phytochemistry and Pharmacology. *J. Ethnopharmacol.* **2023**, *314*, 116630. [CrossRef]
6. Barak, T.H.; Celep, E.; Yesilada, E. Influence of in vitro human digestion on the bioavailability of phenolic content and antioxidant activity of *Viburnum opulus* L. (*European cranberry*) fruit extracts. *Ind. Crop. Prod.* **2019**, *131*, 62–69. [CrossRef]
7. Saeed, M.; Tasleem, M.; Shoaib, A.; Alabdallah, N.M.; Alam, J.; El Asmar, Z.; Jamal, Q.M.S.; Bardakci, F.; Ansari, I.A.; Ansari, M.J.; et al. Investigation of antidiabetic properties of shikonin by targeting aldose reductase enzyme: In silico and in vitro studies. *Biomed. Pharmacother.* **2022**, *150*, 112985. [CrossRef]
8. Wu, M.; Yang, Z.; Zhang, C.; Shi, Y.; Han, W.; Song, S.; Mu, L.; Du, C.; Shi, Y. Inhibition of NLRP3 inflammasome ameliorates podocyte damage by suppressing lipid accumulation in diabetic nephropathy. *Metabolism* **2021**, *118*, 154748. [CrossRef] [PubMed]
9. Nam, J.S.; Cho, M.H.; Lee, G.T.; Park, J.S.; Ahn, C.W.; Cha, B.S.; Lim, S.K.; Kim, K.R.; Ha, H.J.; Lee, H.C. The activation of NF-κB and AP-1 in peripheral blood mononuclear cells isolated from patients with diabetic nephropathy. *Diabetes Res. Clin. Pract.* **2008**, *81*, 25–32. [CrossRef]
10. Abdel-Rahman, E.M.; Alhamad, T.; Reeves, W.B.; Awad, A.S. Management of diabetic nephropathy in the elderly: Special considerations. *J. Nephr Ther.* **2012**, *2*, 124.
11. Barak, T.H.; Celep, E.; İnan, Y.; Yeşilada, E. In vitro human digestion simulation of the bioavailability and antioxidant activity of phenolics from *Sambucus ebulus* L. fruit extracts. *Food Biosci.* **2020**, *37*, 100711. [CrossRef]
12. Yilmaz, M.A. Simultaneous Quantitative Screening of 53 Phytochemicals in 33 Species of Medicinal and Aromatic Plants: A Detailed, Robust and Comprehensive LC–MS/MS Method Validation. *Ind. Crops Prod.* **2020**, *149*, 112347. [CrossRef]

13. Barak, T.H.; Kurt Celep, İ.; Şentürk, T.B.; Bardakci, H.; Celep, E. Anti-Aging Potential Evaluation of *Maclura pomifera* (Rafin.) Schneider 80% Methanol Extract with Quantitative HPTLC Analysis. *Turk. J. Pharm. Sci.* **2022**, *19*, 400–407. [CrossRef]
14. Bardakci, H.; Barak, T.H.; Özdemir, K.; Celep, E. Effect of brewing material and various additives on polyphenolic composition and antioxidant bioactivity of commercial *Tilia platyphyllos* Scop. infusions. *J Res Pharm.* **2020**, *24*, 133–141. [CrossRef]
15. Taşkın, T.; Doğan, M.; Cam, M.E.; Şahin, T.; Şenkardeş, İ. In vitro anti-urease, antioxidant, anticholinesterase, cytotoxic and in vivo anti-inflammatory potential of *Satureja cuneifolia* ten. *Not. Sci. Bio.* **2020**, *12*, 222–232. [CrossRef]
16. İnan, Y.; Kurt-Celep, I.; Akyüz, S.; Barak, T.H.; Celep, E.; Yesilada, E. An Investigation on the Enzyme Inhibitory Activities, Phenolic Profile and Antioxidant Potentials of Salvia Virgata Jacq. *S. Afr. J. Bot.* **2021**, *143*, 350–358. [CrossRef]
17. Zhao, B.T.; Nguyen, D.H.; Le, D.D.; Choi, J.S.; Min, B.S.; Woo, M.H. Protein tyrosine phosphatase 1B inhibitors from natural sources. *Arch. Pharm. Res.* **2018**, *41*, 130–161. [CrossRef]
18. Zhou, X.; Wang, L.L.; Tang, W.J.; Tang, B. Astragaloside IV inhibits protein tyrosine phosphatase 1B and improves insulin resistance in insulin-resistant HepG2 cells and triglyceride accumulation in oleic acid (OA)-treated HepG2 cells. *J. Ethnopharmacol.* **2021**, *268*, 113556. [CrossRef]
19. Kurt-Celep, İ.; Celep, E.; Akyüz, S.; İnan, Y.; Barak, T.H.; Akaydın, G.; Telci, D.; Yesilada, E. *Hypericum olympicum* L. recovers DNA damage and prevents MMP–9 activation induced by UVB in human dermal fibroblasts. *J. Ethnopharmacol.* **2020**, *246*, 112202. [CrossRef]
20. Alaaeldin, R.; Abdel-Rahman, I.A.M.; Hassan, H.A.; Youssef, N.; Allam, A.E.; Abdelwahab, S.F.; Zhao, Q.-L.; Fathy, M. Carpachromene Ameliorates Insulin Resistance in HepG2 Cells via Modulating IR/IRS1/PI3k/Akt/GSK3/FoxO1 Pathway. *Molecules* **2021**, *26*, 7629. [CrossRef]
21. Kurt-Celep, I.; Zheleva-Dimitrova, D.; Sinan, K.I.; Uba, A.I.; Nilofar; Mahomoodally, M.F.; Zengin, G. Uncovering chemical profiles, biological potentials, and protection effect against ECM destruction in H_2O_2 treated HDF cells of the extracts of *Stachys tundjeliensis*. *Arch. Pharm. Res.* **2024**, *357*, 2300528. [CrossRef] [PubMed]
22. Barak, T.H.; Kurt-Celep, I.; Dilek-Tepe, H.; Bardakcı, H.; Akaydın, G.; Yesilada, E.; Celep, E. In vitro assessment of dermatological activity potential of *Achillea clypeolata* Sm. in H_2O_2-treated human dermal fibroblasts. *S. Afr. J. Bot.* **2023**, *160*, 1–8. [CrossRef]
23. Milovanović, V.; Radulović, N.; Todorović, Z.; Stanković, M.; Stojanović, G. Antioxidant, Antimicrobial and Genotoxicity Screening of Hydro-alcoholic Extracts of Five Serbian Equisetum Species. *Plant Foods Hum. Nutr.* **2007**, *62*, 113–119. [CrossRef]
24. Yeganegi, M.; Tabatabaei Yazdi, F.; Mortazavi, S.A.; Asili, J.; Alizadeh Behbahani, B.; Beigbabaei, A. *Equisetum telmateia* extracts: Chemical compositions, antioxidant activity and antimicrobial effect on the growth of some pathogenic strain causing poisoning and infection. *Microb. Pathog.* **2018**, *116*, 62–67. [CrossRef]
25. Correia, H.; González-Paramás, A.; Amaral, M.T.; Santos-Buelga, C.; Batista, M.T. Characterisation of polyphenols by HPLC-PAD-ESI/MS and antioxidant activity in *Equisetum telmateia*. *Phytochem. Anal.* **2005**, *16*, 380–387. [CrossRef]
26. Feng, J.; Wang, J.; Bu, T.; Ge, Z.; Yang, K.; Sun, P.; Cai, M. Structural, in vitro digestion, and fermentation characteristics of lotus leaf flavonoids. *Food Chem.* **2023**, *406*, 135007. [CrossRef]
27. Bortolini, D.G.; Barros, L.; Maciel, G.M.; Brugnari, T.; Modkovski, T.A.; Fachi, M.M.; Pontarolo, R.; Pinela, J.; Ferreira, I.C.; Haminiuk, C.W.I. Bioactive profile of edible nasturtium and rose flowers during simulated gastrointestinal digestion. *Food Chem.* **2022**, *381*, 132267. [CrossRef] [PubMed]
28. Nunes, R.; Pasko, P.; Tyszka-Czochara, M.; Szewczyk, A.; Szlosarczyk, M.; Carvalho, I.S. Antibacterial, Antioxidant and Anti-Proliferative Properties and Zinc Content of Five South Portugal Herbs. *Pharm. Biol.* **2017**, *55*, 114–123. [CrossRef] [PubMed]
29. Hilary, S.; Tomás-Barberán, F.A.; Martinez-Blazquez, J.A.; Kizhakkayil, J.; Souka, U.; Al-Hammadi, S.; Habib, H.; Ibrahim, W.; Platat, C. Polyphenol characterisation of *Phoenix dactylifera* L. (date) seeds using HPLC-mass spectrometry and its bioaccessibility using simulated in-vitro digestion/Caco-2 culture model. *Food Chem.* **2020**, *311*, 125969. [CrossRef]
30. Dixon, R.A.; Xie, D.Y.; Sharma, S.B. Proanthocyanidins-a final frontier in flavonoid research? *New Phytol.* **2005**, *165*, 9–28. [CrossRef]
31. Coughlan, M.T.; Sharma, K. Challenging the dogma of mitochondrial reactive oxygen species overproduction in diabetic kidney disease. *Kidney Int.* **2016**, *90*, 272–279. [CrossRef] [PubMed]
32. Štajner, D.; Popović, B.M.; Čanadanović-Brunet, J.; Boža, P. Free radical scavenging activity of three Equisetum species from *Fruška gora* mountain. *Fitoterapia* **2006**, *77*, 601–660. [CrossRef] [PubMed]
33. Stajner, D.; Popović, B.M.; Čanadanovicć-Brunet, J.; Anacčkov, G. Exploring *Equisetum arvense* L., *Equisetum ramosissimum* L. and *Equisetum telmateia* L. as sources of natural antioxidants. *Phytother. Res.* **2009**, *23*, 546–550. [CrossRef]
34. López, V.; Akerreta, S.; Casanova, E.; García-mina, J.; Cavero, R.; Calvo, M.; López, V.; Akerreta, S.; Casanova, E.; García-mina, J. Screening of Spanish Medicinal Plants for Antioxidant and Antifungal Activities Screening of Spanish Medicinal Plants for Antioxidant and antifungal activities. *Pharm. Biol.* **2008**, *46*, 602–609. [CrossRef]
35. Ge, H.; Chen, Y.; Chen, J.; Tian, J.; Liang, X.; Chen, L. Evaluation of Antioxidant Activities of Ethanol Extract from Ligusticum Subjected to in-Vitro Gastrointestinal Digestion. *Food Chem. Toxicol.* **2018**, *119*, 417–424. [CrossRef]
36. Uzun, E.; Sariyar, G.; Adsersen, A.; Karakoc, B.; Ötük, G.; Oktayoglu, E.; Pirildar, S. Traditional medicine in Sakarya province (Turkey) and antimicrobial activities of selected species. *J. Ethnopharmacol.* **2004**, *95*, 287–296. [CrossRef]
37. Taşkın, T.; Yilmaz, B.; Doğan, A. Antioxidant, Enzyme Inhibitory and Calcium Oxalate Anti-crystallization Activities of *Equisetum telmateia* Ehrn. *Int. J. Sec Met.* **2020**, *7*, 181–191. [CrossRef]

38. Soleimani, S.; Fathiazarbaijani, F.; Nejati, V.; Shojaei, S.H.; Nanghshbandi, N. Effect of *Equisetum arvense* L.(Equisetaceae) in microalbuminuria and creatinine excretion in streptozotocin-induced diabetes in male rats. *Int. J. Pharmacol.* **2007**, *3*, 155–159.
39. Vieira, G.T.; de Oliveira, T.T.; Carneiro, M.A.A.; Cangussu, S.D.; Humberto, G.A.P.; Taylor, J.G.; Humberto, J.L. Antidiabetic effect of *Equisetum giganteum* L. extract on alloxan-diabetic rabbit. *J. Ethnopharmacol.* **2020**, *260*, 112898. [CrossRef]
40. Mata-Torres, G.; Andrade-Cetto, A.; Espinoza-Hernández, F.A.; Cárdenas-Vázquez, R. Hepatic Glucose Output Inhibition by Mexican Plants Used in the Treatment of Type 2 Diabetes. *Front. Pharmacol.* **2020**, *11*, 215. [CrossRef]
41. Barak, T.; Bardakcı, H.; Kurt-Celep, İ.; Özdemir, K.; Celep, E. Evaluation of the influence of in vitro human digestion simulation on the chemical composition and bioactivities of *Ziziphus jujuba* Mill. *Acta Aliment* **2022**, *51*, 105–114. [CrossRef]
42. Combs, A.P. Recent advances in the discovery of competitive protein tyrosine phosphatase 1B inhibitors for the treatment of diabetes, obesity, and cancer. *J. Med. Chem.* **2010**, *53*, 2333–2344. [CrossRef]
43. Khacheba, I.; Djeridane, A.; Yousfi, M. Twenty traditional Algerian plants used in diabetes therapy as strong inhibitors of α-amylase activity. *Int. J. Carbohydrate Chem.* **2014**, *2014*, 287281. [CrossRef]
44. Das, G.; Patra, J.K.; Shin, H.S. Biosynthesis, and potential effect of fern mediated biocompatible silver nanoparticles by cytotoxicity, antidiabetic, antioxidant and antibacterial, studies. *Mater. Sci. Eng. C* **2020**, *114*, 111011. [CrossRef] [PubMed]
45. Boeing, T.; Tafarelo Moreno, K.G.; Gasparotto, A., Jr.; Mota da Silva, L.; de Souza, P. Phytochemistry and pharmacology of the genus Equisetum (Equisetaceae): A narrative review of the species with therapeutic potential for kidney diseases. *Evid.-Based Complement. Altern. Med.* **2021**, *2021*, 6658434. [CrossRef] [PubMed]
46. Alavarce, R.A.S.; Saldanha, L.L.; Almeida, N.L.M.; Porto, V.C.; Dokkedal, A.L.; Lara, V.S. The Beneficial Effect of *Equisetum giganteum* L. against Candida Biofilm Formation: New Approaches to Denture Stomatitis. *Evid.-Based Complement. Altern. Med.* **2015**, *2015*, 939625. [CrossRef]
47. Lemus, I.; Garcia, R.; Erazo, S.; Pena, R.; Parada, M.; Fuenzalida, M. Diuretic activity of an *Equisetum bogotense* tea (Platero herb): Evaluation in healthy volunteers. *J. Ethnopharmacol.* **1996**, *54*, 55–58. [CrossRef]
48. Ding, X.; Jian, T.; Wu, Y.; Zuo, Y.; Li, J.; Lv, H.; Ma, L.; Ren, B.; Zhao, L.; Li, W.; et al. Ellagic acid ameliorates oxidative stress and insulin resistance in high glucose-treated HepG2 cells via miR-223/keap1-Nrf2 pathway. *Biomed. Pharm.* **2019**, *110*, 85–94. [CrossRef]
49. Zhang, X.T.; Yu, C.J.; Liu, J.W.; Zhang, Y.P.; Zhang, C.; Song, C.X.; Wang, F. Qizhi Jiangtang Jiaonang improves insulin signaling and reduces inflammatory cytokine secretion and reactive oxygen species formation in insulin resistant HepG2 cells. *Evid. Based Complement. Altern. Med.* **2015**, *2015*, 518639. [CrossRef]
50. Baker, R.G.; Hayden, M.S.; Ghosh, S. NFκB, Inflammation and Metabolic Disease. *Cell. Metab.* **2011**, *13*, 11–22. [CrossRef]
51. Zhao, X.; An, X.; Yang, C.; Sun, W.; Ji, H.; Lian, F. The crucial role and mechanism of insulin resistance in metabolic disease. *Front. Endocrinol.* **2023**, *14*, 1149239. [CrossRef] [PubMed]
52. Parwani, K.; Mandal, P. Role of advanced glycation end products and insulin resistance in diabetic nephropathy. *Arch. Physiol. Biochem.* **2020**, *129*, 95–107. [CrossRef] [PubMed]
53. Khalid, M.; Petroianu, G.; Adem, A. Advanced Glycation End Products and Diabetes Mellitus: Mechanisms and Perspectives. *Biomolecules* **2022**, *12*, 542. [CrossRef] [PubMed]
54. Mitamura, Y.; Okumura, A.; Harada, C.; Namekata, K.; Nakamura, K.; Tashimo, A.; Harada, T. Activator protein-1 in epiretinal membranes of patients with proliferative diabetic retinopathy. *Diabetologia* **2006**, *49*, 209–211. [CrossRef]
55. Sun, Y.; Wang, J.; Guo, X.; Zhu, N.; Niu, L.; Ding, X.; Xie, Z.; Chen, X.; Yang, F. Oleic Acid and Eicosapentaenoic Acid Reverse Palmitic Acid-induced Insulin Resistance in Human HepG2 Cells via the Reactive Oxygen Species/JUN Pathway. *Genom. Proteom. Bioinform.* **2021**, *19*, 754–771. [CrossRef]

foods

MDPI

Article

Bioaccessibility and Antidiabetic Potential of *xique-xique* and *mandacaru* Fruits in a Simulated Gastrointestinal Tract Model

Fábio Fernandes de Araújo [1], David de Paulo Farias [2], Iramaia Angélica Neri-Numa [2], Glaucia Maria Pastore [2] and Alexandra Christine Helena Frankland Sawaya [1,*]

[1] Faculty of Pharmaceutical Science, University of Campinas, UNICAMP, Campinas 13083-871, SP, Brazil; fabio.fernandesn18@gmail.com

[2] Bioflavors and Bioactive Compounds Laboratory, Department of Food Science, Faculty of Food Engineering, University of Campinas, Rua Monteiro Lobato, 80, Campinas CEP 13083-862, SP, Brazil; glaupast@unicamp.br (G.M.P.)

* Correspondence: alexandra.sawaya@fcf.unicamp.br

Abstract: This study evaluated the influence of gastrointestinal digestion on the bioaccessibility and antidiabetic potential of *xique-xique* (*Pilosocereus gounellei*) and *mandacaru* (*Cereus jamacaru*) fruits. After digestion, the content of total phenolics and flavonoids reduced by 58.3 and 73.51% in *xique-xique* and 48.33 and 88.43% in *mandacaru*. In addition, compounds such as rutin, ρ-coumaric acid, catechin and epicatechin reduced during digestion for both fruits. The antioxidant potential by the ABTS assay increased by 153.3% for *xique-xique* and 273.46% for *mandacaru* in the intestinal phase. However, using the ORAC assay, the antioxidant potential of *xique-xique* reduced from 255.42 to 112.17 μmol TE g^{-1}. The capacity of *xique-xique* fruit to reduce α-amylase activity reduced 23.71-fold after digestion, but the potential to inhibit α-glucosidase increased 17.8-fold. The antiglycation potential reduced in both fruits after the in vitro gastrointestinal digestion. Thus, the bioaccessibility of the phenolic compounds from the fruits, as well as their functional potential, were influenced by the digestive process, as well as by the sample evaluated.

Keywords: unconventional food plants; *Pilosocereus gounellei; Cereus jamacaru;* bioactive compounds; antioxidant potential; antihyperglycemic; antiglycation activity

1. Introduction

The World Health Organization reports that chronic non-communicable diseases (NCDs), including heart disease, cancer, and diabetes, represent a significant public health challenge, being responsible for approximately 74% of all deaths globally [1]. Diabetes has become a significant global concern, as the International Diabetes Federation (IDF) reports that 1 in 11 adults is affected by this disease, amounting to approximately 463 million people. Projections for 2045 estimate that this figure could rise to 700 million. Additionally, it is important to note that about 10% of total health expenditures, approximately USD 760 billion, is currently allocated to the treatment of diabetes [2].

Type 2 diabetes mellitus is a metabolic condition marked by elevated blood sugar levels. This occurs either due to decreased insulin sensitivity or a deficiency in insulin signaling. These issues can result in complications affecting various organs and tissues, including the nerves, eyes, kidneys, and cardiovascular system [3]. Scientific evidence suggests that consuming fruits and vegetables rich in bioactive compounds, such as polyphenols, may be a promising approach for the treatment and prevention of diseases like diabetes [3,4].

Phenolic compounds are secondary metabolites produced by plants that have garnered significant interest from the scientific community for their health benefits. These effects are largely attributed to their ability to sequester or inhibit reactive species, donate electrons to free radicals, and activate antioxidant enzymes, among other mechanisms [3,5,6]. Consuming these compounds in sufficient amounts may offer a range of beneficial effects in preventing various diseases, including diabetes, cancer, obesity, Parkinson's disease, Alzheimer's disease, and cardiovascular and intestinal disorders, among others [5]. While numerous health benefits are associated with the consumption of these compounds, their functional and bioactive properties can be influenced by various factors, particularly their bioaccessibility following gastrointestinal digestion. Bioaccessibility refers to the amount of a specific food component that is released and solubilized in the intestinal lumen during digestion, making it available for absorption. This process is affected by several factors, including the concentrations of substances in food, their release from the food matrix, chemical structure, molecular interactions with other compounds, molecular size, degree of polymerization, and solubility etc. [3,5–8].

Cactus from the Cactaceae family, including *Pilosocereus gounellei* (F.A.C. Weber) and *Cereus jamacaru* DC, commonly known as *xique-xique* and *mandacaru*, play a vital role in the Brazilian semi-arid region [9]. These species have excellent nutritional value, being considered a good source of nutrients such as carbohydrates, fiber and minerals [10]. Due to their richness in nutrients, previous research has demonstrated that these species have significant industrial potential for the development of various products, including cakes, ice creams, yogurts, and juices [11–13]. Additionally, recent studies have revealed that the fruits of species like *xique-xique* and *mandacaru* are rich sources of various bioactive compounds, including phenolic compounds such as chlorogenic acid, catechin, epicatechin, myricetin, quercetin, kaempferol, rutin, and others [14,15]. Furthermore, a recent review paper highlighted that these species may offer numerous health benefits for consumers, exhibiting properties such as antioxidant, in addition to anti-inflammatory, anti-obesity, anti-diabetic, anti-cancer, and anti-nociceptive effects [16]. However, so far, information regarding the effect of the in vitro gastrointestinal digestion process on the bioactivity and bioaccessibility of the phenolic compounds present in these fruits is still scarce.

Studying the effects of the digestive process is important to understand the interactions that may affect their nutritional potential and, therefore, developing strategies to increase their biological effects is essential to encourage the use of underutilized species such as *xique-xique* and *mandacaru* for the preparation of new foods with functional claims. Thus, the objective of this study was to investigate the influences of in vitro gastrointestinal digestion on the profile of phenolic compounds, antioxidant potential, α-amylase and α-glucosidase inhibitory activities and antiglycation properties of *xique-xique* and *mandacaru* fruits.

2. Material and Methods

2.1. Sample Acquisition and Preparation

xique-xique and *mandacaru* fruit were collected in the city of Solânea in the interior of Paraíba, Brazil. The botanical identification of the species was carried out, the exsiccates (Accession n° 208.839 and 208.840) were deposited in the Herbarium-UEC of the Institute of Biology of the State University of Campinas UEC-UNICAMP and the project was registered in the National System for the Management of Genetic Knowledge and Traditional Associate –SisGen (N° ABFC0D8). The collected whole fruits were washed with distilled water and processed in a home juicer. Immediately after processing, they were immediately frozen (−20 °C), freeze-dried and ground in a knife mill. The samples thus obtained were placed in dark packaging and stored at −20 °C to avoid possible changes in chemical composition.

2.2. In Vitro Gastrointestinal Digestion

The digestion process was simulated following the methodology described by Sancho and collaborators [17]. Briefly, 500 mg of each fruit (freeze-dried whole fruits) was homogenized in 9 mL of saline solution (140 mmol L^{-1} NaCl, 5 mmol L^{-1} KCl) and the pH of the mixture was adjusted to 2.0 with 6 mol L^{-1} HCl. An aliquot of 125 μL of porcine pepsin solution (Sigma-Aldrich, 200 mg of 424 U mg^{-1} pepsin in 5 mL of 0.1 mol L^{-1} HCl) was added and the samples were incubated in a water bath with shaking at 130 rpm for 1 h. A sample of 2 mL of this phase was collected and stored in a freezer at −80 °C for later analysis.

After the gastric digestion steps, the pH of the samples was adjusted again to 6.8 by adding 1 mol L^{-1} $NaHCO_3$. Then, 625 µL of pancreatin and bile solution (Sigma-Aldrich, 225 mg of bile extract and 37 mg of pancreatin with activity equivalent to 4x U.S.P, diluted in 18.7 mL of $NaHCO_3$ 0.1 mol L^{-1}) were added and the samples were incubated at 37 °C and 130 rpm for 2 h. All fractions were ultrafiltered using Amicon ultra centrifugal filtration devices (30 kDa, Millipore) and stored at −80 °C until analysis. A sample of saline solution at pH 7.0 (blank) was also subjected to all digestion procedures to eliminate reagent interference. The samples collected after each phase, as well as the undigested samples (suspended in water like a juice), were filtered and used as extracts to perform all analyses.

2.3. *Quantification of Phenolic Compounds*

The identification and quantification of phenolic compounds were performed in Selected Ion Monitoring (SIM) mode (Figures S1–S3—Supplementary Materials) using an Acquity UHPLC chromatograph coupled to an Acquity TQD mass spectrometer (Micromass-Waters Manchester, Manchester, UK) with an ESI source, Waters Acquity C18 BEH column (2.1 mm × 50 mm × 1.7 µm), following the procedures described by Cassola and colleagues [18]. For this analysis, a specific chromatographic method for the samples was developed using the following conditions: phase A composed of acidic water (0.1% formic acid) and phase B of acetonitrile (HPLC grade). The gradient started and remained at 95% A and 5% B for 1 min, then increased to 100% B for 8 min and returned to the initial conditions for 1 min, totaling 10 min. The flow rate was 0.2 mL/min and the injection volume was 7 µL. The results were obtained through calibration curves (concentrations ranging from 0.005 to 100 µg mL^{-1}) constructed with standard compounds of quinic acid, rutin, p-coumaric acid, catechin and epicatechin from Sigma-Aldrich and the values were expressed in mg 100 g^{-1} of lyophilized sample (fd).

2.4. *Determination of Total Phenolic Content*

This assay was conducted following the method proposed by Singleton et al., with minor modifications [19]. In brief, 30 µL of fruit extracts was mixed with 150 µL of 10% Folin Ciocalteau reagent and 120 µL of 7.5% $NaHCO_3$, then incubated for 6 min at 45 °C. The absorbance was measured at 760 nm, and the total phenolic content in the samples was quantified using a standard curve created with gallic acid, with results expressed as mg of gallic acid equivalents (GAE) g^{-1} fd.

2.5. *Determination of Total Flavonoid Content*

The total flavonoid content was determined using the method established by Zhishen et al. [20]. For the reaction system, 30 µL of fruit extracts was combined with 110 µL of ultrapure water. Next, 8 µL of 5% $NaNO_2$ was added, and after 5 min, 8 µL of 10% $AlCl_3$ was introduced, followed by an additional 6 min incubation. Finally, 50 µL of 1 M NaOH and 70 µL of ultrapure water were added. The absorbance of the mixture was measured at 510 nm, with catechin used to prepare the analytical curve; the total flavonoid content was expressed as mg of catechin equivalents (EC) g^{-1} fd.

2.6. *ABTS Assay*

The antioxidant capacity was assessed using the ABTS+ assay, which involved reacting 5 mL of ABTS (7 mM) with 88 µL of 140 mM potassium persulfate [21]. After the formation of the ABTS radical, it was diluted with distilled water to achieve an absorbance value of 0.700 ± 0.02 at 734 nm using a microplate reader. Trolox was employed to prepare the analytical curve, and the results were expressed in µM Trolox equivalents (TE) g^{-1} fd.

2.7. *ORAC Assay*

This assay was performed according to Prior et al., with minor modifications [22]. Samples, standards and reagents were prepared in 75 mM potassium phosphate buffer, pH 7.4. The reaction system consisted of 20 μL of Trolox standard or previously diluted extracts, 120 μL of fluorescein (0.378 μg mL−1, pH 7.4) and 60 μL of AAPH [2,2′-Azobis(2-methylpropionamidine) dihydrochloride] (108 mg/mL). Fluorescence intensity was monitored at 37 °C immediately after the addition of AAPH every 60 s cycle for 80 cycles in a NOVOstar Microplate Reader. Analyses were performed in triplicate and results were expressed in μM TE g^{-1} fd.

2.8. Antidiabetic Potential

2.8.1. α-Amylase Activity

The inhibition of α-amylase activity was evaluated according to the methodology described by Van Quan et al., with modifications [23]. To this end, 20 μL of each extract was pre-incubated for 10 min at 25 °C with 20 μL of 1 U/mL porcine pancreatic α-amylase solution dissolved in 0.1 M sodium phosphate buffer (pH 6.9) containing 0.006 M sodium chloride. Then, 30 μL of starch solution (0.25%) was added and incubated again for 6 min at 25 °C. To stop the reaction, 20 μL of hydrochloric acid (1 M) was pipetted. Next, 120 μL of Lugol's aqueous solution (250 μL of 5% solution) was added to develop a dark blue color. The absorbance was read at 565 nm in a microplate reader. Acarbose was used as a positive control and the inhibitory activity was expressed as IC_{50}.

2.8.2. α-Glucosidase Activity

The inhibition of α-glucosidase activity was evaluated according to Van Quan and collaborators [23]. Briefly, an amount of 20 μL of each extract was mixed with 20 μL of 0.1 M potassium phosphate buffer (pH 7) and 40 μL of α-glucosidase (from Saccharomyces cerevisiae, Sigma-Aldrich, St Louis, MO, USA) (0.5 U/mL in 0.1 M potassium phosphate buffer, pH 7). After 6 min of incubation at 25 °C, a 20 μL aliquot of 5 mM ρ-nitrophenyl-α-D-glucopyranoside substrate (in 0.1 M potassium phosphate buffer, pH 7) was added and the mixture was incubated for another 8 min. The reaction was terminated by the addition of 100 μL of 0.2 M Na_2CO_3, and absorbance was recorded at 405 nm. Acarbose was used as a positive control. The inhibitory activity of the extracts on α-glucosidase was calculated as the percentage of inhibition and the results were expressed as IC_{50} (mg mL^{-1}).

2.8.3. Antiglycation Potential

The antiglycation potential of the extracts was determined according to the protocol described by Sri Harsha et al., with some modifications [24]. Briefly, 200 mM potassium phosphate buffer (pH 7.4) containing 0.02% sodium azide, 50 mg/mL BSA in phosphate buffer, was mixed with 1.25 M fructose in phosphate buffer and the extract. The control consisted of phosphate buffer instead of the antiglycation agent and for the blank, fructose and the antiglycation agent were replaced by phosphate buffer. The mixtures were incubated at 37 °C for 3 days in a water bath in the absence of light. The fluorescence of the mixture was measured in a microplate reader at an excitation/emission wavelength of 350/420 nm, respectively and the results were calculated as percentage of inhibition and expressed as IC_{50} (mg mL^{-1}).

2.9. Statistical Analysis

All experiments conducted in this study were performed in triplicate, and the results were reported as means ± standard deviations. The values of the tested variables were analyzed using analysis of variance (ANOVA), and when significant differences were found, the means of the treatments were compared using Tukey's and Student's t tests, with a significance level of $p \leq 0.05$. Statistical analyses were carried out using Statistica software version 7.0 [25].

3. Results and Discussion

Table 1 shows the total phenolic and total flavonoid content before and during the in vitro gastrointestinal digestion process. The lowest content of total phenolic compounds in the undigested samples was observed for *xique-xique* (11.87 ± 0.67 mg GAE g^{-1} fd), being approximately 1.48 times lower than the value observed for *mandacaru* (17.63 ± 0.08 mg GAE g^{-1} fd). In addition, the greatest reduction in the content of total phenolics in the intestinal phase was also observed in *xique-xique,* which presented a reduction of 58.3%, while the reduction for *mandacaru* was 48.33%. Likewise, flavonoid content was greatly reduced after gastrointestinal digestion. The highest total flavonoid content (16.78 mg CE g^{-1} df) was found in the undigested sample for *mandacaru* ($p \leq 0.05$); however its content reduced approximately 24.97 and 88.43% in the gastric and intestinal phases, respectively. Similarly, the total flavonoid content also reduced approximately 2.97% (gastric phase) and 73.51% (intestinal phase) in *xique-xique* after the in vitro gastrointestinal digestion process.

Table 1. Total phenolics, total flavonoids and antioxidant capacity of *xique-xique* and *mandacaru* fruit after in vitro gastrointestinal digestion.

Fruit	Treatment	Total Phenolic [1]	Total Flavonoids [2]	ABTS [3]	ORAC [3]
xique-xique	Undigested sample	11.87 ± 0.67 Ab	9.25 ± 0.12 Ab	58.64 ± 2.53 Ca	255.42 ± 11.05 Aa
	Gastric phase	9.61 ± 0.39 Bb	8.05 ± 0.16 Bb	86.66 ± 0.99 Bb	104.43 ± 5.85 Bb
	Intestinal phase	4.95 ± 0.24 Cb	2.45 ± 0.05 Ca	148.55 ± 0.91 Aa	112.17 ± 14.35 Bb
mandacaru	Undigested sample	17.63 ± 0.08 Aa	16.78 ± 0.42 Aa	39.80 ± 2.87 Cb	172.28 ± 1.31 Bb
	Gastric phase	17.38 ± 0.04 Aa	12.59 ± 0.16 Ba	103.70 ± 1.70 Ba	168.63 ± 2.04 Ba
	Intestinal phase	9.11 ± 0.52 Ba	1.94 ± 0.04 Cb	148.64 ± 0.48 Aa	174.53 ± 1.73 Aa

Capital letters indicate statistical difference for each treatment in the different digestive phases determined by Tukey's test at 5% significance ($p \leq 0.05$). Lower case letters indicate statistical difference between treatments within the same digestive phase determined by *t* test at 5% significance ($p \leq 0.05$). [1] Results expressed as mg GAE g^{-1} fd. [2] Results expressed as mg CE g^{-1} fd. [3] Results expressed as μM TE g^{-1} fd.

So far, few studies have sought to evaluate the effect of the gastrointestinal digestive process in cactus species, this being the first carried out with *xique-xique* and *mandacaru* fruits. Similar results to those found in this study were reported by Santiago and colleagues when they evaluated the influence of the gastrointestinal digestion process on the content of bioactive compounds in cladodes of *Opuntia ficus-indica*, a species of cactus belonging to the same family as *xique-xique* and *mandacaru*. According to the authors, the content of total flavonoid compounds decreased from 120 ± 0.49 (before digestion) to 58.30 ± 2.28 in the gastric phase and 44.96 ± 0.89 mg g^{-1} dry matter (dm) in the intestinal phase, while the content of total phenolic compounds and total phenolic acids decreased by approximately 55.90 and 44.49%, respectively [26]. In another study, a small reduction in the content of total phenolic and flavonoid compounds was observed after the digestive process in yogurt containing 10% cactus pear extract (*Opuntia oligacantha* C.F. Först). The reported values of total phenolic compounds were 19.9: 17.5 and 16.4 mg GAE 100 mL^{-1} for the undigested samples, gastric and intestinal phases, respectively, representing a reduction of approximately 12.06% in the intestinal phase. Similarly, the reduction in the content of flavonoids was 2.79%, with the reported values being 17.9, 18.2 and 17.4 mg QE 100 mL^{-1} for the undigested samples, gastric and intestinal phases, respectively. [27]. Different results were observed for oat wheat bread enriched with red *pitaya* powder (*Hylocereus polyrhizus*), where the content of total phenolic compounds increased by about 5 and 15% (2.75–3.93 mg GAE g^{-1}) after intestinal digestion [28].

The behavior of the content of total phenolic compounds and total flavonoids was similar to that observed for the profile of individual phenolic compounds after the digestive process for both fruits (Table 2). There was a reduction in the content of phenolic compounds such as rutin, ρ-coumaric acid, catechin and epicatechin in the gastric and intestinal phases for *xique-xique*, with the exception of quinic acid, which demonstrated an increase in its content after the digestive process (1.18-fold in the gastric phase and 1.2 in the intestinal phase). A lower content of individual phenolic compounds was observed in *mandacaru* fruit and, unlike the behavior observed in *xique-xique* fruit, the quinic acid content reduced after the digestive process (reduction of 61.78% in the gastric phase and 49.10% in the intestinal phase). In the gastric phase, a content of 1.40 ± 0.01 mg 100 g^{-1} of catechin was observed for *mandacaru*; however, this value reduced in the intestinal phase.

Table 2. Phenolic compounds identified and quantified in *xique-xique* and *mandacaru* fruit before and after in vitro gastrointestinal digestion.

Fruit	Treatment	Compound				
		Quinic Acid	Rutin	ρ-Coumaric Acid	Catechin	Epicatechin
xique-xique	Undigested sample	1.90 ± 0.01 Bb	9.27 ± 0.27 Aa	0.34 ± 0.09 Ab	n.d.	n.d.
	Gastric phase	2.26 ± 0.11 Ab	0.19 ± 0.07 Ba	n.d.	n.d.	0.44 ± 0.0001
	Intestinal phase	2.29 ± 0.08 Ab	n.d.	n.d.	0.49 ± 0.006	n.d.
mandacaru	Undigested sample	5.60 ± 0.08 Aa	0.10 ± 0.007 Ab	0.67 ± 0.03 Aa	n.d.	n.d.
	Gastric phase	3.46 ± 0.04 Ba	0.0024 ± 0.0002 Bb	0.23 ± 0.03 B	1.40 ± 0.01	n.d.
	Intestinal phase	2.75 ± 0.34 Ca	n.d.	n.d.	n.d.	n.d.

Results expressed as mg 100 g^{-1} fd. Capital letters indicate statistical difference for each treatment in the different digestive phases determined by Tukey's test at 5% significance ($p \leq 0.05$). Lower case letters indicate statistical difference between treatments within the same digestive phase determined by *t* test at 5% significance ($p \leq 0.05$). "n.d.", not detected.

The release of phenolic compounds during the digestive process may vary depending on the sample studied and the bioactive compounds present in the matrix. In whole fruit extracts of cactus berry (*Myrtillocactus geometrizans*), for example, a reduction in the content of some phenolic compounds such as epigallocatechin-glucoside, quercetin-3-*O*-rhamnosyl rutinoside-glucoside, quercetin-3-*O*-rhamnosyl-rhamnosyl-rhamnosyl-glucoside, quercetin-3-*O*-neohesperidosyl-rhamnoside, quercetin-3-rutinoside, and kaempferol-7-*O*-neohesperidoside, among others, was observed [4]. Saniovanni and colleagues, when studying the effect of gastrointestinal digestion on the phenolic compounds of *Vitis vinifera* L. leaves, observed that there was a reduction in the content of several phenolic compounds, including caffeic acid derivatives, flavonols and anthocyanins. In their study, the rutin content, for example, reduced from 1.31 to 1.15 in the gastric phase and 0.96 mg g^{-1} in the intestinal phase. Other compounds such as caftaric acid, quercetin 3-*O*-glucoside, kaempferol 3-*O*-glucoside and delphinidin 3-*O*-glucoside showed a reduction of around 20.02; 38.19; 38.72 and 71.57%, respectively [29]. Similar behavior was also verified in blueberry phenolic compounds after the digestive process, wherein a reduction in the content of flavonoids such as quercetin arabinoside (0.113 to 0.084 mg g^{-1}) and anthocyanins such as delphinidin 3-galactoside (0.107 to 0.004 mg g^{-1}) and malvidin 3-galactoside (0.270 to 0.025 mg g^{-1}) was observed in the intestinal phase; however, the chlorogenic acid content increased by around 18.55%, from 0.291 to 0.345 mg g^{-1} [30]. Compounds such as epicatechin and gallic acid present in wild blackberry fruits reduced by approximately 99.05 and 66.05%, while other compounds such as catechin, quercetin and kaempferol reduced completely in the intestinal phase. On the other hand, some phenolic acids including *p*-coumaric acid and ferulic acid increased their content after the digestive process from 14.97 to 192.00 and 6.95 to 92.67 mg mL^{-1}, respectively [31].

The antioxidant capacity of fruit during the digestive process can be found in Table 1. It can be seen that the antioxidant capacity in the ABTS assay increased for both samples after gastrointestinal digestion. In *xique-xique*, the increase was 47.78% in the gastric phase and 153.3% in the intestinal phase. In *mandacaru*, the increase was 160.55% and 273.46% in the gastric and intestinal phases, respectively. However, statistical analysis demonstrated that there is no difference between the intestinal-phase samples of both fruit ($p \geq 0.05$).

On the other hand, the antioxidant capacity in the ORAC assay significantly reduced during the digestive process in the *xique-xique* sample, with values reducing from 255.42 ±

11.05 to 104.43 ± 5.85 in the gastric phase and 112.17 ± 14.35 μmol TE g^{-1} in the intestinal phase. In *mandacaru*, a reduction in antioxidant capacity was also observed for this assay, although without statistical difference between the three phases evaluated. As can be seen, the behavior of the antioxidant potential of the samples varied depending on the assay used. This behavior may be related to the reduction of some phenolic compounds and to the release of others that could be bound to the cellular matrix and that were released after the digestive process. Phenolic compounds are recognized as natural antioxidants and anti-inflammatory agents, and their beneficial effects are due to their potential to sequester or inhibit reactive species, inactivate free radicals and activate antioxidant enzymes, acting in the prevention and treatment of various chronic non-communicable diseases [5].

The effect of the digestive process on the antioxidant activity of cactus, as well as other native Brazilian fruits, has also been reported. For example, for the ABTS assay, a reduction in the antioxidant potential in raw and cooked cladodes of *Opuntia ficus-indica* was observed after in vitro gastrointestinal digestion [26]. On the other hand, the digestive process resulted in an increase in the antioxidant potential determined by the DPPH and FRAP tests in the gastric and intestinal phases in the oat-wheat bread fortified with red *pitaya* powder [28]. Araujo and colleagues, when studying the effect of in vitro gastrointestinal digestion on *araça-boi* pulp (*Eugenia stipitata*), reported that the antioxidant capacity measured by the DPPH assay reduced from 8.4 to 0.3 μmol TE g^{-1} dw in the intestinal phase, while for the ABTS and ORAC assays an increase of approximately 2.78 and 3.19 times was observed, respectively, after the digestive process [6]. According to the authors, after the digestive process there was a reduction in phenolic acids and a greater release of flavonoids, such as apigenin caffeate hexoside, kaempferol diacetyl dicoumaroylhexoside, quercetin hexopyranosylhexoside and others, directly impacting the increase in antioxidant potential determined by ABTS and ORAC tests [6]. Similarly, the antioxidant capacity in the pulp of *uvaia* (*Eugenia pyriformis*) fruit also increased after gastrointestinal digestion for the ABTS and ORAC assays, with values for ABTS increasing from 44.81 to 77.13 μmol TE g^{-1} dm [7]. In wild blackberry fruit a reduction in the antioxidant potential of the samples occurred during the in vitro gastrointestinal digestion process. In that study, it was possible to observe a reduction of approximately 52.43 and 40.47% in the intestinal phase for the ORAC and ABTS assays, respectively [31]. Similar to the results reported in the present study, there was an increase of approximately 3.06- and 17.32-fold in the antioxidant potential measured by the ABTS assay for hulled barley, after the digestion process [32].

The results regarding the activities of digestive enzymes before and after the digestive process can be seen in Table 3; the IC_{50} of α-amylase for *xique-xique* increased after the gastrointestinal digestion process, and thus, in the intestinal phase, the potential to inhibit this enzyme was reduced by 23.71-fold for this fruit. On the other hand, the potential to inhibit α-amylase increased in *mandacaru* after the digestive process, since the IC_{50} of the sample was reduced in the intestinal phase (77.63 to 46.22). Only undigested samples and the gastric phase of *xique-xique* showed similar potential to arcabose (the therapeutic agent used as a standard to evaluate the inhibition of digestive enzymes) ($p > 0.05$). Regarding the activity of α-glucosidase, except for the undigested *xique-xique* sample, all other samples showed greater potential to inhibit the enzyme when compared to arcabose. Furthermore, there was a 17.8- and 1.68-fold increase in the inhibition potential of α-glucosidase in *xique-xique* and *mandacaru* fruits in the intestinal phase. As previously reported, this increase in the inhibitory potential of digestive enzymes, as well as in the antioxidant potential in the ABTS assay, may be related to the release of bioactive compounds linked to the fiber content or other nutrients present in the fruits. According to Fend and Kong, phenolic compounds are compounds with antioxidant properties that can act competitively or even non-competitively in the inhibition of digestive enzymes, such as α-amylase and α-glucosidase [33].

Table 3. Digestive enzyme inhibition potential of *xique-xique* and *mandacaru* fruit before and after in in vitro gastrointestinal digestion.

Enzyme	Fruit	Treatment	IC_{50} (mg mL^{-1})
α-amylase	*xique-xique*	Undigested sample	30.91 ± 1.03 Eb
		Gastric phase	33.68 ± 3.10 Eb
		Intestinal phase	733.15 ± 93.47 Ba
	mandacaru	Undigested sample	77.63 ± 9.61 Ca
		Gastric phase	1372.54 ± 107.83 Aa
		Intestinal phase	46.22 ± 3.39 Db
	-	Acarbose	36.60 ± 2.80 E
α-glycosidase	*xique-xique*	Undigested sample	160.92 ± 3.13 Aa
		Gastric phase	29.92 ± 2.70 Da
		Intestinal phase	9.04 ± 0.24 Fa
	mandacaru	Undigested sample	41.81± 0.55 Cb
		Gastric phase	18.75 ± 0.30 Eb
		Intestinal phase	8.22 ± 0.04 Fa
	-	Acarbose	121.79 ± 2.78 B

Capital letters indicate statistical difference between treatments in different digestive phases determined by Tukey's test at 5% significance ($p \leq 0.05$). Lower case letters indicate statistical difference within each treatment in specific digestive phases determined by Tukey's test at 5% significance ($p \leq 0.05$).

Recently, the inhibition of enzymes such as α-amylase and α-glucosidase has been identified as a good strategy for controlling hyperglycemia and treating diabetes, since the inhibition of these enzymes is essential to reduce carbohydrate hydrolysis and glucose uptake and to favor the control of postprandial hyperglycemia [3,34]. However, to date, there is little information in the literature regarding the effect of gastrointestinal digestion in cactus on the inhibition of digestive enzymes, as well as on the inhibition of protein glycation. For this reason, our study is extremely important, since it stands out as one of the first to report the antidiabetic potential of species belonging to the Cactaceae family after the digestive process.

Likewise, when compared to the results observed in our study, α-amylase inhibition increased significantly at a concentration of 150 μg mL^{-1} of *Inonotus obliquus* samples after gastric digestion, with the highest inhibitory activity for this enzyme (37.96%) being observed at a concentration of 200 μg mL^{-1}. Similarly, the inhibitory activity of α-glucosidase also increased significantly in the gastric and intestinal phases with the IC_{50} of 60.24 μg mL^{-1} and 52.97 μg mL^{-1}, respectively [35]. In water-soluble polysaccharides from *Opilia amentacea* Roxb fruit, a 69% of α-amylase inhibition in the pre-digested sample at a concentration of 50 μg mL^{-1} was observed, whereas the values of the pancreatic and gastric sample were 72 and 49%, respectively. For α-glucosidase, an inhibition of 70% was observed in the gastric-digested sample and 65% in the pre-digested sample [36]. Different results were found by Gutiérrez and collaborators, depending on the species of oregano studied. For *Hedeoma patens*, for example, the inhibition of α-amylase activity increased from 23.93% in the undigested sample to 25.43 and 52.53% in the gastric and intestinal phases, respectively. On the other hand, there was a reduction in the inhibition of α-glucosidase of approximately 4.80-fold in the gastric phase and 6.68-fold in the intestinal phase. In this study, the authors also observed that certain phenolic compounds disappeared, while others were only found after the digestive process, which certainly influenced the potential of the samples to inhibit digestive enzymes [37].

High blood sugar levels can result in the glycation of proteins and, consequently, the formation and accumulation of advanced glycation end products (AGEs). This process occurs through the binding of carbonyl groups of sugars such as glucose, fructose and ribose to free amino groups of proteins. AGES, in turn, are carbonyl compounds originating from various chemical reactions in Schiff base intermediates that are converted into more stable Amadori products. These can cause changes in proteins, lipids and DNA, resulting in several problems in diabetic patients such as neuropathy, nephropathy, etc. [3,38,39].

Therefore, the search for new natural agents capable of inhibiting protein glycation may be a promising alternative to reduce some diabetic complications [40].

As can be seen in Figure 1, the undigested sample from *mandacaru* fruit showed a higher antiglycation potential (IC_{50} 11.59 $\pm$ 0.001 mg mL^{-1}) than the extract obtained from *xique-xique* fruits (IC_{50} 27.72 $\pm$ 0.01 mg mL^{-1}). After the in vitro gastrointestinal digestion process, it was possible to observe a reduction in the antiglycation potential for both samples in the intestinal phase, with *mandacaru* reducing its potential by approximately 2.50-fold, presenting an IC_{50} of 29.07 $\pm$ 0.80 mg mL^{-1}. Despite there being a reduction in antiglycation potential of *xique-xique* in the intestinal phase (27.72 to 32.40 mg mL^{-1}), there was an increase of approximately 1.65-fold in the gastric phase. These results demonstrate that the antiglycation potential depends on the fruit and the phase studied. Contrary to what was observed for the antioxidant capacity and the potential to inhibit digestive enzymes, the reduction in the antiglycation potential appears to be related to the reduction in the content of phenolic compounds, total flavonoids and compounds such as rutin. Rutin is a flavonol found naturally in plant species that has great bioactive potential. A study previously carried out by Dubey et al. demonstrated that rutin at a concentration of 33 mg ml^{-1} presented antiglycation potential, being able to inhibit 35.55% of fructosamine glycation, 13.45% of protein carbonyls and 80.27% of protein thiols [41]. Thus, the reduction in the concentration of this compound, as well as other phenolic compounds, certainly impacted the antiglycation capacity of the *xique-xique* and *mandacaru* samples after the digestive process.

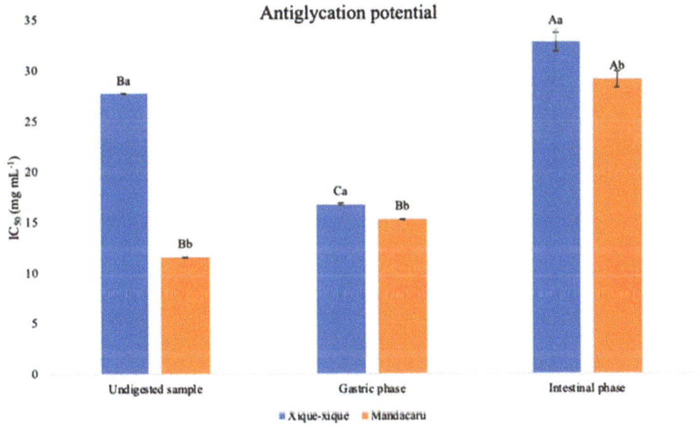

Figure 1. Antiglycation potential of *xique-xique* and *mandacaru* fruit after in vitro gastrointestinal digestion. Capital letters indicate statistical difference for each treatment in the different digestive phases determined by Tukey's test at 5% significance ($p \leq 0.05$). Lower case letters indicate statistical difference between treatments within the same digestive phase determined by *t* test at 5% significance ($p \leq 0.05$).

Different results from ours were observed for the berries of *E. umbellata* and *S. lanceolata*, which showed an increase of 1.74- and 1.85-fold, respectively, after the digestion process [38]. On the other hand, the antiglycation activity of *Vaccinium myrtillus* fruit extract reduced by approximately 26% after gastrointestinal digestion, since the IC_{50} went from 70.41 mg L^{-1} in the undigested sample to 75.15 and 89.04 mg L^{-1} in the gastric and intestinal phases, respectively [42]. Likewise, the BSA glycation potential of phenolic compounds from jabuticaba (*Plinia cauliflora*) peel reduced approximately 4.27-fold after the digestive process. In that study, the IC_{50} went from 0.37 in the undigested sample to 0.59 and 1.58 mg mL^{-1} for the gastric and intestinal phases, respectively [43].

4. Conclusions

In this study, we determined for the first time the bioaccessibility of phenolic compounds from *xique-xique* and *mandacaru* fruits, as well as their antidiabetic properties during the in vitro gastrointestinal digestion process. The content of total phenolics and total flavonoids in *xique-xique* and *mandacaru* fruit reduced after the digestion process. Similarly, there was also a reduction in the content of phenolic compounds such as rutin, ρ-coumaric acid, catechin and epicatechin for both fruits, except for quinic acid, which presented an increase in its content after the digestive process in *xique-xique* fruit. Despite these results, other phenolic compounds may have been released from the plant matrices during the digestion process, which could certainly have influenced the increase in antioxidant potential in the ABTS assay for both fruits and in the ORAC assay for *mandacaru*. Likewise, the antihyperglycemic potential of the samples was positively affected after the digestive process, except for *xique-xique*, which reduced its potential to inhibit α-amylase in the intestinal phase. On the other hand, the antiglycation potential for both samples were reduced after the in vitro gastrointestinal digestion process. Therefore, the functional and antidiabetic potential of the samples depended on the matrix and digestive phase evaluated.

Finally, the results of this study may provide important insights into the impact of the digestion process on the bioactive properties of underutilized fruits from the semi-arid region of Brazil, as well as facilitating their investigation and application in the development of foods with functional claims that may contribute to the maintenance of health.

Supplementary Materials: The following supporting information can be downloaded at: https://www.mdpi.com/article/10.3390/foods13203319/s1, Figure S1. Chromatogram of xique-xique and mandacaru undigested samples in LC/MS analysis in Selected Ion Monitoring (SIM) mode; Figure S2. Chromatogram of samples from the gastric phases of xique-xique and mandacaru in LC/MS analysis in Selected Ion Monitoring (SIM) mode; Figure S3. Chromatogram of samples from the intestinal phases of xique-xique and mandacaru in LC/MS analysis in Selected Ion Monitoring (SIM) mode.

Author Contributions: F.F.d.A.: Conceptualization, data curation, formal analysis, investigation, methodology, software, visualization, writing—original draft, writing—review and editing. D.d.P.F.: Data curation, formal analysis, investigation, methodology. I.A.N.-N.: Supervision, visualization, writing—original draft, writing—review and editing. G.M.P.: Supervision, visualization, funding acquisition, resources, writing—review and editing. A.C.H.F.S.: Supervision, conceptualization, visualization, funding acquisition, resources, writing—review and editing. All authors have read and agreed to the published version of the manuscript.

Funding: This research was funded by São Paulo Research Foundation, FAPESP-Brazil, National Council for Scientific and Technological Development—CNPQ and Coordination for the Improvement of Higher Education Personnel, CAPES-Brazil (Finance Code 001). A.C.H.F. Sawaya thanks the CNPQ for a grant (306100/2021-5). F. F. de Araújo (grant# 2020/15163-6 and 2022/04530-3), and D. P. Farias (grant# 2020/00225-6) thank the FAPESP.

Data Availability Statement: The original contributions presented in the study are included in the article/Supplementary Materials, further inquiries can be directed to the corresponding author.

Conflicts of Interest: The authors declare that they have no conflicts of interest.

References

1. WHO. Noncommunicable Diseases-World Health Organization. Available online: https://www.who.int/health-topics/noncommunicable-diseases#tab=tab_1 (accessed on 30 August 2024).
2. IDF. IDF | International Diabetes Federation. Available online: https://idf.org/ (accessed on 30 August 2024).
3. de Paulo Farias, D.; de Araújo, F.F.; Neri-Numa, I.A.; Pastore, G.M. Antidiabetic Potential of Dietary Polyphenols: A Mechanistic Review. *Food Res. Int.* **2021**, *145*, 110383. [CrossRef] [PubMed]
4. Montiel-Sánchez, M.; García-Cayuela, T.; Gómez-Maqueo, A.; García, H.S.; Cano, M.P. In Vitro Gastrointestinal Stability, Bioaccessibility and Potential Biological Activities of Betalains and Phenolic Compounds in Cactus Berry Fruits (Myrtillocactus Geometrizans). *Food Chem.* **2021**, *342*, 128087. [CrossRef]

5. Araújo, F.F.d.; de Paulo Farias, D.; Neri-Numa, I.A.; Pastore, G.M. Polyphenols and Their Applications: An Approach in Food Chemistry and Innovation Potential. *Food Chem.* **2020**, *338*, 127535. [CrossRef] [PubMed]
6. de Araújo, F.F.; de Paulo Farias, D.; Neri-Numa, I.A.; Dias-Audibert, F.L.; Delafiori, J.; de Souza, F.G.; Catharino, R.R.; do Sacramento, C.K.; Pastore, G.M. Gastrointestinal Bioaccessibility and Bioactivity of Phenolic Compounds from Araçá-Boi Fruit. *LWT* **2021**, *135*, 110230. [CrossRef]
7. de Paulo Farias, D.; de Araújo, F.F.; Neri-Numa, I.A.; Dias-Audibert, F.L.; Delafiori, J.; Catharino, R.R.; Pastore, G.M. Effect of in Vitro Digestion on the Bioaccessibility and Bioactivity of Phenolic Compounds in Fractions of Eugenia Pyriformis Fruit. *Food Res. Int.* **2021**, *150*, 110767. [CrossRef]
8. Ketnawa, S.; Suwannachot, J.; Ogawa, Y. In Vitro Gastrointestinal Digestion of Crisphead Lettuce: Changes in Bioactive Compounds and Antioxidant Potential. *Food Chem.* **2020**, *311*, 125885. [CrossRef]
9. Bezerril, F.F.; de Souza, M.d.F.V.; Lima, M.d.S.; Pacheco, M.T.B.; de Carvalho, P.O.A.A.; Sampaio, K.B.; de Sousa, Y.R.F.; Milani, R.F.; Goldbeck, R.; Borges, G.d.S.C.; et al. Physicochemical Characteristics and Bioactive Compounds of the Xique-Xique (Pilosocereus Gounellei) Cactus from Caatinga Brazilian: Are They Nutritive and Functional? *J. Food Meas. Charact.* **2021**, *15*, 3284–3297. [CrossRef]
10. Fernandes de Araújo, F.; de Paulo Farias, D.; Ranpatabendi, T.; Angélica Neri-Numa, I.; Pastore, G.M.; Sawaya, A.C.H.F. Comparative Assessment of Nutritional Composition, Phenolic Compounds, Antioxidative and Antidiabetic Properties of Pilosocereus Gounellei and Cereus Jamacaru: Two Unexplored Cactus Fruits from the Brazilian Semi-Arid. *CyTA-J. Food* **2024**, *22*, 2383251. [CrossRef]
11. Fidelis, V.R.L.; Pereira, E.M.; Silva, W.P.; Gomes, J.P.; Silva, L.A. Produção de Sorvetes e Iogurtes a Partir Dos Frutos Figo Da Índia e Mandacaru. *Rev. Verde Agroecol. E Desenvolv. Sustentável* **2015**, *10*, 17. [CrossRef]
12. Pinto, M.P. *Desenvolvimento de Sorvete à Base de Polpa de Mandacaru e Xiquexique*; Instituto Federal do Piauí: Teresina, Brazil, 2017.
13. Silva, C.E. *Desenvolvimento, Caracterização e Análise Sensorial de Bolo a Partir Da Farinha de Xiquexique (Pilosocereus Gounellei)*; Instituto Federal do Sertão Pernambucano: Salgueiro, Brazil, 2019.
14. De Assis, P.O.A.; Guerra, G.C.B.; Araújo, D.F.D.S.; De Andrade, L.D.F.L.I.; De Araújo, A.A.; De Araújo, R.F.; De Carvalho, T.G.; De Souza, M.D.F.V.; Borges, G.D.S.C.; Lima, M.D.S.; et al. Intestinal Anti-Inflammatory Activity of Xique–Xique (Pilosocereus Gounellei A. Weber Ex K. Schum. Bly. Ex Rowl) Juice on Acetic Acid-Induced Colitis in Rats. *Food Funct.* **2019**, *10*, 7275–7290. [CrossRef]
15. Da-Silva, J.R.; Correia-Lima, L.; Fernandes, G.; Ribeiro-Filho, N.; Madruga, M.S.; Lima, M.d.S.; Muniz, M.B. Mandacaru Fruit Pulp (Cereus Jamacaru D.C.) as an Adjunct and Its Influence on Beer Properties. *Food Chem.* **2023**, *406*, 135066. [CrossRef] [PubMed]
16. de Araújo, F.F.; de Paulo Farias, D.; Neri-Numa, I.A.; Pastore, G.M. Underutilized Plants of the Cactaceae Family: Nutritional Aspects and Technological Applications. *Food Chem.* **2021**, *362*, 130196. [CrossRef]
17. Sancho, R.A.S.; Souza, J.D.R.P.; de Lima, F.A.; Pastore, G.M. Evaluation of Oligosaccharide Profiles in Selected Cooked Tubers and Roots Subjected to in Vitro Digestion. *LWT-Food Sci. Technol.* **2017**, *76*, 270–277. [CrossRef]
18. Cassola, F.; Reis da Silva, M.H.; Borghi, A.A.; Lusa, M.G.; Sawaya, A.C.H.F.; Garcia, V.L.; Mayer, J.L.S. Morphoanatomical Characteristics, Chemical Profiles, and Antioxidant Activity of Three Species of *Justicia* L. (Acanthaceae) under Different Growth Conditions. *Ind. Crops Prod.* **2019**, *131*, 257–265. [CrossRef]
19. Singleton, V.L.; Orthofer, R.; Lamuela-Raventós, R.M. Analysis of Total Phenols and Other Oxidation Substrates and Antioxidants by Means of Folin-Ciocalteu Reagent. *Methods Enzym. Enzymol.* **1999**, *299*, 152–178. [CrossRef]
20. Zhishen, J.; Mengcheng, T.; Jianming, W. The Determination of Flavonoid Contents in Mulberry and Their Scavenging Effects on Superoxide Radicals. *Food Chem.* **1999**, *64*, 555–559. [CrossRef]
21. Re, R.; Pellegrini, N.; Proteggente, A.; Pannala, A.; Yang, M.; Rice-Evans, C. Antioxidant Activity Applying an Improved ABTS Radical Cation Decolorization Assay. *Free Radic. Biol. Med.* **1999**, *26*, 1231–1237. [CrossRef] [PubMed]
22. Prior, R.L.; Hoang, H.; Gu, L.; Wu, X.; Bacchiocca, M.; Howard, L.; Hampsch-Woodill, M.; Huang, D.; Ou, B.; Jacob, R. Assays for Hydrophilic and Lipophilic Antioxidant Capacity (Oxygen Radical Absorbance Capacity ($ORAC_{FL}$)) of Plasma and Other Biological and Food Samples. *J. Agric. Food Chem.* **2003**, *51*, 3273–3279. [CrossRef]
23. Van Quan, N.; Tran, H.D.; Xuan, T.D.; Ahmad, A.; Dat, T.D.; Khanh, T.D.; Teschke, R. Momilactones A and B Are α-Amylase and α-Glucosidase Inhibitors. *Molecules* **2019**, *24*, 482. [CrossRef]
24. Sri Harsha, P.S.C.; Gardana, C.; Simonetti, P.; Spigno, G.; Lavelli, V. Characterization of Phenolics, in Vitro Reducing Capacity and Anti-Glycation Activity of Red Grape Skins Recovered from Winemaking by-Products. *Bioresour. Technol.* **2013**, *140*, 263–268. [CrossRef]
25. *Statsoft Statistica (Data Analysis Software System)*, TIBCO: Palo Alto, CA, USA, 2004.
26. De Santiago, E.; Pereira-Caro, G.; Moreno-Rojas, J.M.; Cid, C.; De Peña, M.P. Digestibility of (Poly)Phenols and Antioxidant Activity in Raw and Cooked Cactus Cladodes (Opuntia Ficus-Indica). *J. Agric. Food Chem.* **2018**, *66*, 5832–5844. [CrossRef] [PubMed]
27. De Jesús Cenobio-Galindo, A.; Díaz-Monroy, G.; Medina-Pérez, G.; Jesús Franco-Fernández, M.; Ludeña-Urquizo, F.E.; Vieyra-Alberto, R.; Campos-Montiel, R.G. Multiple Emulsions with Extracts of Cactus Pear Added in A Yogurt: Antioxidant Activity, In Vitro Simulated Digestion and Shelf Life. *Foods* **2019**, *8*, 429. [CrossRef] [PubMed]

28. Huang, Y.; He, M.; Kasapis, S.; Brennan, M.; Brennan, C. The Influence of the Fortification of Red Pitaya (Hylocereus Polyrhizus) Powder on the in Vitro Digestion, Physical Parameters, Nutritional Profile, Polyphenols and Antioxidant Activity in the Oat-Wheat Bread. *Int. J. Food Sci. Technol.* **2022**, *57*, 2729–2738. [CrossRef]
29. Saniovanni, E.; Di Lorenzo, C.; Colombo, E.; Colombo, F.; Fumagalli, M.; Frigerio, G.; Restani, P.; DellÁgli, M. The Effect of in Vitro Gastrointestinal Digestion on the Anti-Inflammatory Activity of *Vitis Vinifera* L. Leaves. *Food Funct.* **2015**, *6*, 2453. [CrossRef]
30. Correa-Betanzo, J.; Allen-Vercoe, E.; McDonald, J.; Schroeter, K.; Corredig, M.; Paliyath, G. Stability and Biological Activity of Wild Blueberry (Vaccinium Angustifolium) Polyphenols during Simulated in Vitro Gastrointestinal Digestion. *Food Chem.* **2014**, *165*, 522–531. [CrossRef]
31. Sánchez-Velázquez, O.A.; Mulero, M.; Cuevas-Rodríguez, E.O.; Mondor, M.; Arcand, Y.; Hernández-Álvarez, A.J. In Vitro Gastrointestinal Digestion Impact on Stability, Bioaccessibility and Antioxidant Activity of Polyphenols from Wild and Commercial Blackberries (*Rubus* Spp.). *Food Funct.* **2021**, *12*, 7358–7378. [CrossRef]
32. Zhu, Y.; Yang, S.; Huang, Y.; Huang, J.; Li, Y. Effect of in Vitro Gastrointestinal Digestion on Phenolic Compounds and Antioxidant Properties of Soluble and Insoluble Dietary Fibers Derived from Hulless Barley. *J. Food Sci.* **2021**, *86*, 628–634. [CrossRef]
33. Feng, J.; Kong, F. Enzyme Inhibitory Activities of Phenolic Compounds in Pecan and the Effect on Starch Digestion. *Int. J. Biol. Macromol.* **2022**, *220*, 117–123. [CrossRef]
34. Alqahtani, A.S.; Hidayathulla, S.; Rehman, M.T.; Elgamal, A.A.; Al-Massarani, S.; Razmovski-Naumovski, V.; Alqahtani, M.S.; El Dib, R.A.; Alajmi, M.F. Alpha-Amylase and Alpha-Glucosidase Enzyme Inhibition and Antioxidant Potential of 3-Oxolupenal and Katononic Acid Isolated from Nuxia Oppositifolia. *Biomolecules* **2020**, *10*, 61. [CrossRef]
35. Wang, C.; Li, W.; Chen, Z.; Gao, X.; Yuan, G.; Pan, Y.; Chen, H. Effects of Simulated Gastrointestinal Digestion in Vitro on the Chemical Properties, Antioxidant Activity, α-Amylase and α-Glucosidase Inhibitory Activity of Polysaccharides from Inonotus Obliquus. *Food Res. Int.* **2018**, *103*, 280–288. [CrossRef]
36. Kasipandi, M.; Manikandan, A.; Sreeja, P.S.; Suman, T.; Saikumar, S.; Dhivya, S.; Parimelazhagan, T. Effects of in Vitro Simulated Gastrointestinal Digestion on the Antioxidant, α-Glucosidase and α-Amylase Inhibitory Activities of Water-Soluble Polysaccharides from Opilia Amentacea Roxb Fruit. *LWT* **2019**, *111*, 774–781. [CrossRef]
37. Gutiérrez-Grijalva, E.P.; Antunes-Ricardo, M.; Acosta-Estrada, B.A.; Gutiérrez-Uribe, J.A.; Basilio Heredia, J. Cellular Antioxidant Activity and in Vitro Inhibition of α-Glucosidase, α-Amylase and Pancreatic Lipase of Oregano Polyphenols under Simulated Gastrointestinal Digestion. *Food Res. Int.* **2019**, *116*, 676–686. [CrossRef] [PubMed]
38. Spínola, V.; Pinto, J.; Llorent-Martínez, E.J.; Castilho, P.C. Changes in the Phenolic Compositions of Elaeagnus Umbellata and Sambucus Lanceolata after in Vitro Gastrointestinal Digestion and Evaluation of Their Potential Anti-Diabetic Properties. *Food Res. Int.* **2019**, *122*, 283–294. [CrossRef] [PubMed]
39. Perrone, A.; Giovino, A.; Benny, J.; Martinelli, F. Advanced Glycation End Products (AGEs): Biochemistry, Signaling, Analytical Methods, and Epigenetic Effects. *Oxid. Med. Cell Longev.* **2020**, *2020*, 3818196. [CrossRef] [PubMed]
40. Husain, A.; Alouffi, S.; Khanam, A.; Akasha, R.; Farooqui, A.; Ahmad, S. Therapeutic Efficacy of Natural Product 'C-Phycocyanin' in Alleviating Streptozotocin-Induced Diabetes via the Inhibition of Glycation Reaction in Rats. *Int. J. Mol. Sci.* **2022**, *23*, 14235. [CrossRef]
41. Dubey, S.; Ganeshpurkar, A.; Ganeshpurkar, A.; Bansal, D.; Dubey, N. Glycolytic Enzyme Inhibitory and Antiglycation Potential of Rutin. *Futur. J. Pharm. Sci.* **2017**, *3*, 158–162. [CrossRef]
42. Fraisse, D.; Bred, A.; Felgines, C.; Senejoux, F. Stability and Antiglycoxidant Potential of Bilberry Anthocyanins in Simulated Gastrointestinal Tract Model. *Foods* **2020**, *9*, 1695. [CrossRef]
43. Borges, L.L.R.; Oliveira, L.L.d.; Freitas, V.V.; Leite Júnior, B.R.d.C.; Nascimento, A.L.A.A.; Castro, G.A.D.; Fernandes, S.A.; Stringheta, P.C. Digestive Enzymes Inhibition, Antioxidant and Antiglycation Activities of Phenolic Compounds from Jabuticaba (Plinia Cauliflora) Peel. *Food Biosci.* **2022**, *50*, 102195. [CrossRef]

foods

MDPI

Article

In Vitro Antioxidant, Anti-Inflammatory Activity and Bioaccessibility of Ethanolic Extracts from Mexican *Moringa oleifera* Leaf

Erasmo Herman-Lara [1], Jesús Rodríguez-Miranda [1], Stefany Ávila-Manrique [2], Celia Dorado-López [2], Marisol Villalva [2], Laura Jaime [2], Susana Santoyo [2] and Cecilia E. Martínez-Sánchez [1,*]

[1] Tecnológico Nacional de México Campus, Tuxtepec, Calzada Victor Bravo Ahuja, No. 561, Col. Predio el Paraíso, San Juan Bautista Tuxtepec 68350, Oaxaca, Mexico; erasmo.hy@tuxtepec.tecnm.mx (E.H.-L.); jesus.rm@tuxtepec.tecnm.mx (J.R.-M.)

[2] Institute of Food Science Research (CIAL), Universidad Autónoma de Madrid (CEI, UAM-CSIC), 28049 Madrid, Spain; stefyavilam@gmail.com (S.Á.-M.); cece.dorado@gmail.com (C.D.-L.); marisol.villalva@uam.es (M.V.); laura.jaime@uam.es (L.J.); susana.santoyo@uam.es (S.S.)

* Correspondence: cecilia.ms@tuxtepec.tecnm.mx; Tel.: +52-287-8752170 (ext. 151)

Abstract: This study aimed to assess the antioxidant and anti-inflammatory properties, and bioaccessibility of *Moringa oleifera* ethanolic extracts using pressurized liquid extraction with varying ethanol concentrations (0%, 30%, 50%, 70%, and 100%) in water–ethanol mixtures. Quercetin derivatives and neochlorogenic acid were identified as major compounds via high-performance liquid chromatography with diode array detection. The 70% ethanol extract displayed the highest antioxidant activity and phenolic content, highlighting a strong correlation between phenolics and antioxidant potential. Extracts prepared with 50% and 70% ethanol (30 μg/mL) significantly inhibited TNF-α, IL-1β, and IL-6 cytokine secretion, with the 70% ethanol extract demonstrating robust anti-inflammatory effects. During in vitro digestion (oral, gastric, and intestinal phases), minimal changes were noted in most phenolic compounds' post-oral phase, but reductions occurred after the gastric phase. Substantial decreases in major compounds and antioxidant activity were observed in post-gastric and intestinal phases. Overall, ethanolic extracts of *Moringa oleifera*, particularly those with 70% ethanol, exhibit promising antioxidant and anti-inflammatory properties, suggesting potential for developing therapeutic agents against oxidative stress and inflammation-related disorders. However, it is essential to protect these compounds to prevent their degradation during digestion.

Keywords: anti-inflammatory; antioxidant; *Moringa oleifera*; pressurized fluids extraction

1. Introduction

The Moringa tree (*Moringa oleifera*) is an important part of traditional Mexican horticulture. It is valued for its beauty and its uses in traditional medicine. Originally from the southern Himalayas—India, Bangladesh, Afghanistan, and Pakistan—it is now widely grown for tea, forage, and as a component in traditional medicine with other herbs [1,2]. In recent years, *M. oleifera* has received significant attention for being a rich source of bioactive compounds with diverse biological properties [3–7].

A large number of macronutrients, micronutrients, and secondary metabolites have been recognized in moringa leaves, such as carotenoids, polyphenols, phenolic acids, 10 flavonoids, alkaloids, glucosides, glucosinolates (thioglucoside), isothiocyanates, tannins, saponins, oxalates, and phytates [8,9]. The leaves are a rich source of omega 3, omega 6, and polyunsaturated fatty acids (represented by linoleic and linolenic acid). Palmitic acid occurs as the main saturated fatty acid. Regarding the mineral content, we can find potassium (K), calcium (Ca), and iron (Fe) and, to a lesser extent, magnesium (Mg) [10]. Among the vitamins, moringa leaves contain vitamin A, carotenoids, vitamin C, vitamin

E (tocopherols), and some group B vitamins [9,11]. Dried moringa leaves are a source of polyphenols, such as phenolic acids and flavonoids. The main flavonoids that we find are myricetin, quercetin (also found as isoquercitin, that is, quercetin 3-O-glucoside), and kaempferol (also found as astragalin, that is, kaempferol 3-O-glucoside). However, the amount of these metabolites in *M. oleifera* extracts varies according to the geographical location, soil, sun exposure, and climatic conditions. Furthermore, the method and solvents used for its extraction can also modify the content of plant compounds, mainly phenols and flavonoids [12].

The modern food industry is now focused on developing functional foods and ingredients that aim to reduce disease risks [13]. This change reflects growing consumer interest in healthier dietary choices and the exploration of natural sources of functional ingredients [14]. At the same time, traditional medicine continues to use plants—whether fresh, dried, or in infusions and poultices—to treat a wide range of ailments. Despite these developments, research on Mexican *M. oleifera* using pressurized liquid extraction (PLE) to evaluate its antioxidant and anti-inflammatory activities in vitro is still limited. In vitro digestion models are widely used in research to predict the bioaccessibility of the studied compounds due to several advantages with respect to in vivo models, since they are relatively cheap and simple, faster, do not present ethical restrictions, the conditions can be controlled, sampling is easy, and the results are reproducible [15]. Furthermore, the evaluation of bioaccessibility using this type of model is well correlated with data obtained in animal and human studies [16]. Bioaccessibility has been defined as the amount of an ingested nutrient that is released from a food matrix and is available for intestinal absorption [17]. In the case of phenolic compounds, different food matrices and plant extracts have been subjected to gastrointestinal digestion in order to evaluate their bioaccessibility. Some studies have shown that phenolic compounds can modify their structure during digestion and, therefore, affect their biological activity. One of the most important factors to consider in this digestive process is the change in pH, since this can cause the decomposition and/or transformation of compounds [18]. An example is the case of the study carried out by [19] in which they determined that the carnosic acid present in a rosemary extract was transformed into carnosol due to a possible formation of the epoxy group due to the effect of intestinal pH.

Likewise, interactions with digestive enzymes, bile salts, or other compounds, in addition to other effects (precipitations, dimerizations, or conversions [20–22]) can also modify the chemical structure of phenolic compounds and, therefore, their bioaccessibility, thus affecting their biological activity [23].

Therefore, this study aims to evaluate ethanolic extracts from Mexican *M. oleifera* leaves, with a focus on their potential as natural sources of antioxidant and anti-inflammatory compounds. Additionally, this study aims to assess the in vitro bioaccessibility of the phenolic compounds extracted using pressurized fluids.

While *M. oleifera* has been studied extensively for its bioactive compounds, this study uniquely applies pressurized liquid extraction (PLE) with varying ethanol concentrations to optimize the extraction of phenolic compounds. This systematic approach to varying ethanol concentrations in PLE to assess both the yield and bioactivity of the extracts is innovative and adds depth to the existing body of knowledge on *M. oleifera*. Also, this study provides a thorough investigation of the antioxidant and anti-inflammatory properties of the extracts, specifically highlighting how different ethanol concentrations influence these properties. The in vitro digestion model used to evaluate the bioaccessibility of phenolic compounds adds a critical dimension to the research.

2. Materials and Methods

2.1. Preparation of Samples

M. oleifera leaves were harvested in San Juan Bautista Tuxtepec, Oaxaca, and dried in the shade at temperatures ranging from 27 to 30 °C over three days. After drying, the leaves were ground to achieve a particle size of 0.59 mm and subsequently sieved. Moisture content of the samples was 8.73% ± 0.12. The results for the extracts were reported on a dry weight basis. Analytical reagents and standards (reagent grade and HPLC) were purchased from Sigma-Aldrich, Spain.

2.2. Pressurized Liquid Extraction (PLE)

Extraction of *M. oleifera* samples was conducted using a Dionex system (Dionex Corporation, Sunnyvale, CA, USA) equipped with a solvent controller unit. Five ethanol: water (v/v) solutions with concentrations of 0:100, 30:70, 50:50, 70:30, and 100:0 were employed as extraction solvents. Each sample (1.0 g) was combined with 4.0 g of sea sand in a 10 mL stainless steel cell. The extraction cell was automatically filled with the appropriate solvent under a pressure of 1500 psi. The heating time for the extraction cell was 10 min at 100 °C. Following the heating phase, a static extraction was conducted. After the extraction, the cell was rinsed with a volume of solvent equivalent to 60% of the cell's capacity. To ensure thorough removal of residual solvent, the cell was purged with pressurized nitrogen gas (N_2) for 90 s. Additionally, a complete system rinse was performed between each extraction to prevent cross-contamination and ensure consistency. Following extraction, solvents were evaporated using an IKA RV 10 rotary evaporator (VWR International, Spain), and extracts were freeze-dried (Telstar Lyobeta 15 equipment; Telstar, Madrid, Spain). Powder extracts were kept at −20 °C until analysis [14].

2.3. Determination of Total Phenol Content (TPC) and Total Flavonoid Content (TFC)

TPC was assessed using the Folin–Ciocalteau method [24] with results expressed as mg gallic acid equivalents (GAE) per gram. A 125 mL sample was combined with 500 mL of distilled water and 125 mL of Folin–Ciocalteu reagent, and the mixture was allowed to stand undisturbed for 6 min. Following this, 1.25 mL of a 7% Na_2CO_3 solution and 1 mL of distilled water were added. The resulting mixture was kept in the dark at 25 °C for 90 min without stirring. After the incubation period, a spectrophotometer (Cary 60 UV-Vis, Agilent Technologies Inc., Santa Clara, CA, USA) was used to measure the absorbance at 760 nm. In parallel, a control procedure was conducted using distilled water. Each sample was analyzed in triplicate. Total flavonoid content (TFC) was determined using the method described by Heimler et al. [24]. In a test tube, 1250 µL of deionized water was mixed with 100 µL of the sample or standard (with deionized water serving as the blank) and 75 µL of 5% $NaNO_2$ solution. The mixture was allowed to stand for 6 min, after which 150 µL of 10% $AlCl_3$ solution was added and left to stand for another 5 min. Subsequently, 500 µL of 1 M NaOH was added, and the volume was adjusted to 2.5 mL with 425 µL of deionized water. After a 30 min incubation period, the absorbance of the samples was measured at 510 nm using a spectrophotometer (Agilent Technologies, Cary 60 UV–Vis, USA). The results were expressed as mg of catechin equivalents (mg EC) per gram.

2.4. Determination of Antioxidant Activity (ABTS•+ and DPPH•)

The ABTS•+ radical scavenging activity was determined using the method described by Re et al. [25]. First, 77.6 mg of ABTS•+ reagent was dissolved in 20 mL of distilled water to prepare a 7 mM aqueous solution. Then, 13.2 mg of potassium persulfate (2.45 mM) was added to the ABTS•+ solution in an amber bottle. The mixture was homogenized, covered with aluminum foil, and incubated at room temperature in complete darkness for 12 to 16 h, allowing the formation of the ABTS•+ radical. The ABTS•+ radical solution, stable for up to two days under these conditions, was then diluted with absolute ethanol until an initial absorbance of 0.7 ± 0.05 at 732 nm was achieved. To assess the radical scavenging activity, the ABTS•+ solution was mixed with the samples at various concentrations (10,000, 5000, 2500, 1000, 500, 250, 50, and 25 mg/L) in a 0.1:10 (v/v) ratio. The decrease in absorbance was measured, and the antiradical activity was calculated. The DPPH• (2,2-diphenyl-1-picrylhydrazyl) assay was performed to evaluate the antioxidant activity of the samples. First, 2.4 mg of DPPH• was dissolved in 100 mL of methanol to prepare the DPPH• solution. The samples were initially prepared at a concentration of 10,000 mg/L, followed by serial dilutions to obtain concentrations of 5000, 2500, 1000, and 500 mg/L. For the assay, 975 μL of the DPPH• solution was added to 25 μL of each sample in aluminum-lined Eppendorf tubes. The mixtures were incubated in total darkness for 15 min to ensure consistent reaction conditions. After the incubation period, the absorbance of each sample was measured at 515 nm using a spectrophotometer (Agilent Technologies, Cary 60 UV–Vis, USA). The antioxidant activity of the samples was quantified and expressed as mmol of Trolox equivalents per gram of extract (TEAC value) [25].

2.5. *Analysis of Phenolic Compounds by HPLC-DAD*

The phenolic compounds in the extracts were identified using high-performance liquid chromatography (HPLC). An Infinity 1260 chromatograph (Agilent Technologies, Madrid, Spain) equipped with a photodiode array detector (PAD) was employed. Separation was achieved on a reverse-phase C18 column (ACE Excel 3 Super C18, Symta, Madrid, Spain) with dimensions of 150 mm × 4.6 mm and a particle size of 3 μm. The column was protected by an ACE 3 C18-AR pre-column (7 mm × 13 mm) containing the same stationary phase. The column temperature was maintained at 30 °C, and the injection volume was set to 40 μL. The mobile phase consisted of two solvents: (A) 2% aqueous formic acid and (B) acetonitrile with 2% formic acid. The flow rate was kept constant at 1 mL/min. The gradient elution program was structured as follows: 0 min: 0% B, 1 min: 100% A, 6 min: 85% A, 21 min: 75% A, 26 min: 65% A, 36 min: 50% A, 43 min: 0% A, 48 min: 100% B, 50 min: 100% A, and 55 min: 100% A. Before injection, all samples were diluted in 970 μL of MilliQ ultrapure water, and 370 μL of the internal standard (ethyl gallate) at a concentration of 1 mg/mL was added. The samples were then filtered through a 0.45 μm PVDF filter. Quantification of the phenolic compounds was carried out using external calibration curves with appropriate analytical standards. Detection wavelengths were set according to the compound class: Hydroxybenzoic acids and flavan-3-ols: 280 nm, Hydroxycinnamic acids, and stilbenes: 320 nm and Flavonols: 360 nm.

2.6. *Cell Culture*

To investigate the anti-inflammatory activity of the extracts, we used the THP-1 human monocyte cell line ATCC TIB-202, a human myelomonocytic cell line displaying macrophage-like activity. The cells were cultured and expanded in complete RPMI 1640 medium, which contained 1% HEPES, 100 U/mL penicillin, 100 μg/mL streptomycin, 2 mM L-glutamine, and 10% fetal bovine serum (FBS) (Gibco, Spain). Cultures were maintained at a density of 2–9 × 10^5 cells/mL in a 37 °C incubator with 5% CO_2. Cells from up to 25 passages were used for all experiments. Monocyte differentiation into macrophages was induced by treating the cells with 100 ng/mL phorbol 12-myristate 13-acetate (PMA) (Sigma-Aldrich, Madrid, Spain) for 48 h, as described by Takashiba et al. [26].

2.7. *Cytotoxicity Assays*

The cytotoxicity of the samples was assessed using the MTT assay, which measures mitochondrial respiration based on the reduction of 3-(4,5-dimethylthiazol-2-yl)-2,5-diphenyltetrazolium bromide (MTT) (Sigma-Aldrich, Spain) by succinate dehydrogenase, an enzyme in the mitochondria. This reduction converts MTT into formazan, a colored compound, with the amount of formazan formed being proportional to the number of viable cells. THP-1 cells (5 × 10^5 cells per well) were plated in a 24-well plate. After differentiation into macrophages, the cells were treated with the samples at concentrations of 30, 20, and 10 μg/mL and incubated for 24 h. Following treatment, 0.5 mL of a 0.5 mg/mL MTT solution was added, and the cells were incubated for an additional 3 h at 37 °C with 5% CO_2. To solubilize the formazan, the cells were lysed with a 1:1 mixture of dimethyl sulfoxide (DMSO) (Panreac, Barcelona, Spain) and ethanol. The absorbance of the resulting solution was measured at 540 nm using a plate reader (Sunrise Remote, Tecan, Barcelona, Spain). The optical density of the formazan in the control wells (without sample) was considered as 100% viability, providing a baseline for comparison [27].

2.8. Anti-Inflammatory Activity

Macrophages underwent treatment with various non-cytotoxic sample concentrations and were then stimulated with 0.05 μg/mL bacterial lipopolysaccharide (LPS) (Sigma-Aldrich, Spain) to induce a non-specific inflammatory response. After an incubation period of 24 h at 37 °C with 5% CO_2, samples of the supernatant were collected and stored at −20 °C for subsequent analysis. Macrophages treated with LPS alone were used as positive controls, while non-stimulated macrophages served as the negative controls. The quantification of cytokines released into the culture medium was carried out using a sandwich enzyme-linked immunosorbent assay (ELISA) kit (BD Biosciences, Almaraz, Cáceres, Extremadura, Spain) in accordance with the manufacturer's instructions. Three pro-inflammatory cytokines—IL-6, IL-1β, and TNF-α—were measured. The ELISA was conducted in 96 well plates (Falcon, Corning, VWR, Aurora, OH, USA) that were pre-treated for primary antibody adhesion. The plates were incubated overnight at 4 °C, then washed and equilibrated to room temperature for one hour. Standards for each cytokine and 10 μL of cell supernatant samples, along with their respective controls, were added to the wells. The plate was incubated in the dark for two hours, washed, and then incubated with the secondary antibody and enzyme conjugate for one hour. After washing, an enzyme substrate solution was added, and the reaction was stopped with a stop solution after 30 min. Absorbance was measured at 450 nm with a reference wavelength of 570 nm using a plate reader (Sunrise Remote, Tecan, Männedorf, Zurich, Switzerland) [28].

2.9. In Vitro Gastrointestinal Digestion

For the in vitro digestions, a Titrino plus 877 titrator (Metrohm, Herisau, Switzerland) equipped with a double-jacketed conical vessel and magnetic stirrer was employed and maintained at 37 °C with a constant stream of water. This setup includes an electrode for pH control and a dispenser for titrant solution. The digestion of the moringa extract samples was followed by the method proposed by Soler-Rivas et al. [19] with some adjustments. Initially, a sample solution was prepared at a concentration of 20 mg/mL in a solvent mixture. Specifically, 110 mg of the extract was dissolved in 5.5 mL of a 70:30 ethanol: water mixture.

The oral phase commenced with the addition of 0.1 mL of a salivary solution containing 9.3 mg of human salivary α-amylase type XIII-A in 1 mM calcium chloride at a concentration of 50 U/mg. The mixture was stirred at 37 °C for 2 min. Subsequently, the gastric phase involved adding 25 mL of gastric solution, consisting of 127.3 mg of porcine pepsin from gastric mucosa (536 U/mg), dissolved in 25 mL of acidified water adjusted to pH 2 with 6 M hydrochloric acid. This gastric mixture was incubated under agitation in the dark at 37 °C for one hour. Following the gastric phase, 0.463 mL of a 325 mM calcium chloride solution and 1.389 mL of a 3.25 M sodium chloride solution (Sigma-Aldrich, Madrid, Spain) were added. The pH of the mixture was then adjusted to 7.5 using 0.1 M NaOH to initiate the intestinal digestion phase. This phase began with the addition of 115.7 mg of bile salts and 2.8 mL of a pancreatic solution comprising 9.3 mg of pancreatin (Sigma-Aldrich, Madrid, Spain) in 2.8 mL of 10 mM trizma-maleate buffer. The mixture was stirred in the dark for 2 h, with periodic adjustments of NaOH to maintain a constant pH. Upon completion of digestion, the samples were filtered through a 0.45 μm PVDF filter.

2.10. Statistical Analysis

Data were analyzed using STATGRAPHICS Centurion XVII.II software. One-way analysis of variance (ANOVA) followed by the least significant difference (LSD) test were applied for mean discrimination ($p \leq 0.05$). All experiments were performed in triplicate.

3. Results

3.1. Yield Results of the Hydroalcoholic Extracts Obtained

The extraction yields were initially determined and are presented in Table 1. The results showed a decrease in yield with an increasing ethanol concentration compared to water. Interestingly, there were no significant differences in the extracts obtained with 0% and 30% ethanol, or with 50% and 70% ethanol. However, using 100% ethanol as a solvent resulted in a roughly 50% reduction in the extraction yield. These findings are attributed to the compounds found in moringa leaves, which are more soluble in water than in ethanol.

Table 1. Extraction yield of various extracts in dry basis.

Ethanol (%)	Yield (mg Extract/g)
0	33.73 ± 1.65 [a]
30	33.89 ± 0.40 [a]
50	31.34 ± 0.33 [b]
70	29.27 ± 0.62 [b]
100	16.00 ± 1.33 [c]

Different superscript letters denote significant differences between samples in the same column ($p < 0.05$).

This is likely due to its strong solvent properties that may not efficiently extract all components from the plant material.

3.2. Total Phenol Content (TPC) and Total Flavonoid Content (TFC)

An increasing ethanol concentration is correlated with a higher total phenolic content (TPC) and total flavonoid content (TFC) in the extracts (Table 2). Specifically, the extract using 70% ethanol showed the highest phenolic compound content at 111.30 mg GAE/g. However, there was no significant difference ($p > 0.05$) between the 100% ethanol and ethanol-free extracts, which exhibited lower values at 76.34 and 73.06 mg GAE/g, respectively. This indicates that TPC increased proportionally with the ethanol concentration, peaking at 70% ethanol. Conversely, TPC decreased when the ethanol concentration was raised to 100%, suggesting the reduced solubility of phenolic compounds compared to water–ethanol blends. These results imply that phenolic compounds from *M. oleifera* leaves are more soluble in water–ethanol mixtures than in either water or pure ethanol alone. This finding aligns with previous research; for instance, Wang et al. [29] optimized flavonoid extraction (such as quercetin and kaempferol) from Moringa leaves using pressurized liquid extraction (PLE) with water–ethanol mixtures containing 70% ethanol. Similarly, Rodríguez-Pérez et al. [3] reported the enhanced extraction of phenolic compounds from Moringa leaves using ultrasound-assisted extraction with water–ethanol mixtures containing 50% ethanol. These findings underscore the importance of the ethanol concentration in optimizing the extraction of bioactive compounds from *M. oleifera*, particularly emphasizing the advantages of ethanol–water blends over pure ethanol or water alone.

Table 2. Analysis of total phenolic compounds (TPC), total flavonoids (TFC), and antioxidant activity (ABTS•+ and DPPH•) in the various extracts.

Ethanol (%)	TPC (mg GAE/g Extract)	TFC (mg Q/g Extract)	ABTS•+ (mmol Trolox/g Extract)	DPPH• (mmol Trolox/g Extract)
0	73.06 ± 5.80 [c]	18.80 ± 1.16 [d]	0.36 ± 0.02 [c]	0.26 ± 0.01 [d]
30	96.04 ± 7.58 [b]	24.90 ± 0.71 [c]	0.41 ± 0.01 [b]	0.31 ± 0.01 [c]
50	99.07 ± 5.33 [b]	29.85 ± 1.36 [b]	0.43 ± 0.00 [b]	0.34 ± 0.01 [b]
70	111.30 ± 1.43 [a]	33.20 ± 2.05 [b]	0.52 ± 0.01 [a]	0.46 ± 0.00 [a]
100	76.34 ± 3.94 [c]	51.00 ± 4.79 [a]	0.27 ± 0.02 [d]	0.26 ± 0.00 [d]

Different letters in the same column indicate significant differences ($p < 0.05$) between the extracts.

The maximum total flavonoid content was achieved with a 100% ethanol extraction. The observed increase in flavonoid extraction with the ethanol percentage supports findings from other studies on phenolic compounds and flavonoid extraction from *M. oleifera* leaves [30–32]. Hence, ethanol proves effective for extracting flavonoids, as suggested by Nobossé et al. [33].

3.3. Antioxidant Activities (ABTS•⁺ and DPPH•)

Both ABTS•⁺ and DPPH• assays demonstrated increased values with higher ethanol concentrations (Table 2). However, a decline in antioxidant activity was noted with 100% ethanol extraction. Peak values for both ABTS•+ (0.52 mmol TEAC/g) and DPPH• (0.46 mmol TEAC/g) were recorded with 70% ethanol extracts. These results correlated well with the phenolic compound content in the extracts. The 70% ethanol extract, which exhibited the highest antioxidant activity among the tested methods, also contained the greatest amount of total phenolic compounds. Rocchetti et al. [6] and Saucedo-Pompa et al. [34] have similarly attributed antioxidant activity in plant extracts to the phenolic compound content. Udechukwu et al. [35] emphasized the role of phenolic compounds like chlorogenic acid, quercetin, gallic acid, rutin, and kaempferol in the antioxidant activity of *M. oleifera* leaf extracts. These findings are consistent with studies by Hossain et al. [36], Ahmed et al. [37], and Oyeniran et al. [38] on moringa leaf extracts.

To establish a relationship between the TPC content and antioxidant activity, the DPPH• method was used to measure the antioxidant potential of the extracts. The findings (Table 2) reveal a peak in the antioxidant activity with 70% ethanol. Conversely, increasing the ethanol concentration to 100% markedly decreases the antioxidant activity. Consistent with these results, Braham et al. [31] demonstrated that extracts from moringa leaves using 50% water–ethanol mixtures and 70% ethanol exhibited higher levels of total phenolic compounds and greater antioxidant activity as assessed by the DPPH•method. When the total phenolic compound content of each extract is plotted against its antioxidant activity (Figure 1), a strong linear relationship between these variables emerges, with an R-squared value of 0.91. This result highlights a clear correlation between the antioxidant potential of moringa extracts and their phenolic compound content. This finding is in line with previous studies by authors such as Rocchetti et al. [6] and Saucedo-Pompa et al. [34], who have also linked the antioxidant activity in plant extracts to their phenolic compound concentrations.

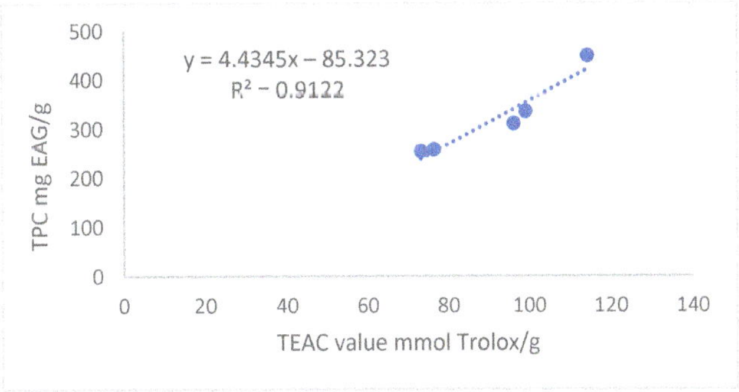

Figure 1. Correlation between TPC and Antioxidant Activity (TEAC Value) of the different extracts.

3.4. Phenolic Composition of the Extracts

The phenolic compounds present in the extracts were chemically characterized using HPLC-PAD. Table 3 shows the percentage composition of each component in the extracts. Among the prominent compounds were two derivatives of the flavonoid quercetin and neochlorogenic acid. One of the quercetin derivatives was identified as quercetin-*O*-hexoside, while the other remained unidentified. Generally, the more polar phenolic compounds (visible at the beginning of the chromatogram in Figure 2 were predominant in extracts using lower ethanol percentages (0 to 30%). Conversely, most phenolic compounds detected in moringa extracts showed increased extraction yields with higher ethanol percentages, peaking between 50% and 70%. Extracts prepared with 100% ethanol contained minimal amounts of polar phenolic compounds, focusing instead on fewer polar varieties. This finding is consistent with studies advocating ethanol or methanol–water blends for optimizing phenolic compound extraction from plant matrices [39–41]. The identified compounds (Table 3) align with previous reports on moringa leaf extracts [3,31,32,42,43].

Table 3. Phenolic composition of the different *Moringa oleifera* extracts.

Peak	Compound	Ethanol (%)				
		0	30	50	70	100
1	Unidentified	--	--	95.5 ± 0.3	90.9 ± 5.8	16.3 ± 0.6
2	Unidentified	125.4 ± 0.1	152.7 ± 0.3	69.8 ± 0.3	60.5 ± 2.9	--
3	Unidentified	--	--	70.7 ± 0.7	71.6 ± 1.7	--
4	Caffeoylquinic acid	124.7 ± 0.1	136.6 ± 2.5	68.9 ± 0.5	61.2 ± 2.1	--
5	Cryptochlorogenic acid derivative	26.9 ± 0.5	21.3 ± 0.4	408.5 ± 3.7	498.7 ± 34.4	502.7 ± 109.5
6	Neochlorogenic acid	1230.5 ± 0.8	1856.1 ± 2.3	1007.7 ± 1.8	875.3 ± 49.6	483.5 ± 30.9
7	Cryptochlorogenic acid	592.7 ± 0.2	797.6 ± 1.8	546.7 ± 0.4	497.2 ± 43.5	276.7 ± 8.2
8	Vicenina	560.5 ± 0.2	635.0 ± 1.1	694.8 ± 0.7	379.2 ± 16.5	200.7 ± 13.3
9	Vitexin	254.9 ± 0.6	383.7 ± 0.4	415.2 ± 0.4	406.3 ± 25.4	208.1 ± 14.8
10	Vitexin derivative	119.4 ± 0.5	124.8 ± 0.4	135.3 ± 0.1	169.7 ± 8.2	220.8 ± 17.1
11	Quercetin-O-hexoside	1314.3 ± 10.8	2093.2 ± 12.9	2308.6 ± 22.9	2627.5 ± 32.9	1657.8 ± 48.8
12	Quercetin derivative	1972.8 ± 0.7	2976.4 ± 4.6	3162.7 ± 3.4	2533.0 ± 20.7	348.2 ± 20.1
13	Kaempferol derivative	274.8 ± 0.1	424.6 ± 0.5	470.3 ± 0.1	524.8 ± 39.1	422.1 ± 13.0
14	Isorhamnetin-3-O-glucoside	60.7 ± 0.6	52.0 ± 0.6	72.8 ± 0.2	157.6 ± 3.3	186.3 ± 4.9
15	Kaempferol derivative	388.7 ± 0.2	583.8 ± 0.9	641.7 ± 0.1	543.3 ± 25.7	220.3 ± 6.9

Data expressed as area under the curve, RT: Retention time.

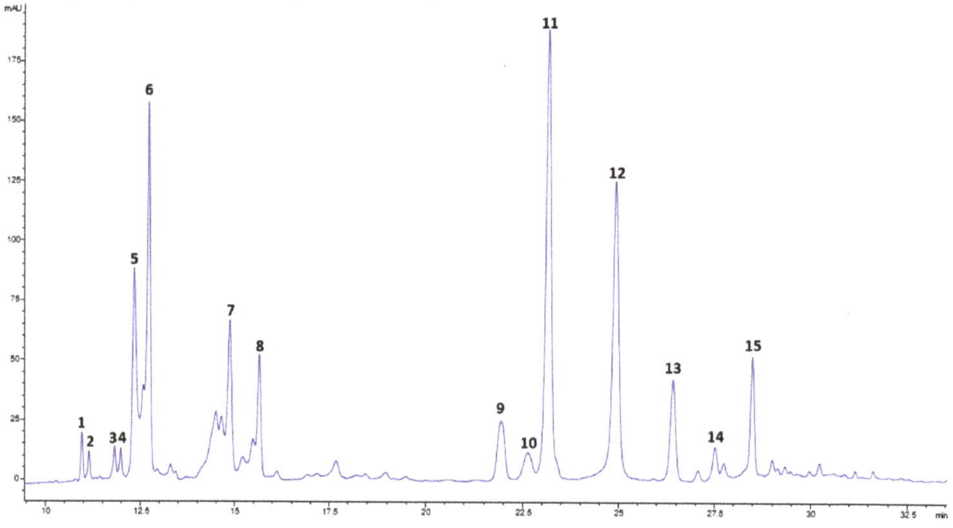

Figure 2. Chromatographic profile of moringa extract obtained with 70% ethanol.

3.5. Evaluation of Cytotoxicity and Anti-Inflammatory Activity of Moringa Extracts

3.5.1. Cytotoxicity Evaluation

Before evaluating the anti-inflammatory activity, cytotoxicity testing was conducted to determine non-toxic extract concentrations for cell cultures. Figure 3 illustrates the results for the highest concentration used (30 µg/mL), indicating no significant differences ($p < 0.05$) between extracts and controls across tested ethanol concentrations, thereby confirming non-cytotoxicity. According to the ISO 10993-5:2009 standard [40], if the relative cell viability for the highest concentration of the test sample is >70% of the control group, then the material shall be considered non-cytotoxic. Consequently, concentrations of 20 and 30 µg/mL were utilized in subsequent in vitro experiments.

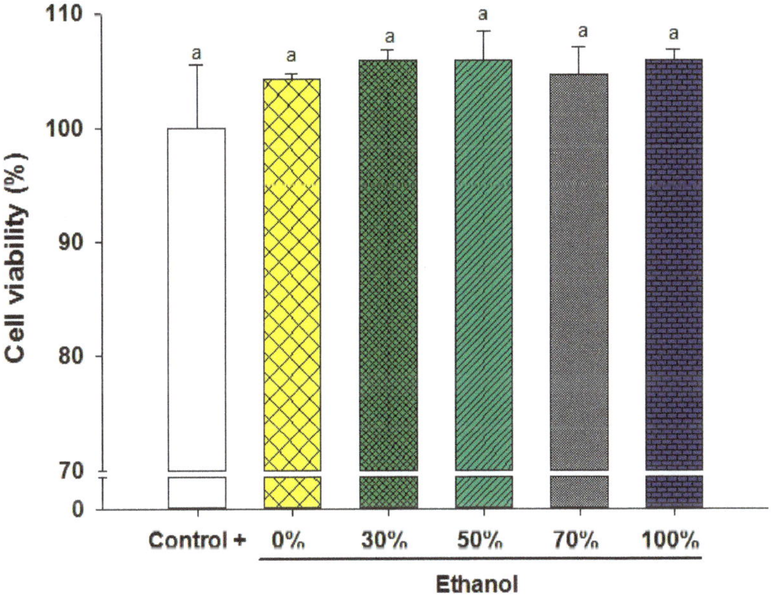

Figure 3. Cytotoxicity of the samples obtained with the different concentrations of ethanol in THP-1 cells differentiated to macrophages after 24 h, with a concentration of 30 µg/mL. Letter between columns indicate significant differences ($p < 0.05$).

3.5.2. Anti-Inflammatory Activity

An anti-inflammatory activity assessment employed the THP-1 monocyte cell line, differentiated into macrophages by PMA induction to simulate inflammatory conditions. Macrophages were stimulated with LPS, with extracts at 20 and 30 µg/mL concentrations added post-cytotoxicity assessment. ELISA quantified pro-inflammatory cytokine secretion (TNF-α, IL-1β, IL-6) in the culture medium. Positive controls included LPS-stimulated macrophages without extracts, while negative controls represented unstimulated cells.

Figure 4a depicts the TNF-α results, showing that at 20 µg/mL, only extracts from 50% and 70% ethanol inhibited cytokine secretion relative to the positive controls, with the 70% ethanol extract demonstrating 40% greater inhibition compared to 50% ethanol. At 30 µg/mL, all extracts except water-based ones inhibited TNF-α release, with the 70% ethanol extract achieving the highest inhibition (60%). For IL-1β (Figure 4b), no inhibition was observed at 20 µg/mL; however, 30 µg/mL concentrations showed 30% inhibition with extracts from 50%, 70%, and 100% ethanol. The IL-6 results (Figure 4c) revealed no reduction in cytokine secretion for aqueous extracts, whereas those from 50% and 70% ethanol displayed potent (70%) inhibition at 20 µg/mL and 80% at 30 µg/mL.

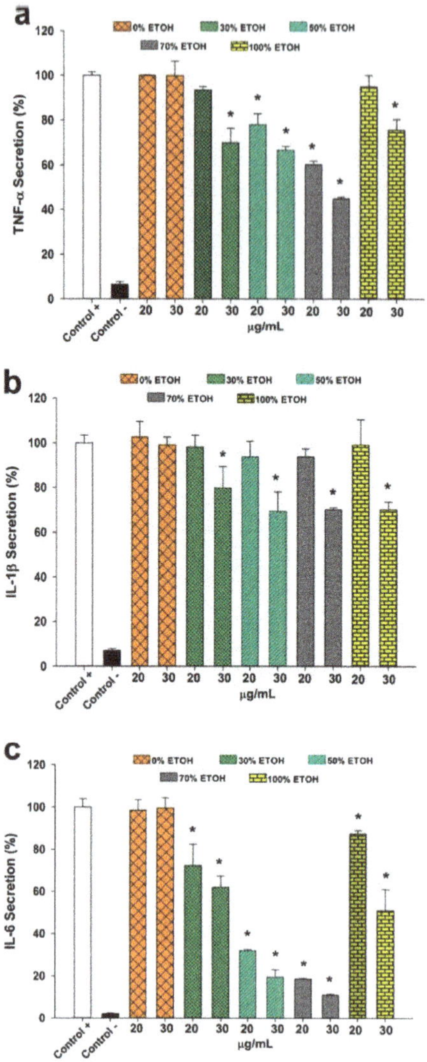

Figure 4. Percentage of secretion of (**a**) TNF-α, (**b**) IL-1β, and (**c**) IL-6 in macrophages after 24 h in the presence of the different extracted moringa extracts and different concentrations of ethanol–water, each with a concentration of 20 and 30 μg/mL. * Indicates the existence of significant differences with the control + ($p < 0.05$).

Moringa extracts obtained with 50% and 70% ethanol demonstrated significant anti-inflammatory activity, effectively inhibiting the secretion of all three pro-inflammatory cytokines at 30 μg/mL concentrations. Notably, the 70% ethanol extract exhibited the highest activity, correlating with its elevated phenolic compound content. Further investigations using high-performance liquid chromatography-mass spectrometry (HPLC-MS) are recommended to delve deeper into the specific compounds responsible for this anti-inflammatory effect. Martínez-González et al. [4] also suggested that the in vivo anti-inflammatory activity of moringa leaf extracts could stem from compounds like chlorogenic acid and kaempferol-3-glucoside. Studies by Koheil et al. [44], Onsare et al. [45], and Prabakaran et al. [46] highlighted the anti-inflammatory, anti-rheumatic, anti-arthritic, and anti-cancer activities of ethanolic moringa leaf and seed extracts. These activities are attributed to phenolic compounds such as hydroxycinnamic acids, flavanols, hydroxybenzoic acids, and other phenolics [47]. Additionally, Ferreira et al. [48] identified primary antioxidant compounds like myricetin and quercetin, renowned for their diverse therapeutic applications. Zhang et al. [49] further demonstrated positive correlations between the polyphenol content and antioxidant and anti-inflammatory activities across 14 Chinese medicinal plants, gauging TNF-α secretion and NO production in murine macrophages.

3.6. Effect of In Vitro Digestion on the Phenolic Composition of Moringa Extract

The moringa extract obtained through Pressurized Liquid Extraction (PLE) using 70% ethanol is rich in phenolic compounds and has significant antioxidant properties. For these beneficial phenolic compounds to effectively unleash their potential in a living organism, they must be bioavailable, meaning they must successfully traverse the digestive system and be absorbed in the intestine. To investigate this, the moringa extract was subjected to an in vitro digestion simulation to assess how the process, including digestive enzymes and pH variations at different stages, could impact the extract's phenolic compounds and antioxidant activity. The extraction's in vitro digestion procedure followed the methodology outlined in Section 2. This protocol had been refined by the research group before. In addition, a control digestion was carried out, substituting the extract with a 70% ethanol–water mixture. Samples were collected after the oral, gastric, and intestinal phases in each digestion cycle to analyze how each phase of digestion affects the phenolic compounds of the extract. Figure 5 displays the chromatographic profiles of the original extract and after oral, gastric, and intestinal digestion phases. These results do not show differences between the profiles of the original extract and those obtained after the oral phase of digestion. After the gastric and intestinal phases, there are notable differences between the obtained profiles.

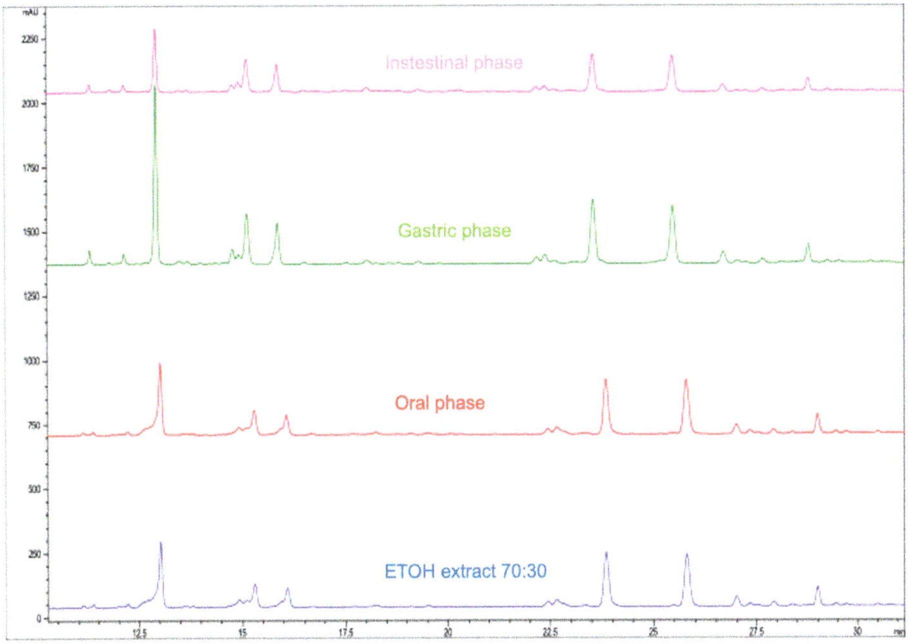

Figure 5. Shows the chromatographic profiles of the original extract after oral, gastric, and intestinal digestion phases.

Table 4 illustrates the identification and quantification of various phenolic compounds after the different phases of digestion. Relative to the original extract, most of the quantified phenolic compounds show no significant differences after the oral phase. After the gastric phase, an increase in neochlorogenic acid is notable compared to its amount in the original extract. However, only minimal differences are observed for the other phenolic compounds after the gastric phase.

Table 4. Phenolic composition of the extract before and after oral, gastric, and intestinal digestion phases.

Phenolic Compounds	Content (mg/g Extract)			
	Extract	Oral	Gastric	Intestinal
Caffeoylquinic acid	0.28 ± 0.03^{b}	0.33 ± 0.03^{b}	0.52 ± 0.01^{a}	0.47 ± 0.01^{a}
Neochlorogenic acid	7.69 ± 0.02^{b}	8.29 ± 0.18^{ab}	9.58 ± 0.13^{a}	5.28 ± 0.66^{c}
Chlorogenic acid	3.10 ± 0.02^{b}	3.46 ± 0.15^{ab}	3.93 ± 0.11^{a}	3.68 ± 0.47^{a}
Vicenine	2.88 ± 0.38^{c}	2.88 ± 0.03^{c}	4.47 ± 0.26^{a}	3.80 ± 0.01^{b}
Luteolin glycoside	0.09 ± 0.02^{a}	0.11 ± 0.02^{a}	0.13 ± 0.02^{a}	0.13 ± 0.03^{a}
Luteolin glycoside derivative	0.02 ± 0.01^{b}	0.03 ± 0.01^{ab}	0.05 ± 0.01^{a}	0.01 ± 0.01^{b}
Vitexin isomer	0.54 ± 0.04^{b}	0.58 ± 0.01^{b}	0.69 ± 0.08^{a}	0.53 ± 0.03^{b}
Vitexin	0.73 ± 0.04^{a}	0.77 ± 0.01^{a}	0.72 ± 0.22^{a}	0.65 ± 0.04^{b}
Quercetin-*O*-hexoxide	13.87 ± 0.47^{a}	14.48 ± 0.14^{a}	13.44 ± 0.79^{a}	10.92 ± 0.60^{b}
Isoramnetin-*O*-hexoxide	0.25 ± 0.09^{b}	0.21 ± 0.30^{b}	0.52 ± 0.26^{a}	0.32 ± 0.16^{b}
Quercetin derivative	14.54 ± 0.68^{a}	14.52 ± 0.18^{a}	12.97 ± 0.97^{b}	12.12 ± 1.58^{b}
Kaempferol derivative	2.89 ± 0.14^{a}	3.01 ± 0.01^{a}	2.93 ± 0.15^{a}	2.47 ± 0.15^{b}
Isorhammetin-*O*-hexoside	0.72 ± 0.01^{b}	0.82 ± 0.00^{a}	0.58 ± 0.10^{c}	0.34 ± 0.09^{d}
Quercetin derivative	0.52 ± 0.05^{a}	0.56 ± 0.01^{a}	0.53 ± 0.03^{a}	0.42 ± 0.05^{b}
Isorhammetin-*O*-hexoside	0.88 ± 0.05^{a}	0.90 ± 0.06^{a}	0.84 ± 0.05^{a}	0.74 ± 0.06^{b}
Kaempferol-*O*- hexoside	4.35 ± 0.18^{a}	4.65 ± 0.11^{a}	3.51 ± 0.61^{b}	3.85 ± 0.29^{b}
Quercetin derivative	0.33 ± 0.02^{a}	0.36 ± 0.03^{a}	0.30 ± 0.02^{a}	0.25 ± 0.05^{b}
Kaempferol derivative	0.37 ± 0.03^{a}	0.39 ± 0.04^{a}	0.33 ± 0.04^{b}	0.38 ± 0.02^{a}

Different superscript letters denote significant differences between the compounds after the different phases of digestion ($p < 0.05$).

In this context, it can be stated that the gastric enzymes and acidic pH during this phase are unlikely to have a significant impact on most of the identified phenolic compounds, except for neochlorogenic acid. The observed increase in neochlorogenic acid may be attributed to transformations or degradation processes experienced by other compounds within the same family. However, after the intestinal phase, a reduction in most of the studied phenolic compounds is observed. This reduction is likely due to the action of intestinal enzymes and/or the basic pH, which can affect almost all the phenolic compounds to varying extents.

Most phenolic compounds, including quercetin-*O*-hexoside and its derivative, experience losses of approximately 17–20% after the intestinal phase. However, it is worth noting that neochlorogenic acid shows better stability, with around 69% of this compound being retained after the intestinal phase. The varying chemical structures of these compounds result in different sensitivities to changes in pH and the actions of digestive enzymes during the in vitro gastrointestinal digestion process. This observation has been highlighted by several authors [50,51]. In the majority of phenolic compounds, both quercetin-*O*-hexoside and its derivative experience losses of about 17–20% after the intestinal phase. However, it is important to note that neochlorogenic acid shows a higher percentage of recovery, retaining about 69% of the compound after the intestinal phase. This difference in stability during the in vitro gastrointestinal digestion process can be attributed to the distinct chemical structures of these compounds.

Guo et al. [52] studied the stability of various phenolic compounds during simulated digestion. They found that compounds with specific structural characteristics, such as hydroxyl groups and glycosylation, were more stable compared to others. In a separate study, Reboul et al. [53] examined the digestion of phenolic compounds in fruits and vegetables, noting that stability varied depending on the specific compound and the food matrix. Significant variations in the release and bioaccessibility of phenolic compounds during digestion were observed. Ribeiro et al. [54] focused on the stability of phenolic compounds in beverages during simulated digestion and noted that compounds with ester bonds and higher molecular weights were more susceptible to degradation. Serra et al. [55] investigated the impact of digestion on the stability of phenolic compounds in olive oil and observed significant changes in the composition and bioavailability of these compounds during the digestion process. These studies emphasize the importance of considering the chemical structure of phenolic compounds when assessing their stability during gastrointestinal digestion.

Influence of In Vitro Digestion on Total Phenolic Compounds and Antioxidant Activity of Moringa Extract

This study also investigated how digestion affects the total phenolic compounds and antioxidant activity of the extract at each phase, as depicted in Table 5. The results show that the amount of total phenolic compounds in the moringa extract remains relatively unchanged after the oral and gastric phases compared to the original extract. However, there is a decrease in the quantity of total phenolic compounds after the intestinal phase. These findings are consistent with the HPLC-PAD analysis, which also demonstrates a reduction in phenolic compounds following the intestinal phase.

Table 5. Total phenolic compounds and antioxidant activity of the extract during different phases of gastrointestinal digestion.

Samples	TPC (mg GAE/g)	DPPH• (mmol Trolox/g)
Moringa extract	76.21 ± 2.37 ^{a}	0.308 ± 0.009 ^{a}
Oral phase	78.81 ± 2.75 ^{a}	0.314 ± 0.018 ^{a}
Gastric phase	78.87 ± 2.17 ^{a}	0.246 ± 0.008 ^{c}
Intestinal phase	73.23 ± 1.98 ^{b}	0.283 ± 0.019 ^{b}

Different superscript letters denote significant differences between samples in the same column ($p < 0.05$).

In terms of the sample's antioxidant activity, it remained stable during the oral phase but experienced a significant decrease following the gastric and intestinal phases. Particularly noteworthy is the low antioxidant activity observed after the gastric phase, which is not consistent with the abundance of total phenolic compounds in the extract. This discrepancy might be attributed to a potential interaction between phenolic compounds and gastric enzymes, rendering them undetectable by HPLC analysis. Similar phenomena have been documented by other researchers [56], especially in the case of luteolin-7-O-glucoside. The sample's antioxidant activity remained stable during the oral phase but showed a significant decrease after the gastric and intestinal phases. It is worth noting that the antioxidant activity was particularly low after the gastric phase, despite the high abundance of total phenolic compounds in the extract. This inconsistency may be due to a potential interaction between phenolic compounds and gastric enzymes, which makes them undetectable by HPLC analysis.

4. Conclusions

This study on *M. oleifera* ethanolic extracts reveals several practical applications due to their potent antioxidant and anti-inflammatory properties. The identification of key compounds, particularly in 70% ethanol extracts, underscores their potential in nutraceuticals and functional foods aimed at managing oxidative stress and inflammation, both of which are linked to chronic diseases like cardiovascular conditions, diabetes, and cancer. This research provides a solid scientific foundation for incorporating *M. oleifera* into health-promoting products, reinforcing its role in preventive healthcare. This study's findings on the strong anti-inflammatory effects of these extracts suggest that they could serve as natural sources for developing plant-based anti-inflammatory drugs or supplements, which are increasingly favored due to their lower risk of side-effects compared to synthetic alternatives. Moreover, the antioxidant properties of Moringa extracts present opportunities in the cosmetics industry, particularly in products designed to protect the skin from oxidative damage and aging. These extracts could be integrated into skincare formulations to help mitigate the effects of environmental stressors on the skin. Additionally, the potential use of Moringa extracts as natural preservatives in food products could extend the shelf life by preventing oxidation and spoilage, offering a natural alternative to synthetic preservatives. This study also explored the stability of these bioactive compounds during in vitro digestion, a critical factor for their efficacy when consumed. Understanding how these compounds behave throughout digestion can inform the development of delivery systems that protect and enhance their bioavailability, ensuring that consumers gain maximum health benefits from *M. oleifera* products.

Author Contributions: Conceptualization, M.V., L.J. and S.S.; Formal analysis, S.Á.-M., C.D.-L. and C.E.M.-S.; Investigation, S.Á.-M., C.D.-L. and C.E.M.-S.; Methodology, J.R.-M., M.V., L.J., S.S. and C.E.M.-S.; Resources, S.S.; Software, C.E.M.-S. and E.H.-L.; Supervision, M.V. and L.J.; Validation, J.R.-M., M.V. and S.S.; Visualization, E.H.-L.; Writing—original draft, S.Á.-M., C.D.-L. and C.E.M.-S.; Writing—review and editing, S.S, L.J. and C.E.M.-S. All authors have read and agreed to the published version of the manuscript.

Funding: The authors appreciate the financial support of the National Council for Science and Technology of Mexico (CONACyT) through Grant 26410 for a sabbatical stay at the Autonomous University of Madrid and to the Technological Institute of Mexico through Grant 8035.20-P.

Institutional Review Board Statement: No ethical review or approval were required for this study as participation was voluntary, and all data collected were anonymous.

Informed Consent Statement: Informed consent was obtained from all the participants involved in this study.

Data Availability Statement: The data presented in this study are available on request from the corresponding author.

Conflicts of Interest: The authors declare no conflicts of interest.

References

1. Guzmán-Maldonado, S.H.; López-Manzano, M.J.; Madera-Santana, T.J.; Núñez-Colín, C.A.; Grijalva-Verdugo, C.P.; Villa-Lerma, A.G.; Rodríguez-Núñez, J.R. Nutritional characterization of *Moringa oleifera* leaves, seeds, husks and flowers from two regions of Mexico. *Agron. Colomb.* **2020**, *38*, 287–297. [CrossRef]
2. Sreeja, M.; Jayasri, P.; Keerthi, N.; Yeshashwini, J.; Praveen, J. *Moringa oleifera*: A review on nutritive importance and its potential use as nutraceutical plant. *J. Med. Plants Stud.* **2021**, *9*, 15–17.
3. Rodríguez-Pérez, C.; Quirantes-Piné, R.; Fernández-Gutiérrez, A.; Segura-Carretero, A. Optimization of extraction method to obtain a phenolic compounds-rich extract from *Moringa oleifera* Lam leaves. *Ind. Crops Prod.* **2015**, *66*, 246–254. [CrossRef]
4. Martínez-González, C.L.; Martínez, L.E.; Martínez-Ortiz, J.; González-Trujano, M.E.; Déciga-Campos, M.; Ventura-Martínez, R.; Díaz-Reval, I. *Moringa oleifera*, a species with potential analgesic and anti-inflammatory activities. *Biomed. Pharmacother.* **2017**, *87*, 482–488. [CrossRef] [PubMed]

5. Nuapia, Y.; Cukrowska, E.; Tutu, H.; Chimuka, L. Statistical comparison of two modeling methods on pressurized hot water extraction of vitamin C and phenolic compounds from *Moringa oleifera* leaves. *S. Afr. J. Bot.* **2020**, *129*, 9–16. [CrossRef]
6. Rocchetti, G.; Blasi, F.; Montesano, D.; Ghisoni, S.; Marcotullio, M.C.; Sabatini, S.; Cossignani, L.; Lucini, L. Impact of conventional/non-conventional extraction methods on the untargeted phenolic profile of *Moringa oleifera* leaves. *Food Res. Int.* **2019**, *115*, 319–327. [CrossRef] [PubMed]
7. Abou-Zeid, S.M.; Ahmed, A.I.; Awad, A.; Mohammed, W.A.; Metwally, M.M.; Almeer, R.; Khalil, S.R. *Moringa oleifera* ethanolic extract attenuates tilmicosin-induced renal damage in male rats via suppression of oxidative stress, inflammatory injury, and intermediate filament proteins mRNA expression. *Biomed. Pharmacother.* **2021**, *133*, 110997. [CrossRef]
8. Martín, C.; Martín, G.; García, A.; Fernández, T.; Hernández, E.; Puls, J. Potenciales aplicaciones de *Moringa oleifera*. Una revisión crítica. *Pastos y Forrajes* **2013**, *36*, 137–149.
9. Leone, A.; Spada, A.; Battezzati, A.; Schiraldi, A.; Aristil, J.; Bertoli, S. Cultivation, genetic, ethnopharmacology, phytochemistry and pharmacology of *Moringa oleifera* leaves: An overview. *Int. J. Mol. Sci.* **2015**, *16*, 12791–12835. [CrossRef]
10. Saini, R.K.; Sivanesan, I.; Keum, Y.S. Phytochemicals of *Moringa oleifera*: A review of their nutritional, therapeutic and industrial significance. *3 Biotech.* **2016**, *6*, 203. [CrossRef]
11. Vergara-Jimenez, M.; Almatrafi, M.M.; Fernandez, M.L. Bioactive components in *Moringa oleifera* leaves protect against chronic disease. *Antioxidants* **2017**, *6*, 91. [CrossRef] [PubMed]
12. Anwar, F.; Latif, S.; Ashraf, M.; Gilani, A.H. *Moringa oleifera*: A food plant with multiple medicinal uses. *Phytother. Res.* **2007**, *21*, 17–25. [CrossRef] [PubMed]
13. Jaime, L.; Santoyo, S. The health benefits of the bioactive compounds in foods. *Foods* **2021**, *10*, 325. [CrossRef]
14. Nieto, J.A.; Santoyo, S.; Prodanov, M.; Reglero, G.; Jaime, L. Valorization of grape stems as a source of phenolic antioxidants by using a sustainable extraction methodology. *Foods* **2020**, *9*, 604. [CrossRef] [PubMed]
15. Minekus, M.; Alminger, M.; Alvito, P.; Ballance, S.; Bohn, T.; Bourlieu, C. Standardized static in vitro digestion method suitable for food -an international consensus. *Food Funct.* **2014**, *5*, 1113. [CrossRef] [PubMed]
16. Biehler, E.; Bohn, T. Methods for assessing aspects of carotenoid bioavailability. *Curr. Nut. Food Sci.* **2010**, *6*, 44–69. [CrossRef]
17. Parada, J.; Aguilera, J.M. Food microstructure affects the bioavailability of several nutrients. *J. Food Sci.* **2007**, *72*, R21–R32. [CrossRef]
18. Noguer, M.; Cereza, A.B.; Rentzsch, M.; Winterhalter, P.; Troncoso, A.M.; García-Parrilla, M.C. Simulated Digestion and Antioxidant Activity of Red Wine Fractions Separated by High Speed Countercurrent Chromatography. *J. Agric. Food Chem.* **2008**, *56*, 8879–8884. [CrossRef]
19. Soler-Rivas, C.; Marín, F.R.; Santoyo, S.; García Risco, M.R.; Señoráns, F.J.; Reglero, G. Testing and Enhancing the in vitro Bioaccessibility and Bioavailability of *Rosmarinus officinalis* Extracts with a High Level of Antioxidant Abietanes. *J. Agric. Food Chem.* **2010**, *58*, 1144–1152. [CrossRef]
20. Zhu, Q.Y.; Holt, R.R.; Lazarus, S.A.; Ensunsa, J.L.; Hammerstone, J.F.; Schmitz, H.H.; Keen, C.L. Stability of the Flavan-3-Ols Epicatechin and Catechin and Related Dimeric Procyanidins Derived from Cocoa. *J. Agric. Food Chem.* **2002**, *50*, 1700–1705. [CrossRef]
21. Neilson, A.P.; Hopf, A.S.; Cooper, B.R.; Pereira, M.A.; Bomser, J.A.; Ferruzzi, M.G. Catechin degradation with concurrent formation of homo-and heterocatechin dimers during in vitro digestion. *J. Agric. Food Chem.* **2007**, *55*, 8941–8949. [CrossRef] [PubMed]
22. Siracusa, L.; Kulisic-Bilusic, T.; Politeo, O.; Krause, I.; Dejanovic, B.; Ruberto, G. Phenolic Composition and Antioxidant Activity of Aqueous Infusions from *Capparis spinosa* L. and *Crithmum maritimum* L. before and after Submission to a Two-Step in vitro Digestion Model. *J. Agric. Food Chem.* **2011**, *59*, 12453–12459. [CrossRef]
23. Tarko, T.; Duda-Chodak, A.; Zając, N. Digestion and absorption of phenolic compounds assessed by in vitro simulation methods. A Review. *Rocz. Państwowego Zakładu Hig.* **2013**, *64*, 79–84.
24. Heimler, D.; Vignolini, P.; Dini, M.G.; Romani, A. Rapid test to assess the antioxidant activity of *Phaseolus vulgaris* L. dry beans. *J. Agric. Food Chem.* **2005**, *53*, 3053–3056. [CrossRef]
25. Re, R.; Pellegrini, N.; Proteggente, A.; Pannala, A.; Yang, M.; Rice-Evans, C. Antioxidant activity applying an improved ABTS radical cation decolorization assay. *Free Radic. Biol. Med.* **1999**, *26*, 1231–1237. [CrossRef] [PubMed]
26. Takashiba, S.; Van Dyke, T.E.; Amar, S.; Murayama, Y.; Soskolne, A.W.; Shapira, L. Differentiation of monocytes to macrophages primes cells for lipopolysaccharide stimulation via accumulation of cytoplasmic nuclear factor kappa B. *Infect. Immun.* **1999**, *67*, 5573–5578. [CrossRef]
27. Mosmann, T. Rapid colorimetric assay for cellular growth and survival-application to proliferation and cytotoxicity assays. *J. Inmmunol. Meth.* **1983**, *65*, 55–63. [CrossRef]
28. Kim, J.A.; Kong, C.S.; Pyun, S.Y.; Kim, S.K. Phosphorylated glucosamine inhibits the inflammatory response in LPS-stimulated PMA-differentiated THP-1 cells. *Carb. Res.* **2010**, *345*, 1851–1855. [CrossRef]
29. Wang, Y.; Gao, Y.; Ding, H.; Liu, S.; Han, X.; Gui, J.; Liu, D. Subcritical ethanol extraction of flavonoids from *Moringa oleifera* leaf and evaluatin of antioxidant activity. *Food Chem.* **2015**, *208*, 152–158.

30. Vongsak, B.; Sithisarn, P.; Mangmool, S.; Thongpradltchote, S.; Wongkrajang, Y.; Gritsanapan, W. Maximizing total phenolics, total flavonoids contents and antioxidant activity of *Moringa oleifera* leaf extract by the appropriate extraction method. *Ind. Crops Prod.* **2013**, *44*, 566–571. [CrossRef]
31. Braham, F.; Carvalho, D.O.; Almeida, C.M.R.; Zaidi, F.; Magalhães, J.M.C.S.; Guido, L.F.; Gonçalves, M.P. Online HPLC-DPPH screening method for evaluation of radical scavenging phenols extracted from *Moringa oleifera* leaves. *S. Afr. J. Bot.* **2020**, *129*, 146–154. [CrossRef]
32. Saleem, A.; Saleem, M.; Akhtar, M.F. Antioxidant, anti-inflammatory and antiarthritic potential of *Moringa oleifera* Lam: An ethnomedicinal plant of Moringaceae family. *S. Afr. J. Bot.* **2020**, *128*, 246–256. [CrossRef]
33. Nobossé, P.; Fombang, E.N.; Mbofung, C.M. Effects of age and extraction solvent on phytochemical content and antioxidant activity of fresh *Moringa oleifera* L. leaves. *Food Sci. Nutr.* **2018**, *6*, 2188–2198. [CrossRef]
34. Saucedo-Pompa, S.; Torres-Castillo, J.A.; Castro-López, C.; Rojas, R.; Sánchez-Alejo, E.J.; Ngangyo-Heya, M.; Martínez-Ávila, G.C.G. Moringa plants: Bioactive compounds and promising applications in food products. *Food Res. Int.* **2018**, *111*, 438–450. [CrossRef] [PubMed]
35. Udechukwu, M.C.; Abbey, L.; Nwodo, U.; Udenigwe, C.C. Potential of *Moringa oleifera* seeds and leaves as functional food ingredients for human health promotion. *J. Food Nutr. Res.* **2018**, *57*, 1–14.
36. Hossain, M.A.; Disha, N.K.; Shourove, J.H.; Dey, P. Determination of antioxidant activity and total tannin from drumstick (*Moringa oleifera* lam.) leaves using different solvent extraction methods. *Turk. J. Agric. Food Sci. Technol.* **2020**, *8*, 2749–2755. [CrossRef]
37. Ahmed, K.S.; Jahan, I.A.; Jahan, F.; Hossain, H. Antioxidant activities and simultaneous HPLC-DAD profiling of polyphenolic compounds from *Moringa oleifera* Lam. Leaves grown in Bangladesh. *Food Res.* **2021**, *5*, 401–408. [CrossRef] [PubMed]
38. Oyeniran, O.H.; Ademiluyi, A.O.; Oboh, G. Comparative study of the phenolic profile, antioxidant properties, and inhibitory effects of Moringa (*Moringa oleifera* Lam.) and Almond (*Terminalia catappa* Linn.) leaves on acetylcholinesterase and monoamine oxidase activities in the head region of Fruitfly (*Drosophila melanogaster* Meigen) in vitro. *J. Food Biochem.* **2021**, *45*, e13401.
39. Hossain, M.B.; Barry-Ryan, C.; Martin-Diana, A.B.; Brunton, N.P. Optimization of accelerated solvent extraction of antioxidant compounds from rosemary (*Rosmarinus officinalis* L.), marjoram (*Origanum* majorana L.), and oregano (*Origanum vulgare* L.) using response surface methodology. *Food Chem.* **2011**, *126*, 339–346. [CrossRef]
40. Zhao, B.; Deng, J.; Li, H.; He, Y.; Lan, T.; Wu, D.; Gong, H.; Zhang, Y.; Chen, Z. Optimization of phenolic compound extraction from Chinese *Moringa oleifera* leaves and antioxidant activities. *J. Food Qual.* **2019**, *2019*, 5346279. [CrossRef]
41. Bennour, N.; Mighri, H.; Eljani, H.; Zammouri, T.; Akrout, A. Effect of solvent evaporation method on phenolic compounds and the antioxidant activity of *Moringa oleifera* cultivated in Southern Tunisia. *S. Afr. J. Bot.* **2020**, *129*, 181–190. [CrossRef]
42. Pollini, L.; Tringaniello, C.; Ianni, F.; Blasi, F.; Manes, J.; Cossignani, L. Impact of ultrasound extraction parameters on the antioxidant properties of *Moringa oleifera* leaves. *Antioxidants* **2020**, *9*, 277. [CrossRef] [PubMed]
43. Lin, X.; Wu, L.; Wang, X.; Yao, L.; Wang, L. Ultrasonic-assisted extraction for flavonoid compounds content and antioxidant activities of India *Moringa oleifera* L. leaves: Simultaneous optimization, HPLC characterization and comparison with other methods. *J. Appl. Res. Med. Aromat. Plants* **2021**, *20*, 100284. [CrossRef]
44. Koheil, M.A.; Hussein, M.A.; Othman, S.M.; El-Haddad, A. Anti-inflammatory and antioxidant activities of *Moringa peregrina* seeds. *Free Rad. Antiox.* **2011**, *1*, 49–61. [CrossRef]
45. Onsare, J.G.; Kaur, H.; Arora, D.S. Antimicrobial activity of *Moringa oleifera* from different locations against some human pathogens. *Acad. J. Med. Plants* **2013**, *1*, 80–91.
46. Prabakaran, M.; Kim, S.H.; Sasireka, A.; Chandrasekaran, M.; Chung, I.M. Polyphenol composition and antimicrobial activity of various solvent extracts from different plant parts of *Moringa oleifera*. *Food Biosci.* **2018**, *26*, 23–29. [CrossRef]
47. Charoensin, S. Antioxidant and anticancer activities of *Moringa oleifera* leaves. *J. Med. Plants Res.* **2014**, *8*, 318–325.
48. Ferreira, P.M.P.; Farias, D.F.J.; Oliveira, T.A.; Carvalho, A.F.U. *Moringa oleifera*: Bioactive compounds and nutritional potential. *J. Nutr.* **2008**, *21*, 431–437. [CrossRef]
49. Zhang, L.; Ravipati, A.S.S.; Koyyalamudi, R.S.; Jeong, C.; Reddy, N.; Smith, P.T.; Bartlett, J.; Shanmugam, K.; Münch, G.; Wu, M.J. Antioxidant and anti-inflammatory activities of selected medicinal plants containing phenolic and flavonoid compounds. *J. Agric. Food Chem.* **2011**, *59*, 12361–12367. [CrossRef]
50. Goulas, V.; Hadjisolomou, A. Dynamic changes in targeted phenolic compounds and antioxidant potency of carob fruit (*Ceratonia siliqua* L.) products during in vitro digestion. *LWT-Food Sci. Technol.* **2019**, *101*, 269–275. [CrossRef]
51. Lima, K.; Silva, O.; Figueira, M.E.; Pires, C.; Cruz, D.; Gomes, S.; Mauricio, E.M.; Duarte, M.P. Influence of the in vitro gastrointestinal digestion on the antioxidant activity of *Artemisia gorgorum* Webb and *Hyptis pectinata* (L.) Poit. Infusions from Cape Verde. *Food Res. Int.* **2019**, *115*, 150–159. [CrossRef]
52. Guo, J.; Giusti, M.M.; De Santiago, E. Stability of various phenolic compounds during simulated digestion. *J. Agric. Food Chem.* **2018**, *66*, 12609–12618.
53. Reboul, E.; Gonçalves, A.; Comera, C.; Bott, R.; Nowicki, M.; Landrier, J.F.; Jourdheuil-Rahmani, D. Digestion of phenolic compounds from different fruits and vegetables during simulated digestion: Impact on phenolic metabolites availability and lipid peroxidation. *Food Chem.* **2017**, *228*, 186–192.
54. Ribeiro, D.; Freitas, V.; Silva, A.M.; Carvalho, F.; Fernandes, E.; Mateus, N. Influence of ester bonds on the stability of phenolic compounds in beverages during simulated digestion. *J. Agric. Food Chem.* **2016**, *64*, 3287–3296.

55. Serra, A.; Rubió, L.; Borràs, X.; Macià, A.; Romero, M.P. Digestion stability of phenolic compounds in virgin olive oil: A review. *Comp. Rev. Food Sci. Food Saf.* **2014**, *13*, 180–198.
56. Gutiérrez-Grijalva, E.P.; Picos-Salas, M.; Leyva-López, N.; Criollo-Mendoza, M.S.; Vázquez-Olivo, G.; Basilio Heredia, J. Flavonoids and Phenolic Acids from Oregano: Occurrence, Biological Activity and Health Benefits. *Plants* **2017**, *7*, 2. [CrossRef] [PubMed]

foods

MDPI

Article

Nanosuspensions Loaded with Acetogenins: Physical Stability During In Vitro Digestion, Genotoxicity and Cytotoxicity

Brandon Alexis López-Romero [1], Carmen Magdalena Gurrola-Díaz [2], Belinda Vargas-Guerrero [2], María de Lourdes García Magaña [1], Efigenia Montalvo-González [1,*] and Gabriela Aguilar-Hernández [3,*]

[1] Laboratorio Integral de Investigación en Alimentos, Tecnológico Nacional de México/Instituto Tecnológico de Tepic, Av. Tecnológico 2595, Lagos del Country, Tepic 63175, Nayarit, Mexico; brallopezro@ittepic.edu.mx (B.A.L.-R.); mgarciam@ittepic.edu.mx (M.d.L.G.M.)

[2] Departamento de Biología Molecular y Genómica, Centro Universitario de Ciencias de la Salud, Universidad de Guadalajara, Sierra Mojada 950, Guadalajara 44340, Jalisco, Mexico; carmenhpv@yahoo.de (C.M.G.-D.); belinda.vargas@academicos.udg.mx (B.V.-G.)

[3] División de Ciencias Agropecuarias e Ingenierías, Centro Universitario de los Altos, Universidad de Guadalajara, Av. Rafael Casillas Aceves 1200, Guadalajara 47600, Jalisco, Mexico

* Correspondence: emontalvo@ittepic.edu.mx (E.M.-G.); gaby.mca2017@gmail.com (G.A.-H.)

Abstract

This study assesses the stability, in vitro bioaccessibility and potential bioavailability, and in vivo genotoxicity and toxicity of polyethylene glycol–soy lecithin (PEGSL-ACG-NSps) or β-cyclodextrin–soy lecithin (βCDSL-ACG-NSps) nanosuspensions (NSps). Both formulations exhibited initial particle sizes below 130 nm and PDI values below 0.3. Under simulated gastrointestinal conditions, PEGSL-ACG-NSps preserved structural integrity, with only a moderate size increase (~239 nm) in the intestinal phase and controlled release of acetogenins (ACGs); in contrast, βCDSL-ACG-NSps destabilized considerably (size > 500 nm) and released ACGs rapidly. Consistently, βCDSL-ACG-NSps achieved higher in vitro bioaccessibility and a potential bioavailability (up to 95% from post-digestion recovery). In contrast, PEGSL-ACG-NSps displayed a more gradual release profile (up to 55%). In vivo toxicity tests in mice showed no significant genotoxic or cytotoxic effects for either formulation, even at high doses. These findings suggest that selecting appropriate food-grade stabilizing polymers is crucial for optimizing NSps for the oral delivery of ACGs as therapeutic agents.

Keywords: acetogenins; nanosuspensions; physical stability; in vitro digestion; polymers; in vivo toxicity

1. Introduction

Acetogenins (ACGs) are secondary metabolites produced through the acetate-polyketide pathway and are primarily extracted from plants of the Annonaceae family. These compounds are molecules with 35–37 carbon atoms connected to one, two, or three tetrahydrofuran or tetrahydropyran rings in their central region, along with several oxygenated groups (–OH) and an α-β-unsaturated or saturated γ-lactone [1]. ACGs exhibit a wide range of biological activities, including antibacterial, antifungal, and antitumor effects, which have generated significant interest for their potential therapeutic applications [2,3]. Their primary mechanism involves inhibiting mitochondrial complex I (NADH: ubiquinone oxidoreductase), thereby decreasing ATP production and inducing cell apoptosis [3,4].

However, despite their high therapeutic potential, ACGs are limited by their very lipophilic structure, which reduces their solubility in water, stability under physiological conditions, and oral bioavailability [5–7]. In addition to pharmacokinetic challenges, some in vivo studies in rats have shown that certain ACGs can be toxic, especially when administered systemically at high doses and under conditions that promote their accumulation in sensitive tissues. Champy et al. [8] reported that annonacin, a monotetrahydrofuran acetogenin, induced neurotoxicity in rats when administered intravenously at doses of 3.8–7.6 mg/kg/day for 28 days. This exposure resulted in inhibition of mitochondrial complex I in brain tissue, reduced cerebral ATP levels, and pronounced neurodegeneration, characterized by significant losses of dopaminergic, cholinergic, and GABAergic neurons, as well as marked astrocytic and microglial proliferation. Consequently, Chen et al. [9] reported that repeated administration of bullatacin (300 μg/kg) over three weeks caused liver and kidney toxicity by increasing calcium levels, reactive oxygen species production, Bax expression, and the Bax/Bcl-2 ratio in rats. These findings highlight the risks linked to ACGs' use, particularly when systemic exposure and accumulation in the central nervous system are involved.

To overcome these obstacles and reduce systemic toxicity while maintaining antitumor effectiveness, new and promising drug delivery strategies, such as nanosuspensions (NSps), have been developed. These colloidal systems consist of nanometer-sized (<200 nm) particles of the active ingredient dispersed in water and stabilized by polymeric matrices, thereby improving the dispersibility and bioavailability of hydrophobic drugs [10].

In this context, several studies have demonstrated that using hydrophilic polymers, such as polyethylene glycol (PEG) and modified cyclodextrins (βCD), in combination with soy lecithin (SL) enables the formation of NSps capable of encapsulating ACGs. This improves their pharmacokinetic profile, reduces systemic toxicity, and provides self-assembly properties that are resistant to gastrointestinal conditions [11–15]. A study by Hong et al. [11] revealed that ACG NSps stabilized with hydroxypropyl–beta-cyclodextrin–soy lecithin showed sustained in vitro release, good stability in plasma and simulated gastrointestinal fluid, and significant antitumor efficacy both orally (49.74%) and intravenously (70.31%). Similarly, Hong et al. [12] prepared ACG NSps using mPEG2000–PCL2000 as a stabilizer, achieving particle sizes of 123.2 nm, high ACG loading, efficient in vitro cytotoxicity (4T1, MCF-7, and HeLa lines), good stability in physiological media, sustained release, and improved antitumor efficacy (74.83%) in mice bearing the 4T1 tumor compared to free ACGs (45.53%).

Although NSps show significant potential as delivery systems for ACGs, few studies have simultaneously examined their physicochemical stability in a gastrointestinal environment, bioaccessibility, bioavailability, genotoxicity, and cytotoxicity.

From a therapeutic perspective, ACGs are promising compounds with bioactive potential and health-promoting properties. However, their practical application in functional foods or therapeutic products is limited by poor water solubility, low gastrointestinal tract stability, and safety concerns arising from uncontrolled exposure. In this context, nanosuspension-based delivery systems made with food-grade polymers and surfactants are a viable strategy to enhance the oral delivery, gastrointestinal behavior, and safety of lipophilic food bioactives. Therefore, this study aimed to evaluate two acetogenin-loaded NSps stabilized with polyethylene glycol–soy lecithin (PEGSL-ACG-NSps) and β-cyclodextrin–soy lecithin (βCDSL-ACG-NSps). The focus was on their physical stability during simulated gastrointestinal digestion, in vitro bioaccessibility and potential bioavailability, and in vivo cytotoxicity and genotoxicity. The formulations were developed using food-grade excipients, including soy lecithin, PEG, and β-CD, and tested under gastrointestinal conditions relevant to oral consumption. In this sense, the present study contributes to evaluating whether these NSPs meet the preliminary criteria that justify their

https://doi.org/10.3390/foods15050937

future exploration as potential therapeutic agents, while recognizing the need for further studies to confirm their safety and applicability.

2. Materials and Methods

2.1. Materials

The ACGs used in this study were obtained from the endosperm of A. muricata seeds at the Bromatology and Nutrition Laboratory of the National Technological Institute of Mexico in Tepic, following the methodology described by López-Romero et al. [2]. Briefly, ACGs were extracted from the defatted endosperm by thermosonication (50 °C, 50 min, 100% amplitude, and 0.5 s pulse cycles) using an ultrasonic system (UP400S, Hielscher Ultrasonic, Teltow, Germany). The optimized extract was purified by open-column chromatography on silica gel (60 mesh) and then further purified on silica gel (230–400 mesh). TLC was used to analyze the resulting fractions, and ACG-rich fractions were characterized by HPLC-DAD. Also, structural elucidation of the ACGs was performed by HPLC-MS and NMR, confirming purities greater than 95% [16]. The isolated ACGs mainly consisted of pseudoannonacin (60%), annonacin (6%), and other ACGs (34%) such as bullatacin, squamostatin-D, squamocin, isodesacetyluvarcicin, and desacetyluvaricin [2]. Polyethylene glycol 6000 (PEG), (2)-hydroxypropyl-β-cyclodextrin (βCD), soy lecithin (SL), pancreatic lipase, porcine pepsin, gastric mucin, bile salts, and cyclophosphamide (1000 mg/mL) were purchased from Sigma-Aldrich Co., Ltd. (St. Louis, MO, USA). Acetonitrile, HPLC-grade water, and all analytical-grade reagents such as sodium hydroxide (NaOH), hydrochloric acid (HCl), sodium chloride (NaCl), ammonium nitrate (NH_4NO_3), potassium phosphate (KH_2PO_4), potassium chloride (KCl), potassium citrate ($C_6H_5K_3O_7$), urea ($CO(NH_2)_2$), and calcium chloride ($CaCl_2$) were purchased from Jalmek Scientific S.A., Guadalajara, Jalisco, Mexico.

2.2. Preparation of Nanosuspensions Loaded with Acetogenins

Nanosuspensions (NSps) loaded with acetogenins (ACG) were prepared following the protocols established by Montalvo-González et al. [17] and López-Romero et al. [18]. In summary, PEG (0.20% *w*/*v*) or βCD (16% *w*/*v*), dissolved in sterile water (10 mL), was added dropwise to 800 μL of a surfactant solution (10–15 mg/mL SL in ethanol). The mixture was stirred at 500 rpm for 30 min to form amphiphilic complexes (PEGSL and βCDSL). Amphiphilic complexes are self-assembled supramolecular structures formed by the association of hydrophilic and hydrophobic domains in aqueous media [12]. During this process, the hydrophilic portion of PEG or βCDSL interacts with water. In contrast, the hydrophobic portion of SL organizes toward the core of the complex, enabling the subsequent association with ACGs.

Subsequently, the ethanol was evaporated using a rotary evaporator (Yamato RE300, Yamato, Tokyo, Japan). Next, 160 μL of an ACG solution (62.5 mg/mL in ethanol) was added dropwise, and the mixture was stirred at 500 rpm for 30 min. Finally, the ethanol was evaporated, yielding ACG-loaded nanosuspensions (PEGSL-ACG-NSps and βCDSL-ACG-NSps) at a final concentration of 1.024 mg/mL. The resulting systems are obtained as stable nanosuspensions, consisting of colloidal particles in the nanometer range dispersed in an aqueous continuous phase [10]. Particle size depends on the concentration of amphiphilic complexes and ACGs, as well as on the dispersion conditions [19]. The particle size of the nanosuspensions ranged from 113.32 nm for PEGSL-ACG-NSps to 133.77 nm for βCDSL-ACG-NSps, with a polydispersity index (PDI) between 0.19 and 0.20, indicating a homogeneous size distribution and good colloidal stability. Encapsulation efficiencies of 83.33% and 56.09% for PEGSL-ACG-NSps and βCDSL-ACG-NSps, respectively, have been reported in these previous studies [19].

https://doi.org/10.3390/foods15050937

2.3. Simulated In Vitro Gastrointestinal Digestion Study

The simulated in vitro digestion of PEGSL-ACG-NSps and βCDSL-ACG-NSps was performed in three sequential phases (mouth, gastrointestinal, and small intestine) following the methodology described by Brodkorb et al. [20] and Bravo-Alfaro et al. [21], with some modifications.

During the mouth phase, 7 mL of PEGSL-ACG-NSps or βCDSL-ACG-NSps was mixed with 7 mL of simulated salivary fluid (see Table 1) and 0.3 g/L of gastric mucin. The pH was adjusted to 6.8, and the mixture was incubated for 10 min at 37 °C with agitation at 100 rpm.

Table 1. Composition of simulated salivary fluid.

Compound	Formula	Concentration (g/L)
Sodium chloride	NaCl	0.157
Ammonium nitrate	NH_4NO_3	0.328
Potassium phosphate	KH_2PO_4	0.489
Potassium chloride	KCl	0.020
Potassium citrate	$C_6H_5K_3O_7$	0.032
Urea	$CO(NH_2)_2$	0.019
Calcium chloride	$CaCl_2$	0.014

For the gastrointestinal phase, 14 mL of simulated gastric juice (containing 0.0034 g NaCl, 0.7 mL HCl, and 0.32 g pepsin per 100 mL of distilled water, with a pH of 2.0) was added to the oral bolus. The pH was then adjusted to 2.5 using 1 M NaOH, and the mixture was incubated for 2 h at 37 °C with shaking at 100 rpm.

During the small intestine phase, the pH was adjusted to 7.0 using 1 M NaOH, and then 3.6 mL of bile salt solution (46 mg/mL in phosphate buffer, pH 7.0) and 2.2 mL of pancreatic lipase (15 mg/mL in 100 mM phosphate buffer, pH 7.0) were added. The mixture was incubated for 2 h at 37 °C with shaking at 100 rpm.

The stability analysis was conducted to evaluate the resistance of NSps to simulated human digestion conditions. During each digestion phase, 1 mL aliquots were collected to determine particle size and PDI and then stored at −20 °C until further analysis.

In Vitro Bioaccessibility and Potential Bioavailability of Nanosuspensions During In Vitro Digestion: Analysis by High-Performance Liquid Chromatography (HPLC).

The collected sample (2 mL) at each stage of the digestion assay was mixed with 2 mL of chloroform to extract the ACGs, which are highly soluble in this solvent. The mixture was homogenized at 10,000 rpm for 1 min (Ultraturrax Kaibrite FS-2A, Shanghai, China) to promote effective disruption of the nanosuspensions and facilitate transfer of ACGs into chloroform, followed by centrifugation at 8000 rpm for 10 min at 4 °C. The chloroform was then evaporated to dryness. The residue was reconstituted in 1 mL of HPLC-grade methanol.

The resulting solutions were then filtered through nylon filters (0.22 μm), and 10 μL was injected into an HPLC system (Agilent Technologies 1260 Infinity, Agilent Technologies, Waldbronn, CA, USA) equipped with a diode array detector and a C18 column (5 μm, 4.6 mm × 250 mm; Thermo Scientific, Sunnyvale, CA, USA). Separation was carried out using a gradient elution of acetonitrile (eluent A) and water (eluent B) as follows: 60% A (0–30 min), 70% A (35–40 min), and 60% A (41–60 min), at a flow rate of 0.8 mL/min. Peak areas were recorded at 210 nm and quantified using calibration curves for annonacin (0.093–1.50 mg/mL, R^2 = 0.9991) and pseudoannonacin (0.13–2.10 mg/mL, R^2 = 0.9998) [16].

https://doi.org/10.3390/foods15050937

In vitro bioaccessibility and potential bioavailability were calculated according to Zhao et al. [22] using Equations (1) and (2).

$$In\ vitro\ Bioaccessibility\ (\%) = \left(\frac{\text{ACGs in micellar fraction}}{\text{ACGs in digested fraction}}\right) \times 100 \quad (1)$$

$$In\ vitro\ potential\ Bioavailability\ (\%) = \left(\frac{\text{ACGs in digested fraction}}{\text{ACGs before digestion}}\right) \times 100 \quad (2)$$

where the micellar fraction refers to the ACGs solubilized in mixed micelles, which are colloidal structures formed during intestinal digestion through the interaction of bile salts, phospholipids, and lipid digestion products. The digested fraction, in turn, represents the total amount of ACGs recovered after digestion, including both micellized compounds (ACGs incorporated into and solubilized in mixed micelles, making them potentially available for intestinal absorption) and those still associated with the nanoparticles present in the NSps.

2.4. *In Vivo Genotoxicity and Cytotoxicity Study*

2.4.1. Animal Handling and Care

The male C57BL/6 mice (6–8 weeks old, 20–25 g) used in this study were sourced from the Animal Facility at the University Health Sciences Center, University of Guadalajara. Animals were acclimated under standard housing conditions in polycarbonate cages at 24 ± 2 °C, 50–60% relative humidity, and a 12 h light/dark cycle, with ad libitum access to food and water. The protocol was approved by the Research Ethics Committee of the University of Guadalajara (No. Protocol CI-03523). All procedures adhered to the Guide for the Care and Use of Laboratory Animals and the CONCEA Euthanasia Practice Guidelines.

2.4.2. Formation of Experimental Groups

Animals (36 mice) were randomly assigned to nine experimental groups, with four mice in each group. The number of animals per group was determined by the nature of the in vivo micronucleus assay, which allows the evaluation of a large number of erythrocytes per animal, providing high analytical sensitivity. This experimental design was selected in accordance with internationally accepted guidelines for genotoxicity testing, emphasizing ethical principles and the reduction in animal use (3Rs), while maintaining sufficient power to detect biologically relevant genotoxic effects when appropriate positive controls are included [23].

Group 1 served as the positive control and received cyclophosphamide (10 mg/kg of body weight). Groups 2 and 3 served as negative controls and received the PEGSL or βCDSL complex, respectively. Groups 4, 5, and 6 were treated with PEGSL-ACG-NSps at doses of 400, 800, and 1600 µg/kg of body weight, respectively. Groups 7, 8, and 9 received βCDSL-ACG-NSps at the same doses.

All treatments were administered via orogastric tube once daily for five consecutive days. At the end of the experiment, the animals were euthanized using an overdose of tiletamine–zolazepam anesthesia (Zoletil® 50, Virbac, Carros, France).

2.4.3. Micronucleus Test

Cytotoxic and genotoxic effects were assessed using the micronucleus assay in bone marrow. Before administering the treatment, peripheral blood smears were collected by making a small cut at the tip of the tail at 0, 24, 48, 72, and 96 h. The samples were fixed with absolute ethanol for 10 min, stained with acridine orange, and examined under an epifluorescence microscope (Olympus CX3, Olympus, Tokyo, Japan).

 https://doi.org/10.3390/foods15050937

Recent genotoxic damage, induced 24 h after treatment, was assessed by counting micronucleated polychromatic erythrocytes (MNPCE) per 1000 total polychromatic erythrocytes (TPE). Accumulated genotoxicity was measured by the number of micronucleated normochromatic erythrocytes (MNE) per 10,000 total erythrocytes (TE). At the same time, cytotoxicity was evaluated as the reduction in the number of polychromatic erythrocytes (PCE) per 1000 TE [24–26].

2.4.4. Ethical Considerations

The study was conducted in accordance with Mexican Official Standards NOM-062-ZOO-1999, NOM-033-SAG/ZOO-2014, and NOM-087-ECOL-SSA1-2002 [27–29]. All experimental procedures were reviewed and approved by the appropriate ethics committee and followed NOM-062-ZOO-1999 for laboratory animal care and use, NOM-033-SAG/ZOO-2014 for humane euthanasia, and NOM-087-ECOL-SSA1-2002 for the management of bio-infectious waste.

2.5. Statistical Analysis

The data obtained from the in vitro digestion study were expressed as mean ± standard deviation and analyzed using one-way analysis of variance (ANOVA), followed by Fisher's least significant difference (LSD) post hoc test for multiple comparisons between formulations and controls. For the in vivo study, data normality (MNE, PCEMN, and PCE) was evaluated using the Kolmogorov–Smirnov test before parametric analysis. Intragroup comparisons were made relative to baseline values, while intergroup comparisons were conducted relative to the positive control (cyclophosphamide, 10 mg/kg) and the negative controls (PEGSL and βCDSL complex). Repeated measures ANOVA followed by Fisher's LSD test was applied, and differences were considered statistically significant ($p < 0.05$). All statistical analyses were performed using STATISTICA software (v.10, StatSoft Inc., Tulsa, OK, USA).

3. Results and Discussion

3.1. Analysis of the Stability of Nanosuspensions During the In Vitro Digestion Study

Figures 1 and 2 show the results obtained for particle size and PDI of the PEGSL-ACG-NSps (Figure 1A,B) and βCDSL-ACG-NSps (Figure 2A,B) formulations during each stage of the simulated digestion study.

Initially, PEGSL-ACG-NSPs and βCDSL-ACG-NSPs had average sizes of ~100 nm and ~128 nm, respectively, with PDI values of ~0.22, indicating a homogeneous size distribution. At the start of digestion, in the buccal and gastric phases, particle size remained constant ($p > 0.05$). Although a moderate increase in PDI (~0.32) was observed for PEGSL-ACG-NSps, the values remained within acceptable ranges for colloidal systems [5].

These results suggest good colloidal stability of NSps under adverse gastrointestinal conditions, such as bile salts, digestive enzymes, or acidic pH, which can alter particle stability, leading to possible aggregation or structural collapse of the systems [21]. Furthermore, Figure 1A shows that the PEGSL-ACG-NSps formulation exhibited a gradual increase in particle size, from 130.8 nm at the beginning of the intestinal phase to a maximum rise of 239.25 nm at the end of digestion. The gradual increase in size in PEGSL-ACG-NSps is due to the hydration of the PEG coating, which forms a flexible steric barrier that protects the particles from digestive conditions, allowing only moderate swelling without compromising their structural integrity. These results indicate a stable system, ideal for controlled release of active compounds [7].

 https://doi.org/10.3390/foods15050937

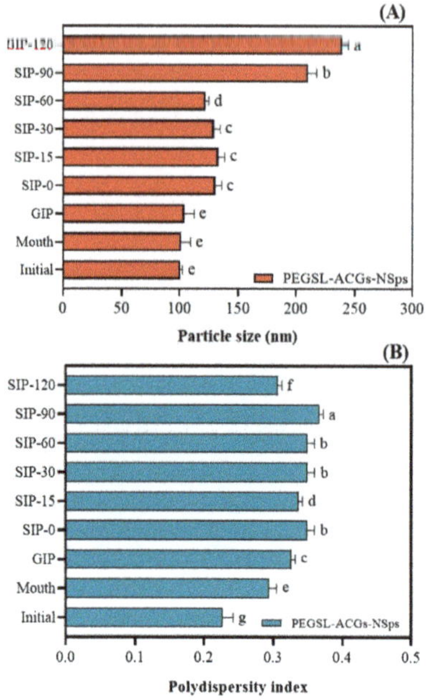

Figure 1. Particle size (**A**) and polydispersity index (**B**) of PEGSL-ACG-NSps during each stage of the in vitro digestion study. SIP, small intestine phase; GIP, gastrointestinal phase. Lowercase letters indicate significant differences between the various stages of the in vitro digestion study ($p < 0.05$).

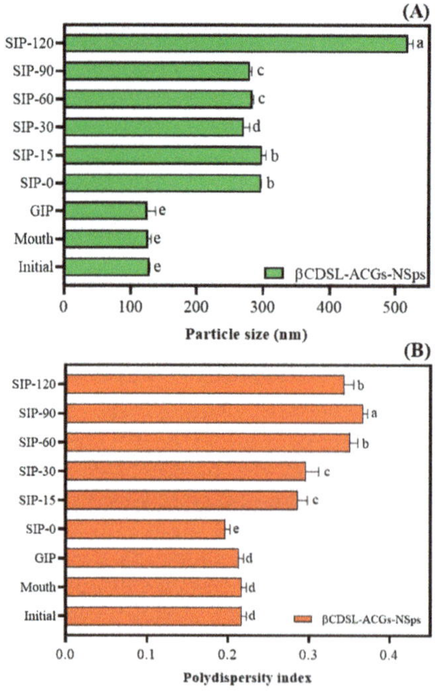

Figure 2. Particle size (**A**) and polydispersity index (**B**) of βCDSL-ACG-NSps during each stage of the in vitro digestion study. SIP, small intestine phase; GIP, gastrointestinal phase. Lowercase letters indicate significant differences between the various stages of the in vitro digestion study ($p < 0.05$).

https://doi.org/10.3390/foods15050937

In contrast, βCDSL-ACG-NSps (Figure 2A) showed a dramatic increase at the beginning of this intestinal phase (297.22 nm), reaching 518.23 nm at 120 min. This behavior suggests that βCD does not provide as adequate steric protection against digestive conditions as PEG, potentially leading to destabilization of the colloidal system and premature or uncontrolled release of ACGs. These results suggest that PEG formulations may be suitable and effective for the controlled release of ACGs during gastrointestinal transit.

Particle size is a critical parameter in lipid nanotransport systems, as it directly influences their stability, encapsulation efficiency, release profile, distribution in the body, interaction with mucous membranes, and, especially, cellular uptake. The latter process, essential for in vivo efficacy, occurs mainly through endocytosis mechanisms, such as pinocytosis and phagocytosis. In general, particles smaller than 1 µm are internalized by cells via different pinocytosis pathways (mediated or not by clathrin or caveolin), which depend on the particle's size and uniformity (PDI). Therefore, characterizing the size and distribution of NSps is essential for predicting their biological behavior and optimizing their therapeutic efficacy [5].

3.2. In Vitro Bioaccessibility and Potential Bioavailability of Nanosuspensions During In Vitro Digestion

The results obtained (Figures 3 and 4) show significant differences ($p < 0.05$) between the PEGSL-ACG-NSps and βCDSL-ACG-NSps formulations in terms of release, in vitro bioaccessibility, and potential bioavailability during simulated digestion. The PEGSL-ACG-NSps formulation exhibited lower bioaccessibility (25–29%) and gradual potential bioavailability (25–55%), with ACGs released 60 min after the intestinal phase. In contrast, the βCDSL-ACG-NSps formulation showed faster release ($p < 0.05$), with higher bioaccessibility (15–50%) and high potential bioavailability (35–95%) of ACGs from the first 30 min of the small intestine phase.

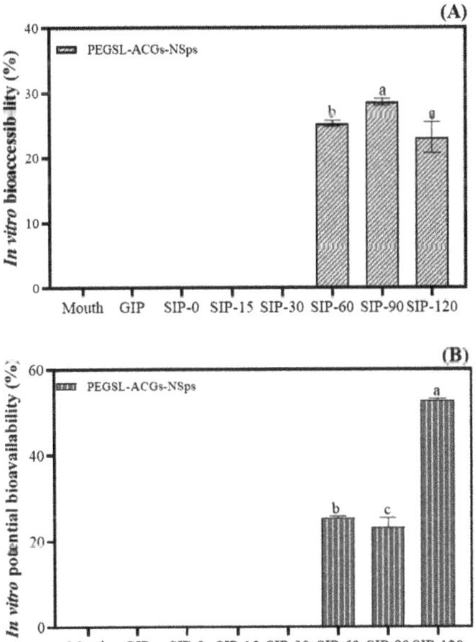

Figure 3. Percentage of in vitro bioaccessibility (**A**) and potential bioavailability (**B**) of pseudoannonacin present in PEGSL-ACG-NSps during each stage of the in vitro digestion study. SIP, small intestine phase; GIP, gastrointestinal phase. Lowercase letters indicate significant differences between the various stages of the in vitro digestion study ($p < 0.05$).

 https://doi.org/10.3390/foods15050937

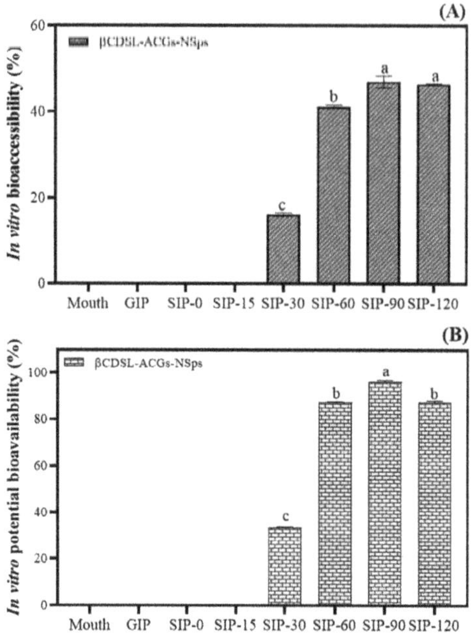

Figure 4. Percentage of in vitro bioaccessibility (**A**) and potential bioavailability (**B**) of pseudoannonacin present in βCDSL-ACG-NSps during each stage of the in vitro digestion study. SIP, small intestine phase; GIP, gastrointestinal phase. Lowercase letters indicate significant differences between the various stages of the in vitro digestion study ($p < 0.05$).

In this context, evaluating in vitro bioaccessibility and potential bioavailability is essential for predicting the gastrointestinal behavior of NSps after oral administration. These determinations allow us to estimate how much of the ACGs could be released under physiological digestive conditions and resist degradation caused by gastrointestinal factors (pH, acidity, enzymatic activity, bile salts), and what fraction is actually available for intestinal absorption.

Additionally, HPLC-DAD analysis performed at each stage of the simulated gastrointestinal digestion did not show significant changes in ACGs concentration or chromatographic profiles (Figure 5), suggesting that the acetogenins did not undergo relevant chemical degradation under the evaluated conditions. In addition, determining in vitro potential bioavailability provides critical information for the design of optimal formulations, such as selecting appropriate coatings (e.g., PEG or βCD) and anticipating in vivo behavior, thereby reducing the time and costs associated with animal testing.

These results reflect a controlled release profile for PEGSL-ACG-NSps, due to the protection provided by the flexible steric layer formed by PEG, which acts to protect the particles from digestive conditions and delay the release of ACG, as mentioned by Wiecinski et al. [30], who demonstrated that PEG 5000 improves the structural stability of nanoparticles, slowing their degradation under gastric conditions and improving their biodegradability. Furthermore, Hong et al. [11] and Kesisoglou et al. [31] confirmed that the use of high-molecular-weight PEG improves dispersion and colloidal stability and promotes slower, prolonged release by forming a physical barrier around the nanoparticles, which is key for applications requiring controlled drug release.

https://doi.org/10.3390/foods15050937

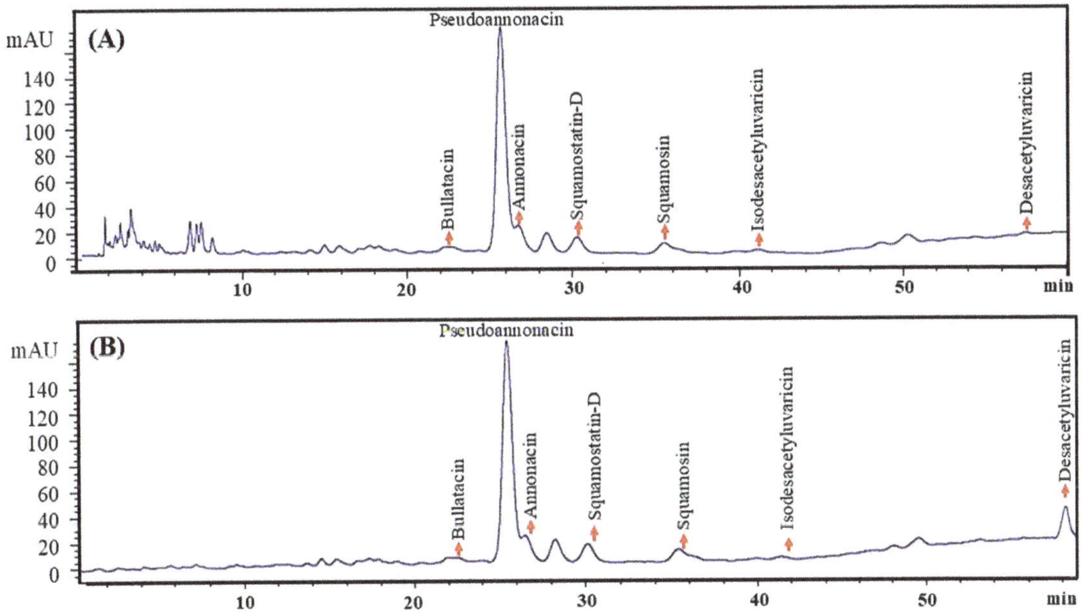

Figure 5. Representative HPLC-DAD chromatographic profile of ACGs into nanosuspensions, after simulated gastrointestinal digestion (**A**) and isolated ACGs (**B**).

In contrast, the βCDSL formulation, with its lower structural protection capacity against gastric conditions, allowed for a more rapid release of the ACGs. This characteristic could favor rapid, albeit less controlled, absorption, which is appropriate in applications requiring immediate therapeutic action, such as skin wounds at risk of infection [32]. Therefore, although the βCDSL formulation may not be ideal for sustained-release therapies, its kinetic profile positions it as a promising alternative for rapid, effective intervention.

In a study conducted by Hong et al. [13], it was reported that NSps loaded with ACGs and stabilized with βCD and soy lecithin, evaluated under controlled laboratory conditions at pH 7.4, exhibited more sustained ACGs release than in the present study. In contrast, the faster release profile reported here may be attributable to several factors, including differences in formulation composition, experimental design, and the gastrointestinal conditions under which the NSps were exposed. These conditions may involve more aggressive physicochemical and enzymatic environments, which can compromise the stability of βCD/ACGs complexes and promote accelerated desorption of the active compounds during the gastric and intestinal phases. Consequently, the results obtained in this study provide data that could be considered for future in vivo conditions.

Taken together, these results demonstrate that the choice of coating has a decisive impact on the stability and release profile of NSps, highlighting the effectiveness of PEG in designing colloidal systems with sustained release and greater gastrointestinal resistance.

3.3. *In Vivo Genotoxicity and Cytotoxicity Study*

The results obtained from this study demonstrate that neither the amphiphilic complexes nor the PEGSL-ACG-NSps and βCDSL-ACG-NSps at any of the concentrations evaluated caused significant genotoxic effects (Figure 6) in the in vivo model used ($p < 0.05$).

https://doi.org/10.3390/foods15050937

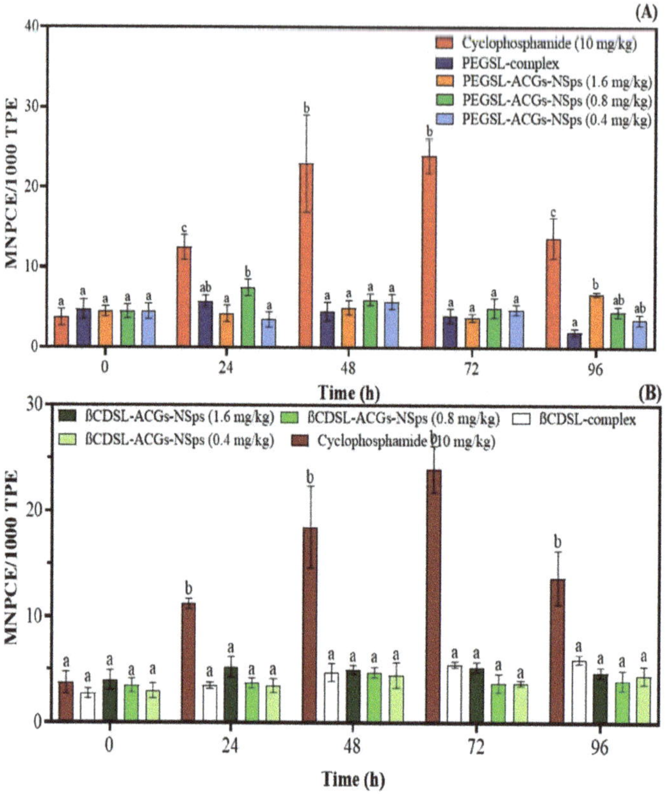

Figure 6. Genotoxicity of PEGSL-ACG-NSps (**A**) and βCDSL-ACG-NSps (**B**) at different concentrations produced after 24, 48, 72, and 96 h of treatment. The PEGSL-complex is polyethylene glycol complexed with soy lecithin; the βCDSL complex is beta-cyclodextrin complexed with soy lecithin. The different letters indicate significant differences among treatments within each evaluated time period ($p < 0.05$).

The frequency of micronucleated polychromatic erythrocytes (MNPCE), a reliable biomarker for assessing chromosomal damage [33], remained constant (3 MNPCE) up to 72 h after administration. However, a slight increase (7 MNPCE) was observed in the group treated with PEGSL-ACG-NSps at 1.6 mg/kg. Nevertheless, these results are lower than those observed with cyclophosphamide ($p < 0.05$), which showed a gradual increase of 3–23 MNPCE at 72 h, confirming its well-established genotoxic capacity and suggesting ongoing DNA damage or a deterioration of repair mechanisms [26,34].

Similarly, analysis of cumulative genotoxic damage (Figure 7) showed no evidence of progressive accumulation of normochromatic micronuclei (MNE) in groups treated with NSps, even after exposure for up to 96 h. These results indicate that NSps do not cause cumulative genotoxic stress or DNA integrity alterations over the exposure period. In contrast, the group treated with cyclophosphamide (10 mg/kg) showed a significant increase ($p < 0.05$) in MNE formation, reaching up to 46 MNE at 72 h, which corroborates its cumulative genotoxic effect on DNA.

https://doi.org/10.3390/foods15050937

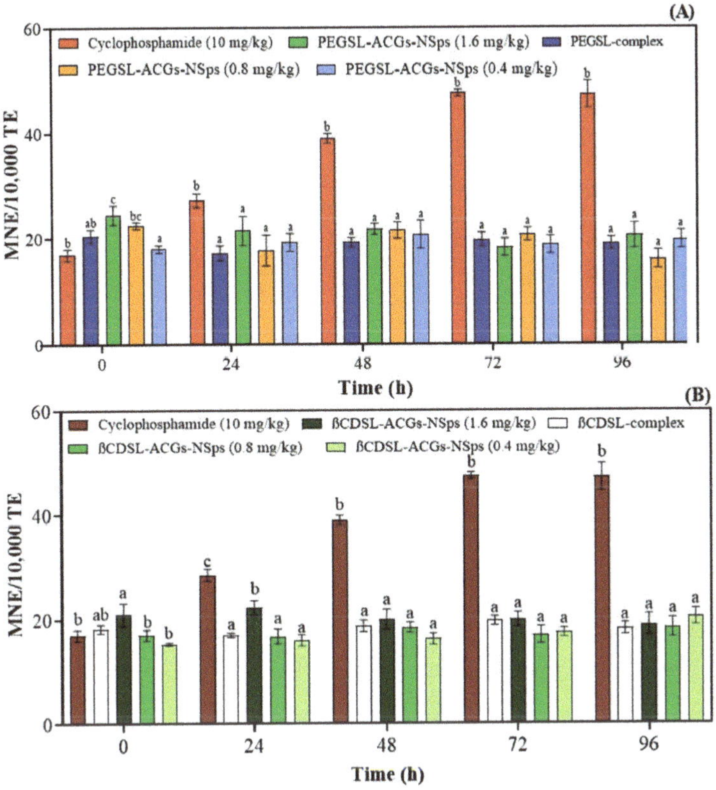

Figure 7. Cumulative genotoxic damage of PEGSL-ACG-NSps (**A**) and βCDSL-ACG-NSps (**B**) at different concentrations produced after 24, 48, 72, and 96 h of treatment. The PEG-SL complex is polyethylene glycol complexed with soy lecithin; the βCDSL complex is beta-cyclodextrin complexed with soy lecithin. The different letters indicate significant differences among treatments within each evaluated time period ($p < 0.05$).

Finally, Figure 8 shows that the amphiphilic complexes did not cause genetic damage. Figure 8A shows that the number of polychromatic erythrocytes (PCE), an indicator of bone marrow damage, remained constant for PEGSL-ACG-NSps at any of the concentrations evaluated. In contrast, βCDSL-ACG-NSps at 1.6 mg/kg showed a slight increase from 45 PCE at 24 h to 58 PCE at 96 h. Other indicators of cytotoxicity did not accompany this variation and therefore do not suggest a cytotoxic effect under the evaluated conditions. In contrast, cyclophosphamide showed a gradual decrease in the number of PCE (from 53 to 25) from 48 h, suggesting possible mitotic suppression or severe cytotoxic damage to bone marrow cells [34].

In contrast to these findings, the genotoxic potential of ACGs administered in their free form has been reported previously. García-Aguirre et al. [35] evaluated acetogenin isomers isolated from *Annona cherimolia* in a murine model using male NIH mice, administering doses of 1–5 mg/kg. Their results demonstrated a significant, dose-dependent increase in genotoxicity, with a peak effect at 48 h, comparable to that induced by daunorubicin. This discrepancy underscores the importance of the amphiphilic complexes employed in the present NSps, as these may modulate the biodistribution, cellular interactions, and biological activity of ACGs, thereby reducing their genotoxic potential.

https://doi.org/10.3390/foods15050937

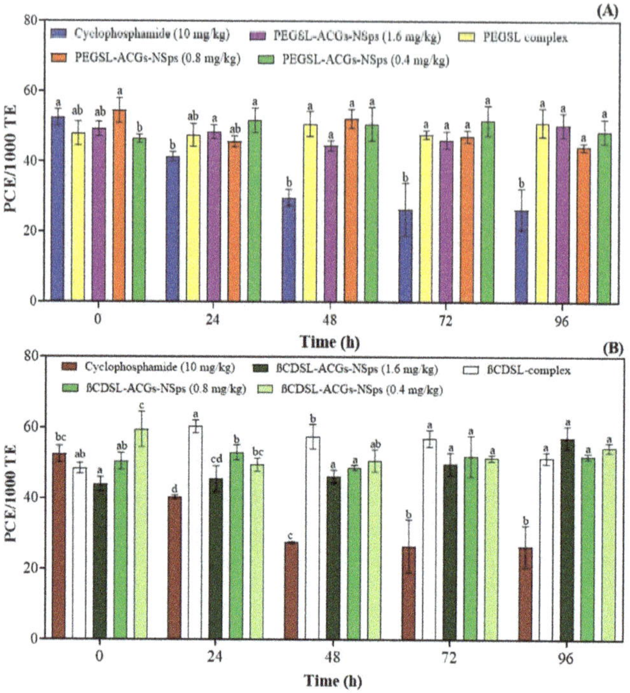

Figure 8. Cytotoxicity of PEGSL-ACG-NSps (**A**) and βCDSL-ACG-NSps (**B**) at different concentrations produced after 24, 48, 72, and 96 h of treatment. The PEG-SL complex is polyethylene glycol complexed with soy lecithin; the βCDSL complex is beta-cyclodextrin complexed with soy lecithin. The different letters indicate significant differences among treatments within each evaluated time period ($p < 0.05$).

Evaluation of genotoxicity and cytotoxicity is a key aspect of characterizing NSps for potential use in pharmaceuticals, as their nanometric size can induce unwanted interactions with cellular components, including DNA, causing toxic damage in the short or long term. Therefore, it is necessary to evaluate the biological safety of any nanosystem before its application. In this study, the toxicity of NSps was assessed using the micronucleus assay. This sensitive and reliable tool allows the evaluation of DNA damage induced by physical and chemical agents at early stages, by quantifying complete or fragmented chromosomes that are not incorporated into the nucleus during mitosis due to aneugenic or clastogenic effects. This method can be applied to any nucleated cell and requires small sample volumes, making it an ideal approach for toxicological studies [36]. Overall, these findings indicate that both PEGSL-ACG-NSps and βCDSL-ACG-NSps did not induce significant genotoxic or cytotoxic effects in the micronucleus assay at the tested doses following short-term exposure.

Despite the comprehensive characterization performed in this study, some limitations should be considered. The gastrointestinal behavior and potential bioavailability of ACGs were evaluated using an in vitro digestion model, which, although widely accepted, cannot fully replicate the complexity of in vivo digestion and absorption processes. In addition, the in vivo experiments were conducted at short-term exposure levels, which do not allow conclusions regarding chronic toxicity or long-term safety. Furthermore, pharmacokinetic parameters were not evaluated, and no specific biomarkers of oxidative stress were included, limiting the mechanistic interpretation of the observed biological responses. However, based on the results obtained in this first stage, we will conduct

 https://doi.org/10.3390/foods15050937

the second in vivo study to assess the pharmacokinetic profile, long-term toxicity, and incorporation of relevant biomarkers of oxidative stress.

4. Conclusions

This study demonstrates that both NSps are effective strategies for transporting ACGs without undergoing any changes during in vitro digestion. The βCDSL-ACG-NSps had 95% in vitro potential bioavailability, while PEGSL-ACG-NSps had only 55%. Finally, none of the formulations or their individual components induced detectable genotoxic or cytotoxic effects under the short-term conditions evaluated.

Author Contributions: Writing—original draft, G.A.-H. and E.M.-G.; Conceptualization, G.A.-H., B.A.L.-R. and E.M.-G.; methodology, G.A.-H., B.A.L.-R., C.M.G.-D. and B.V.-G.; validation, G.A.-H., B.A.L.-R. and E.M.-G.; formal analysis, G.A.-H. and B.A.L.-R.; investigation, M.d.L.G.M., G.A.-H. and B.A.L.-R.; writing—review and editing, E.M.-G., M.d.L.G.M., C.M.G.-D. and B.V.-G. All authors have read and agreed to the published version of the manuscript.

Funding: This research was funded by the Secretaría de Ciencia, Humanidades, Tecnología e Innovación de México (SECIHTI), grant number CBF-2025-I-58.

Institutional Review Board Statement: The experimental protocol was approved by the Research Ethics Committee of the University of Guadalajara (No. Protocol CI-03523, Approved on 22 May 2023).

Informed Consent Statement: Not applicable.

Data Availability Statement: The original contributions presented in this study are included in the article. Further inquiries can be directed to the corresponding authors.

Acknowledgments: The authors thank SECIHTI-Mexico for the postdoctoral grants to Gabriela Aguilar-Hernández (No. 590664) and Brandon A. López-Romero (No. 958281). This work was partially supported by Grammarly (v.1.2.229), which was employed to assist with language editing. The final manuscript was thoroughly reviewed and revised by the authors.

Conflicts of Interest: The authors declare no conflicts of interest.

Abbreviations

The following abbreviations are used in this manuscript:

NSps	Nanosuspensions
ACGs	Acetogenins
PEGSL-ACG-NSps	Nanosuspensions loaded with acetogenins and stabilized with either polyethylene glycol–soy lecithin
βCDSL-ACGsNSps	Nanosuspensions loaded with acetogenins and stabilized with either β-cyclodextrin

References

1. Bermejo, A.; Figadere, B.; Zafra-Polo, M.C.; Barrachina, I.; Estornell, E.; Cortés, D. Acetogenins from Annonaceae: Recent progress in isolation, synthesis and mechanisms of action. *Nat. Prod. Rep.* **2005**, *22*, 269–303. [CrossRef]
2. López-Romero, B.A.; Luna-Bárcenas, G.; García-Magaña, M.L.; Anaya-Esparza, L.M.; Zepeda-Vallejo, L.G.; López-García, U.M.; Ortiz-Basurto, R.I.; Aguilar-Hernández, G.; Pérez-Larios, A.; Montalvo-González, E. Extraction of acetogenins using thermosonication-assisted extraction from *Annona muricata* seeds and their antifungal activity. *Molecules* **2022**, *27*, 6045. [CrossRef] [PubMed]
3. Aguilar-Hernández, G.; López-Romero, B.A.; Pérez-Larios, A.; Ruvalcaba-Gómez, J.M.; Castellanos-Huerta, I.; Téllez-Isaías, G.; Petrone-García, V.M.; Anaya-Esparza, L.M.; Montalvo-González, E. Antibacterial activity of crude extract and purified acetogenins from *Annona muricata* seeds. *Appl. Sci.* **2023**, *13*, 558. [CrossRef]
4. Neske, A.; Ruiz-Hidalgo, J.; Cabedo, N.; Cortes, D. Acetogenins from the Annonaceae family. Their potential biological applications. *Phytochemistry* **2020**, *174*, 112332. [CrossRef]

 https://doi.org/10.3390/foods15050937

5. Danaei, M.; Dehghankhold, M.; Ataei, S.; Hasanzadeh-Davarani, F.; Javanmard, R.; Dokhani, A.; Khorasani, S.; Mozafari, M.R. Impact of particle size and polydispersity index on the clinical applications of lipidic nanocarrier systems. *Pharmaceutics* **2018**, *10*, 57. [CrossRef] [PubMed]
6. Lynnerup, J.T.; Eriksen, J.B.; Bauer Brandl, A.; Holsæter, A.M.; Brandl, M. Oral bioavailability enhancement by nanosuspensions. *Eur. J. Pharm. Sci.* **2023**, *184*, 106417. [CrossRef]
7. Mishra, J.N.; Sharma, U.; Chauhan, M.; Sharma, S.; Kumar, H.; Chen, Y.; Singh, P. Polyethylene glycol assisted synthesis of ZnO nanoparticles: Correlating structural modifications with temperature and pH dependent electrochemical behavior. *Int. J. Hydrogen Energy* **2025**, *164*, 150840. [CrossRef]
8. Champy, P.; Höglinger, G.U.; Féger, J.; Gleye, C.; Hocquemiller, R.; Laurens, A.; Guérineau, V.; Laprévôte, O.; Medja, F.; Lombès, A.; et al. Annonacin, a lipophilic inhibitor of mitochondrial complex I, induces nigral and striatal neurodegeneration in rats. *J. Neurochem.* **2004**, *88*, 63–69. [CrossRef]
9. Chen, Y.; Chen, J.W.; Zhai, J.H.; Wang, Y.; Wang, S.L.; Li, X. Antitumor activity and toxicity relationship of annonaceous acetogenins. *Food Chem. Toxicol.* **2013**, *58*, 394–400. [CrossRef]
10. Chamundeeswari, M.; Jeslin, J.; Verma, M.L. Nanocarriers for drug delivery applications. *Environ. Chem. Lett.* **2018**, *17*, 849–865. [CrossRef]
11. Hong, J.; Li, Y.; Li, Y.; Xiao, Y.; Kuang, H.; Wang, X. Annonaceous acetogenins nanosuspensions stabilized by PCL–PEG block polymer. *Int. J. Nanomed.* **2016**, *11*, 3239–3253. [CrossRef] [PubMed]
12. Hong, J.; Li, Y.; Xiao, Y.; Li, Y.; Guo, Y.; Kuang, H.; Wang, X. Annonaceous acetogenins nanosuspensions based on a self-assembly stabilizer. *Colloids Surf. B Biointerfaces* **2016**, *145*, 319–327. [CrossRef]
13. Hong, J.; Sun, Z.; Li, Y.; Guo, Y.; Liao, Y.; Liu, M.; Wang, X. Folate-modified Annonaceous acetogenins NSps. *Int. J. Nanomed.* **2017**, *12*, 5053–5067. [CrossRef]
14. Kim, S.H.; Lee, E.S.; Lee, K.T.; Hong, S.T. Stability and antioxidant activity of curcumin nanosuspensions. *CyTA J. Food* **2021**, *19*, 40–48. [CrossRef]
15. Ma, Y.; Cong, Z.; Gao, P.; Wang, Y. Nanosuspensions technology for natural products. *Eur. J. Pharm. Sci.* **2023**, *185*, 106425. [CrossRef] [PubMed]
16. Aguilar-Hernández, G.; Zepeda-Vallejo, L.G.; García-Magaña, M.L.; López-García, U.M.; Aguilera-Aguirre, S.; Montalvo-González, E. Acetogenin content from soursop fruit (*Annona muricata* L.) seeds: Evaluation of different extraction conditions. *Biotecnia* **2022**, *24*, 12–19. [CrossRef]
17. Montalvo-González, E.; Tellez-Isaías, G.; López-Romero, B.A.; Zamudio-Ojeda, A.; Pérez-Ladrón de Guevara, H.; Castellanos-Huerta, I.; Pérez-Larios, A.; Hernández-Aguilar, G. Polyethylene glycol/soy lecithin-nanosuspensions as carriers of acetogenins: Optimization, characterization, antibacterial and hemolytic activity. *Drug Deliv. Transl. Res.* **2024**, *91*, 105228. [CrossRef]
18. López-Romero, B.A.; Aguilar-Hernández, G.; Hargis, B.M.; García-Magaña, M.L.; López-García, U.M.; Ortiz-Basurto, R.I.; Zamudio-Ojeda, A.; Latorre, J.D.; Montalvo-González, E. Characterization of an acetogenin-carrying nanosuspension and its effect on bacteria of interest in the poultry industry. *Microorganisms* **2025**, *13*, 18. [CrossRef]
19. López-Romero, B.A.; Téllez-Isaías, G.; Luna-Bárcenas, G.; Zamudio-Ojeda, A.; Velázquez-Juárez, G.; de Lourdes García-Magaña, M.; Aguilar-Hernández, G.; Montalvo-González, E. Nanosuspensions as carriers of *Annona muricata* acetogenins: Antibacterial activity against *Enterococcus faecalis* and *Listeria monocytogenes*. *CyTA—J. Food* **2023**, *21*, 771–780. [CrossRef]
20. Brodkorb, A.; Egger, L.; Alminger, M.; Alvito, P.; Assunção, R.; Ballance, S.; Bohn, T.; Bourlieu-Lacanal, C.; Boutrou, R.; Carrière, F.; et al. INFOGEST static in vitro simulation of gastrointestinal food digestion. *Nat. Protoc.* **2019**, *14*, 991–1014. [CrossRef] [PubMed]
21. Bravo-Alfaro, D.A.; Ochoa-Rodríguez, L.R.; Villaseñor-Ortega, F.; Luna-Barcenas, G.; García, H.S. Self-nanoemulsifying drug delivery system (SNEDDS) improves the oral bioavailability of betulinic acid. *J. Mol. Liq.* **2022**, *364*, 119946. [CrossRef]
22. Zhao, Q.; Wang, Z.; Wang, X.; Yan, X.; Guo, Q.; Yue, Y.; Yue, T.; Yuan, Y. The bioaccessibility, bioavailability, bioactivity, and prebiotic effects of phenolic compounds from raw and solid-fermented mulberry leaves during in vitro digestion and colonic fermentation. *Food Res. Int.* **2023**, *165*, 112493. [CrossRef] [PubMed]
23. OECD (Organisation for Economic Cooperation and Development). *OECD Guideline for the Testing of Chemicals: Mammalian Erythrocyte Micronucleus Test*; OECD: Paris, France, 2016; pp. 1–10. Available online: https://www.oecd.org/en/publications/test-no-474-mammalian-erythrocyte-micronucleus-test_9789264071285-en.html (accessed on 20 February 2026).
24. Hayashi, M.; MacGregor, J.T.; Gatehouse, D.G. In vivo rodent erythrocyte micronucleus assay. II. *Environ. Mol. Mutagen.* **2000**, *35*, 234–252. [CrossRef]
25. Zúñiga-González, G.; Gómez-Meda, B.C.; Zamora-Pérez, A. Induction of micronuclei in vaginal cells from treated rats. *Environ. Mol. Mutagen.* **2003**, *42*, 306–310. [CrossRef]
26. Afrin, S.; Rahman, Y.; Sarwar, T.; Ansari, M.O.; Ahmed, S.; Rizvi, M.A.; Shadab, G.G.H.A.; Tabish, M. Cytotoxic and genotoxic evaluation of dipyridamole and its alternative therapeutic potential in cancer therapy: An in vitro and in vivo approach. *J. Mol. Struct.* **2022**, *1249*, 131626. [CrossRef]

https://doi.org/10.3390/foods15050937

27. *NOM-062-ZOO-1999*; Especificaciones Técnicas para la Producción, Cuidado y Uso de los Animales de Laboratorio. Diario Oficial de la Federación: Ciudad de México, Mexico, 2001.
28. *NOM-033-SAG/ZOO-2014*; Métodos para dar Muerte a los Animales Domésticos y Silvestres. Diario Oficial de la Federación: Ciudad de México, Mexico, 2014.
29. *NOM-087-ECOL-SSA1-2002*; Protección Ambiental—Salud Ambiental—Residuos Peligrosos Biológico-Infecciosos—Clasificación y Especificaciones de Manejo. Diario Oficial de la Federación: Ciudad de México, Mexico, 2003.
30. Wiecinski, P.N.; Metz, K.M.; Mangham, A.N.; Jacobson, K.H.; Hamers, R.J.; Pedersen, J.A. Gastrointestinal biodurability of engineered nanoparticles: Development of an in vitro assay. *Nanotoxicology* **2009**, *3*, 202–214. [CrossRef]
31. Kesisoglou, F.; Panmai, S.; Wu, Y. Nanosizing—Oral formulation development and biopharmaceutical evaluation. *AAPS J.* **2007**, *14*, 677–685. [CrossRef]
32. Trejo-Caballero, M.E.; Elizalde-Mata, A.; González-Reyna, M.A.; Martín-Martínez, M.J.; Bonilla-Peregrino, R.; Padilla-Vaca, F.; España-Sánchez, B.L.; Estevez, M. Alginate-based composites incorporating acetogenins. *Chem. Eng. J.* **2025**, *522*, 167354. [CrossRef]
33. Sommer, S.; Buraczewska, I.; Kruszewski, M. Micronucleus assay: The state of art, and future directions. *Int. J. Mol. Sci.* **2020**, *21*, 1534. [CrossRef] [PubMed]
34. García-Martínez, Y.; Razo-Estrada, A.C.; Pérez-Pastén Borja, R.; Galván-Colorado, C.; Chamorro-Cevallos, G.; Chanona-Pérez, J.J.; López-Canales, O.A.; Islas-Flores, H.; Pérez-Gutiérrez, S.; Cordero-Martínez, J.; et al. Protective effects of phycobiliproteins from *Arthrospira maxima* against cyclophosphamide-induced embryotoxicity and genotoxicity. *Pharm. Prod.* **2025**, *18*, 101. [CrossRef]
35. García-Aguirre, K.K.; Zepeda-Vallejo, L.G.; Ramón-Gallegos, E.; Álvarez-González, I.; Madrigal-Bujaidar, E. Genotoxic and cytotoxic effects produced by acetogenins obtained from *Annona cherimolia* Mill. *Biol. Pharm. Bull.* **2008**, *31*, 2346–2349. [CrossRef] [PubMed]
36. Zapata, L.M.; Bock, B.C.; Orozco, L.Y.; Palacio, J.A. Micronucleus test and comet assay in *Trachemys callirostris*. *Ecotoxicol. Environ. Saf.* **2016**, *127*, 108–116. [CrossRef] [PubMed]

 https://doi.org/10.3390/foods15050937

foods

MDPI

Review

Therapeutic and Preventive Potential of Plant-Derived Antioxidant Nutraceuticals

Antia G. Pereira [1,2], Javier Echave [1,3], Ana O. S. Jorge [1,4], Rafael Nogueira-Marques [1], Ezgi Nur Yuksek [1], Paula Barciela [1], Ana Perez-Vazquez [1], Franklin Chamorro [1], Maria B. P. P. Oliveira [4], Maria Carpena [1,*] and Miguel A. Prieto [1,*]

1 Universidade de Vigo, Nutrition and Food Group (NuFoG), Department of Analytical Chemistry and Food Science, Instituto de Agroecoloxía e Alimentación (IAA)—CITEXVI, 36310 Vigo, Spain; antia.gonzalez.pereira@uvigo.es (A.G.P.); javier.echave@uvigo.es (J.E.); anolijorge@gmail.com (A.O.S.J.); nogueirarafael29@gmail.com (R.N.-M.); ezginur.yuksek@uvigo.es (E.N.Y.); paula.barciela@uvigo.es (P.B.); ana.perez.vazquez@uvigo.es (A.P.-V.); franklin.noel.chamorro@uvigo.es (F.C.)

2 Investigaciones Agroalimentarias Research Group, Galicia Sur Health Research Institute (IIS Galicia Sur), SERGAS-UVIGO, 36312 Vigo, Spain

3 Centro de Investigação de Montanha (CIMO), LA SusTEC, Instituto Politécnico de Bragança, Campus de Santa Apolónia, 5300-253 Bragança, Portugal

4 REQUIMTE/LAQV, Department of Chemical Sciences, Faculty of Pharmacy, University of Porto, R. Jorge Viterbo Ferreira 228, 4050-313 Porto, Portugal; beatoliv@ff.up.pt

* Correspondence: mcarpena@uvigo.es (M.C.); mprieto@uvigo.es (M.A.P.)

Abstract: Oxidative stress and its relation to the onset of several chronic diseases have been increasingly highlighted in recent years. In parallel, there has been an increasing interest in the antioxidant properties of phytochemicals. Phytochemicals are products of plant secondary metabolism, including structural polysaccharides, unsaturated fatty acids, pigments (chlorophylls, carotenoids, and anthocyanins), or phenolic compounds. Phytochemicals can be obtained from lower and higher plants, their fruits, and even from macro- or microalgae. Their diverse structural features are linked to different beneficial effects through various molecular mechanisms, contributing to disease prevention. Beyond antioxidant activity, many phytochemicals also display anti-inflammatory, antidiabetic, anti-obesity, and neuroprotective effects, which can be intertwined. Beyond these, other natural antioxidants can also be obtained from animal, fungal, and bacterial sources. Thus, a wide range of antioxidants have the potential to be used as nutraceuticals with chemopreventive effects on the onset of various diseases related to antioxidant stress. Given their enormous structural and sourcing diversity, the present work provides an updated insight into the therapeutic and preventive potential of plant-derived antioxidants and nutraceuticals.

Keywords: phytochemicals; natural antioxidants; chemopreventive; nutraceuticals

1. Introduction

Phytochemicals are secondary metabolites synthesized by plants, encompassing a wide variety of chemical groups. These include phenolic compounds (PCs), fatty acids (FAs), pigments, polysaccharides (PSs), terpenoids, alkaloids, phytosterols, and saponins, among others. Their broad functional potential has positioned them as essential components in various industrial and health-related processes, including medicine, agriculture, food, textile, and cosmetic industries [1]. In particular, dietary phytochemicals have garnered increasing scientific attention in recent decades, largely due to their associated health benefits, such as cardioprotective, neuroprotective, anti-inflammatory, and gut-health-promoting effects, among others [2]. Yet, the major capacity of phytochemicals is to act as ameliorators of oxidative stress. Oxidative stress is the result of excessive levels of reactive oxygen species (ROS), which are produced by normal oxidative metabolism and exposure to the environment. The term ROS refers to molecules, such as superoxide radical (O_2^{2-}), hydroxyl radicals (OH^-), and hydrogen peroxide (H_2O_2), among others. These are produced by the sequential reduction of molecular oxygen in metabolic reactions. Oxidative stress may ultimately cause damage to biomolecules and can deregulate and even disrupt physiological redox signaling, causing changes in the rate of metabolic reduction reactions, an increased rate of DNA mutations, and deregulation of cell proliferation [3]. They are also one of the major markers of cell death, which, in turn, triggers inflammation and intertwines both processes. The impact that oxidative stress has on altering normal physiological performance is known to be a factor leading to the development of inflammatory, cardiovascular, neurodegenerative, and metabolic disorders, which, in turn, may result in cancer development [4,5]. Thus, despite the presence of endogenous antioxidant systems, the intake and use of phytochemicals as significant antioxidants has been proven as a suitable alternative to improve oxidative stress status [2].

Key contributors to this activity include PCs, unsaturated fatty acids (UFAs), carotenoids, and polysaccharides, which are commonly isolated from plant-based matrices [6]. The growing recognition of these bioactivities has not only influenced dietary recommendations and favored increased consumption of fruits and vegetables, but it has also catalyzed the expansion of the nutraceutical sector. This industry now focuses on formulating products enriched with specific phytochemicals, such as purified phenolics, aimed at delivering targeted health benefits [7]. However, in recent decades, the interest in dietary phytochemicals has risen because they are thought to provide many health benefits, such as cardioprotection, neuroprotection, anti-inflammatory effects, and improved gut health, among others [2]. These effects are mainly attributed to the antioxidant properties of various groups of phytochemicals, with phenolic compounds, UFAs, and polysaccharides being chief among the key contributors [6]. Thus, diets rich in vegetables and fruits that may act as sources of phytochemicals have attracted more interest from researchers, industry, and society, with a focus on more conscious and healthy diets, as extensive research in the last decades has suggested that they are an overall healthy alternative [8,9].

To support the formulation of effective phytochemical-enriched products, it is essential to optimize extraction methodologies that maximize recovery while preserving compound integrity. Among the wide array of phytochemicals, UFAs are a particularly valuable group because of their well-documented health benefits. However, despite advances in extraction technologies, most industrial-scale studies and applications continue to rely on conventional techniques, such as solvent extraction or mechanical pressing. These approaches, while widely implemented, often present limitations in terms of selectivity, efficiency, and environmental impact [10]. UFAs, for example, can be feasibly obtained from oily crops, such as olive, sunflower, rapeseed, sesame, palm, and peanut [11], although their composition and concentration vary depending on the plant source and extraction technique. Beyond these conventional crops, several underutilized fruit seeds, including those of pepper, grape, and avocado, have also been identified as promising UFA sources [12,13]. Additionally, microalgae have gained increasing attention in recent years as sustainable and efficient producers of UFAs with high nutritional value [14].

In addition to UFAs, pigments represent another important class of phytochemicals with significant health-promoting properties and broad industrial applicability, primarily due to their antioxidant activity and their function as natural colorants [15]. Among these, carotenoids stand out as the principal group of pigments used in industrial sectors. Carotenoids can be distinguished between carotenes, which include carotene and lycopene, and xanthophylls, with some main examples being lutein, zeaxanthin, and lycopene [15]. Moreover, they can be further distinguished between provitamin A carotenoids, which include carotenes and β-cryptoxanthin, and non-provitamin A carotenoids, including mainly xanthophylls, like lycopene, lutein, and zeaxanthin [16]. Intuitively, carotenoids can be obtained from yellow, orange, and red-colored fruits and vegetables, with higher concentrations in their peels. In recent years, both seaweeds and microalgae have become feasible and cost-effective sources of carotenoids, with some major examples being fucoxanthin from brown seaweeds and astaxanthin from red microalgae [17]. In this sense, carotenoids have been highlighted as relevant phytochemicals with various potential health benefits, ranging from their nutritional value as precursors of vitamin A to health-related properties, such as antioxidant, metabolic modulation, anti-inflammatory, and neuroprotective [18].

Another important class of phytochemicals is PSs, which play a critical role in both plant biology and human health. Among them, cellulose, hemicellulose, and pectin are the major PSs present in plants, which constitute dietary fiber [19]. These structural PSs may be found more abundantly in the "harder" parts of plants, such as seeds, leaves, and peels, and are of high interest as promoters of healthy gut microbiota [20]. Conversely, storage polysaccharides (that act as stored energy), which include starches and other PSs (i.e., fructans), are more abundant in bulbs, tubers, and pulps [21].

Lastly, PCs are antioxidant phytochemicals of paramount importance that have been exhaustively studied in the last two decades and have a ubiquitous distribution in the plant kingdom. PCs are mainly synthesized in plants in response to biotic (predators, infections) or abiotic (drought, salinity) factors [3]. Their diverse structural complexity and degree of polymerization also determine their bioavailability, activities, and occurrence. Some well-known examples of PCs and sources include flavanols in grapes or tea, flavonols in citrus, and anthocyanins in berries [3,22].

Given their vast variety, sources, and beneficial effects related to health, natural antioxidants have been increasingly used as nutraceuticals. Although these diverse molecules have been generally observed to display null or usually lower toxicity than synthetic or conventional chemicals, they may also display some undesired effects. For instance, it is known that PCs establish synergies in their antioxidant effects when various diverse molecules are present; however, when a single compound is present at high concentrations, PCs may act as prooxidant molecules, a fact that has also been observed with other antioxidant molecules [2]. Moreover, as with every other nutrient and chemical, an excessive intake can lead to adverse effects on health. For example, excessive intake of β-carotene has been associated with an increase in serum triglycerides, cholesterol, and bone loss [23]. On the other hand, ROS and other reactive species that are produced as a result of physical activity are needed for endogenous responses that enable muscle growth, which antioxidant supplementation may partially hurdle [24]. Nonetheless, beyond potential hepatotoxicity if the daily consumption is consistently very high, negative side effects of these types of natural compounds are limited, and mainly attributed to other elements, especially metals [25,26].

As research continues to uncover the full extent of their health benefits, these compounds are increasingly recognized for their potential to act as antioxidants, support disease prevention, enhance immune function, and improve overall well-being. As interest in these compounds continues to grow, there is a shift towards more economically viable production methods, including chemical and microbial synthesis, which offer greater consistency and scalability [27]. Additionally, the extraction of phytochemicals from natural sources, particularly through the use of agri-food waste from fruits and vegetables, is emerging as a sustainable and cost-effective approach [28]. Although these methods may be cost-effective, chemical extraction of phytochemicals from their natural sources is becoming increasingly widespread, especially using fruit and vegetable agri-food waste as a source of phytochemicals [28]. The aim of the present work is to discuss the role of antioxidants in plants, their significance in the mitigation of oxidative damage, and their implications for plant resilience and human overall health through other beneficial effects on health, such as anti-inflammatory, antidiabetic, anti-obesity, and neuroprotection. Although this work focuses on UFAs, PCs, PSs, and carotenoids, various other natural compounds with antioxidant properties, as well as significant benefits to health and potential use as nutraceuticals, are also explored.

2. Chemical Classes of Antioxidant Phytochemicals: Structural Significance

2.1. Unsaturated Fatty Acids

FAs are hydrophobic aliphatic carbon chains of variable length that contain a carboxyl (-COOH) group at the end of the chain [29]. FAs are classified by the number of carbons in the chain, but they mainly depend on the presence of double bonds between the carbons (Figure 1). FAs with no double bonds present are called saturated FAs (SMFAs), while those with double bonds are called UFAs, which are further classified based on whether they have a single bond (monounsaturated FAs, MFAs) or two or more bonds (polyunsaturated FAs, PUFAs) [30]. In addition, fatty acids can also be distinguished depending on their stereochemistry, presenting *cis* or *trans* isomers, with *cis* being more prevalent in plants and *trans* more prevalent in animal fats [31]. Depending on the position of the double bonds in the chain, UFAs can be further distinguished as omega (ω). In the case of plant organisms, fatty acids may usually present a length in the range of 12 to 22 carbons, with 16 and 18 carbons being generally the most abundant [32]. Among MFAs, plant organisms generally accumulate palmitic (C16:0) and stearic (C18:0) acids. Moreover, plants are the main source of UFAs in nature and, thus, in the human diet, with some major representatives being oleic acid (C18:1) and linoleic acid (C18:2) [11].

2.2. Phenolic Compounds

As mentioned, PCs are produced by plants primarily in response to environmental stress factors, such as drought, extreme temperatures, UV exposure, pollution, and pathogen attacks [33], and their biosynthesis follows pathways like the pentose phosphate, shikimate, and phenylpropanoid routes [34]. Chemically, PCs are defined by the presence of one or more hydroxyl groups attached to an aromatic ring, though their complexity varies across different subclasses. These compounds encompass a broad range of molecules, including phenolic acids, flavonoids, stilbenes, lignans, and related structures, like chalcones, humulones, and alcohols [33,35]. These compounds are classified according to their structure, with more than 8000 variations currently identified. The classification is typically based on the number of phenolic rings and the nature of the linkages between them. The major groups include phenolic acids, flavonoids, stilbenes, and lignans [36].

Phenolic acids are generally the most ubiquitous and the smallest PCs in terms of molecular size and weight, and they are divided into two main categories: hydroxybenzoic acids (e.g., gallic acid) and hydroxycinnamic acids (e.g., caffeic, ferulic, coumaric acids) [36,37]. Flavonoids, another key category, have a C6–C3–C6 backbone and are further categorized into subgroups (Figure 1), such as flavonols (e.g., quercetin, kaempferol, morin, myricetin), flavones (e.g., luteolin, acacetin, chrysin, baicalein, wogonin), isoflavones (e.g., genistein), anthocyanins, flavanones (e.g., naringenin, hesperetin), flavanols, and chalcones (e.g., xanthohumol isoxanthohumol, isobavachalcone, licochalcone A) [37]. In plant cells, flavonoids are typically stored as glycosides, and their antioxidant potential is influenced by the degree of glycosylation. Aglycones, like quercetin and myricetin, are generally more potent antioxidants than their glycoside forms, but with hampered bioavailability [38].

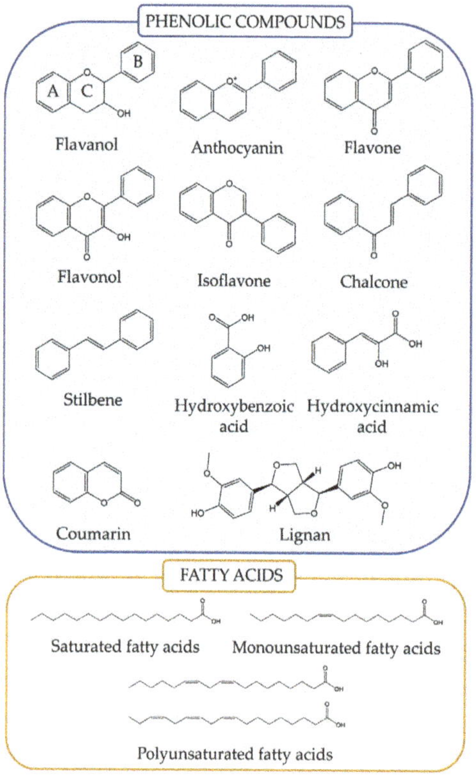

Figure 1. Chemical structure and examples of phenolic compounds and fatty acids.

Tannins, another class of phenolics, are mainly categorized into hydrolysable tannins (gallotannins, ellagitannins) and condensed tannins and phlorotannins (based on phloroglucinol, which is only found in brown seaweeds). These compounds are known for their protein-binding capabilities, contributing to the astringent taste of certain fruits and their antimicrobial activities [36,39].

Additionally, some PCs, like curcumin (diferuloylmethane), display a broad spectrum of biological functions, including antioxidant, anti-inflammatory, antimicrobial, antitumor, and liver-protective effects [40]. The structure of PCs is closely linked to their bioactivities, with different chemical arrangements providing specific biological functions [41,42].

Given their broad spectrum of bioactivities, PCs are increasingly being incorporated into nutraceuticals and functional foods aimed at improving human health and preventing chronic diseases [43]. Furthermore, these compounds contribute to the sensory qualities of many plant-based foods and beverages, particularly in terms of flavor and color [44]. In the context of human nutrition, PCs are known for their strong antioxidant properties and represent a significant portion of the antioxidant compounds in plants [45]. In terms of dietary applications, phenolic acids are commonly found in functional foods because of their beneficial health effects, such as reducing inflammation, preventing allergic reactions, and offering cardiovascular protection [46]. As research continues to elucidate the specific mechanisms underlying their health benefits, phenolic compounds hold promise as key ingredients in the development of novel therapeutic agents and functional food products [46,47]. Additionally, phenolic compounds have extensive applications in the food, cosmetic, and pharmaceutical industries, where they are used for their antioxidant and preservative properties, as well as their ability to improve the stability and shelf life of products [2]. They also represent a significant portion of the antioxidant content in plant-based foods [45].

2.3. *Carotenoids*

Given their nature and metabolic function as reducing agents, plant pigments are used as colorants and antioxidants. Carotenoids represent a class of pigments naturally produced by plants, fungi, algae, and specific bacteria [48]. These compounds are responsible for the vibrant red, yellow, and orange colors of various fruits and vegetables [49]. Structurally, carotenoids are classified as tetraterpenoids, composed of a 40-carbon polyene chain that provides their antioxidant properties (Figure 2) [50]. The conjugated double bond system present in the structure of carotenoids facilitates electron delocalization, thereby increasing their ability to neutralize free radicals and singlet oxygen [51]. Additionally, their antioxidant efficiency is modulated by the degree of unsaturation and the presence of functional groups, like hydroxyl (-OH) or keto oxo (C=O) groups [52].

Based on their structure, carotenoids can be divided into two primary categories: carotenes, which are purely hydrocarbons, and xanthophylls, which contain oxygenated groups [53]. The functional groups in xanthophylls generally improve their solubility and interaction with cellular membranes, potentially influencing their antioxidant effectiveness [54]. In fact, the presence of additional oxygen-containing functional groups often makes xanthophylls more potent antioxidants compared to carotenes [55].

Plants, algae, and fungi can produce different isomers of carotenes, with α-carotene and β-carotene being the most prevalent in plants (Table 1). Other isomeric forms include γ-carotene, δ-carotene, ε-carotene, and ζ-carotene. Both α-carotene and β-carotene serve as precursors to vitamin A [56], although β-carotene exhibits greater antioxidant activity compared to α-carotene [57]. Among xanthophylls, the key molecules include fucoxanthin, astaxanthin, lutein, zeaxanthin, and β-cryptoxanthin [53]. These compounds are recognized for their strong antioxidant properties, which play a role in their anticancer, anti-inflammatory, and neuroprotective effects, as well as their ability to enhance immune responses [53,58]. Owing to these attributes, carotenoids hold considerable potential for therapeutic use in managing OS-related diseases [59]. The variation in their efficacy stems from differences in their structural characteristics, such as terminal functional groups and chain length [53]. For instance, lutein and zeaxanthin, commonly found in leafy greens, are known to accumulate in the retina's macular region, where they filter harmful blue light and serve as powerful antioxidants [58].

Additionally, several studies support the inclusion of carotenoids in functional foods and dietary supplements, linking their intake to a reduced risk of chronic diseases, including cardiovascular disorders, specific cancers, and age-related macular degeneration [49]. Their role as precursors to vitamin A further enhances their significance in promoting visual health and supporting the immune system [60]. The numerous benefits of these compounds have favored the inclusion of these molecules as ingredients in the development of nutraceuticals aimed at improving health and preventing diseases.

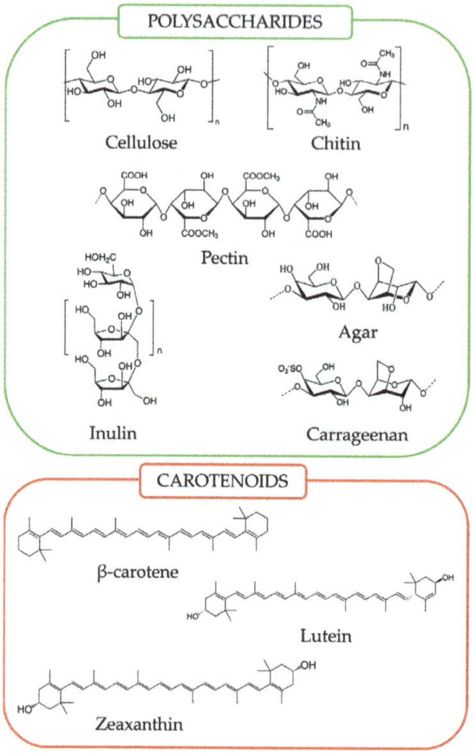

Figure 2. Chemical structure of some major examples of polysaccharides and carotenoids with beneficial health effects.

2.4. Polysaccharides

PSs are complex polymers composed of monosaccharide units connected by glycosidic bonds, which are widely distributed in plants, fungi, and algae [61]. These natural polymers have attracted significant attention due to their potential as nutraceuticals, particularly for their antioxidant, immunomodulatory, and anti-inflammatory activities [62,63]. Additionally, PSs prevent ROS-induced tissue damage through both free radical scavenging (Table 1) and modulation of the immune response [64,65]. However, under specific conditions, certain PSs may also exhibit prooxidant behavior, adding complexity to their antioxidant profile [66,67].

Overall, the antioxidant capacity of PSs is multifactorial, owing not only to their size but also to their structure and functional groups [68]. Structural features, such as monosaccharide composition, glycosidic linkages, molecular weight, and branching degree, contribute to their antioxidant activity [69]. Moreover, PSs with higher molecular weights, increased branching, and specific functional groups, such as sulfate or carboxyl groups, often demonstrate stronger antioxidant potential [70,71]. Moreover, the method of extraction plays a crucial role in determining their functional properties [72,73]. However, despite the established antioxidant potential of PSs, there is limited research on highly purified forms; thus, the possible influence of other bioactive components, which may be present in combination with PSs, must be taken into account [68].

PSs are classified based on their structure, origin, and function. Structurally, they can be divided into hetero-PSs (e.g., hyaluronic acid, chondroitin sulfate, heparin) and homo-PSs (e.g., fructans, galactan, glycogen) [74]. Based on their source, they are categorized into several groups: plant-based (e.g., pectin, cellulose, starch), microbial-based (e.g., curdlan, dextran, bacterial cellulose), fungi-derived (e.g., *β*-glucans, chitin derivatives), animal-derived (e.g., chitosan, chitin, heparin), and marine-derived (e.g., agar, alginate, fucoidans) [75]. PSs like pectin, which is rich in galacturonic acid, exhibit significant antioxidant properties. This bioactivity is influenced by their uronic acid content and degree of polymerization [76]. Moreover, sulfated PSs from marine algae are particularly effective in neutralizing free radicals and chelating metal ions, enhancing their antioxidant performance [77]. Ongoing research continues to shed light on the mechanisms by which these PSs exert their health-promoting effects, opening new avenues for their use in preventive and therapeutic nutrition [78]. Moreover, beyond their role as prebiotics, PSs are also widely utilized in industry for various functions, including as thickeners, emulsifiers, stabilizers, gelling agents, and controlled-release agents, broadening their applications [74,78].

Table 1. Concentration and antioxidant activity of antioxidant phytochemicals in principal sources: data from chemical assays.

Class	Compound	Main Source	Concentration	Assay	AA	Ref.
UFAs *						
ω-3	ALA	Olive, sunflower, linseed, rapeseed, fruit and vegetable seeds, other oily crops	5.5–61.5%	ROS	16.86 mM	[79–82]
	EPA	Seaweed, microalgae, fish oil	6.6–22.5%	ROS	150 μM	[83,84]
	DHA	Seaweed, microalgae, fish oil	1–6.6%	ROS	100 μM	[83,85,86]
ω-6	LA	Olive, sunflower, linseed, rapeseed, nuts, fruit and vegetable seeds, other oily crops	16.5–62.5%	ROS	39.5 mM	[79,82,87–89]
ω-7	PA	Olive, nuts, macadamia nuts, microalgae	0.6–50.1	-	-	[90–94]
ω-9	OA	Microalgae, linseed, rapeseed, nuts, fruit and vegetable seeds, other oily crops	1.4–79.6%	SOD	53.1 mM	[82,83,91,95–97]
Carotenoids						
Carotenes	*α*-carotene	Carrots, pumpkins	13.44–30.11 mg/kg fw	ROS	40.6 μmol TE/g dw	[98,99]
	β-carotene	Carrots, red peppers, oranges, potatoes, green vegetables	41.60–71.2 mg/kg fw	ROS	7.2 μmol TE/g dw	[98,99]
Xanthophylls	Fucoxanthin	Brown algae	0.02–18.60 mg/g dw	ROS	201 μg/mL	[53,100]
	Astaxanthin	*Haematococcus pluvialis*	3.8%	ROS	1.33 mM	[101]
	Lutein	Microalgae, algae, vegetables (i.e., kale, spinach)	0.7–5% dw	ROS	1.8–22 μg/mL	[102]
	Zeaxanthin	Red and brown seaweed, red/orange vegetables/fruits	0.49–1230 μg/g dw	ROS	2.2 μg/mL	[53,103]
	β-cryptoxanthin	Algae, red/orange vegetables/fruits	409–1103 μg/g dw	ROS	38.30 μg/mL	[104,105]
Polysaccharides						
HE	Hyaluronic acid	*Streptococcus* spp., *Tremella fuciformis*	1300 μg/mL	ROS	69.2–78.4%	[106,107]
	Chondroitin sulfate	Bacteria and cartilage		MCC	3.33 mg/mL	[108,109]
	Heparin	Marine organism, Asteraceae plants	-	EA	2.20 mg/mL	[110,111]
HO	Fructan	Prokaryotes, lower and higher plants	0.9–1.8 g/100 g in different wheat cultivars	EA, MCC	0.12 mg/mL	[21,112]
	Galactan	Seaweeds, seeds of some plants	-	SOD, GSH-Px	9 μM	[113]
Plant	Pectin	Cell walls of terrestrial plants	Citrus peels 30% fw, oranges 0.5–3.5% fw, carrots 1.4% fw	ROS	161.94 ppm	[114,115]
	Cellulose	Cell walls of terrestrial plants	40–50% fw	ROS	80.9 ppm	[116,117]
	Starch	Cereals, pseudocereals, legumes, root tubers	60–75% fw	ROS	97 μg/mL	[118,119]
Microbial	Curdlan	*Agrobacterium* sp., *Rhizobium* sp.	34.04 mg/g	ROS	82% DPPH, 72% ABTS	[120,121]
	Dextran	Lactic acid bacteria	580 mg/100 mL dw	ROS	97 μg/mL	[122,123]
	Cellulose	*Acetobacter* spp., *Sarcina* spp., *Agrobacterium* spp.	60.7% dw	ROS	80.9 ppm	[116,124]
Fungi	β-glucans	Fungal cell walls	31% dw	ROS, EA	161–4019 μg/mL	[125,126]
	Chitosan	Cell wall of filamentous fungi	20–45% dw	ROS	0.022 mg/mL	[127,128]
Marine	Fucoidan	Brown seaweed	20% dw	ROS	0.058 mg/mL	[129,130]
	Alginate	Brown seaweed	20–60% dw	ROS	121.4–346.3 mol/g	[131,132]
	Cellulose	Green algae	1.5–34%dw	ROS	0.15–0.39 mg/mL	[124,133,134]

Table 1. *Cont.*

Class	Compound	Main Source	Concentration	Assay	AA	Ref.
			UFAs *			
			Phenolic compounds			
Total PCs	-	*Phoenix dactylifera* var Bunarinja	34.90 mg/100g fw	ROS	0.875 µg/mL	[135]
Phenolic acids	Caffeic acid	Green coffee	6.56 µg/mL	ROS	6.31 µg/mL	[136]
	Chicoric acid	*Echinacea purpurea*	56.03 mg/g dw	ROS	15 µg/mL	[137]
	Ferulic acid	Rice bran	8.71 mg/g	ROS	9.9 µg/mL	[138,139]
Flavonoids	Myricetin	Green tea	0.40–0.79 mg/g	ROS	4.68 µg/mL	[140]
	Apigenin	*Gentiana veitchiorum*	37.50 mg/L	ROS	8.26 mg/mL	[141]
Total tannin	-	Ginger	35.08 mg/g	ROS	1 mg/mL	[142]
	-	Garlic	7.44 mg/g	ROS	3.7 mg/mL	[142]
	-	*Myristica fragrans*	14.03% *w/w*	ROS	89.98 µg/mL	[143]

Abbreviations: UFAs: unsaturated fatty acids; EFAs: Essential Fatty Acids; OA: oleic acid; DHA: docosahexaenoic acid; EPA: Eicosapentaenoic Acid; HEs: Heteropolysaccharides; HOs: Homopolysaccharides; PCs: phenolic compounds; dw: Dry Weight; fw: Fresh Weight; MCC: metal cation chelation; EAs: enzymatic antioxidants; ROS: reactive oxygen species; AA: antioxidant activity (IC_{50}); ns: Not Specified. * UFA concentrations are presented as % of total fatty acids.

3. Chemopreventive and Therapeutic Properties of Antioxidant Phytochemicals

Diets rich in plant-based foods are generally associated with reduced cancer risk and all-cause mortality due to their related health benefits, including antioxidant, anti-inflammatory, antidiabetic, anti-obesity, and, potentially, anticancer effects (Figure 3). This suggests that dietary phytochemicals from various plant species hold potential for cancer chemoprevention [144]. This section will explore in vivo and in vitro studies on the biological properties of phytochemicals and other antioxidants from common and uncommon sources, highlighting their effects and mechanisms described (Table 2) [145].

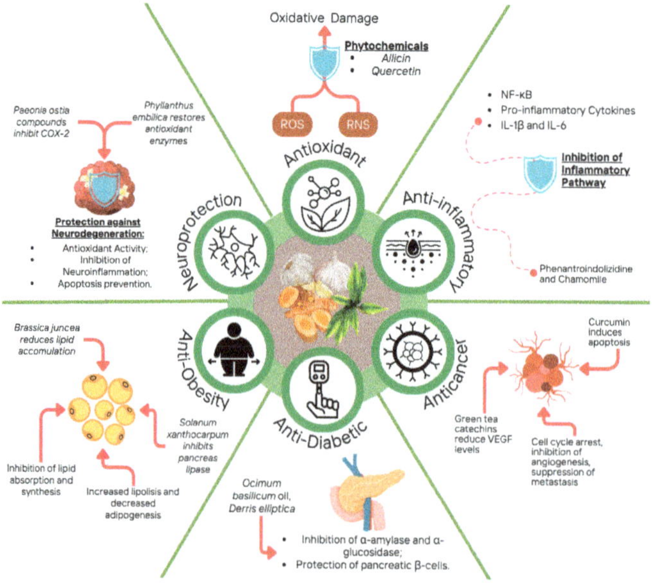

Figure 3. Overall view of the therapeutic properties of antioxidant phytochemicals, as illustrated with some of the examples mentioned.

Table 2. Chemopreventive activities of plant extracts and their phytochemicals: in silico, in vivo, and in vitro examples.

Plant Species	Extract	Phytochemicals	Activity	Analysis Method	Assay	Mechanism of Action	Results	Ref.
Antioxidant activity								
Garlic (*Allium sativum* L.)	ns	Allicin (purity > 90%)	Antitumor (cholangiocarcinoma)	CCK-8, FC, WB	In vitro/in vivo	STAT3 inhibition via SHP-1 upregulation	Suppressed proliferation, invasion, EMT, and tumor growth	[146]
ns	HS-1793	Resveratrol analogue	Antitumor (murine breast cancer)	LYM assays, DNA damage, Treg/TAM	In vivo	Inhibits immune suppression by Treg/TAM	Enhanced LYM proliferation, reduced Tregs, and decreased IL-10/TGF-β	[147]
Green tea *Curcuma longa* L.	ns	Catechins, curcumin	Antitumor (OSCC)	Histology, immunofluorescence, FC	In vivo	AP induction and anti-angiogenesis	Reduced tumor growth, increased apoptosis	[144]
Melilotus officinalis L.	ns	DC (coumarin derivative)	Anti-proliferative, gonad-safe	DC injection in BALB/c mice ovarian apoptosis, meiotic spindle	In vitro/In vivo	Cell cycle alteration without tubulin disruption	DC suppressed cell proliferation and increased AP in Vero and MCF-7 cells	[148,149]
Polyalthia longifolia L.	ME	Tetranorditerpene	Anticancer (prostate, leukemia cells)	Proteomic analysis	In vitro	ER stress activation, apoptosis	Inhibited tumor cell growth	[150]
Fagonia cretica L.	AqE	ns	Cytotoxic, induces cell cycle arrest	siRNA knockdown, MTT and FC, comet assay, WB	In vitro	Induction of DNA damage, activation of p53 and FOXO3a	Induced cell cycle arrest and AP in two phenotypic breast cancer cell lines	[151]
Onobrychis argyrea L.	ME (leaves)	Quinic acid, isoquercitrin, epicatechin, routine	Antioxidant, antidiabetic, anticancer	LC-MS/MS, DPPH, iron reduction, enzyme inhibition, XTT, FC	In vitro	ME induces apoptosis in HT-29 cells by disrupting mitochondrial membranes and activating caspases	Strong antioxidant and cytotoxic effect	[152]
Elephantopus mollis Kunth.	ME	3,4-di-*O*-caffeoyl quinic acid	Cytotoxicity, glucosidase inhibition	DPPH, FRAP, metal chelation, β-carotene cytotoxicity	In vitro	Induces cell death in NCI-H23 cells by triggering apoptosis	High antioxidant capacity, induced apoptosis	[153]
Thymus vulgaris L.	MetE, EAE, ChE, BolE, AqE, PEE	Polyphenols, tannins, flavonoids, sterols/triterpenes	Antioxidant	DPPH, ABTS, ferrous ion chelation, CVT	In vitro	Radical scavenging, metal chelation	Strong antioxidant capacity correlated with phenol and flavonoid content	[154]
Markhamia lutea L.	Leaf extract	Flavonoids (*O*-glycosides)	Antioxidant, anti-AChE, BChE, Aβ-42	DPPH, ORAC, iron reduction, FRAP	In silicoin vitro	Inhibits AChE, BuChE, and Aβ-amyloid-42	DPPH: 35.69 µg/mL, ORAC: 16,694.4 µM TE/mg, and iron chelation: 70.7 µM EDTA eq/mg	[155]
Hertia cheirifolia L.	Organic and EtOAc fraction	Total phenolics (100–250 mg GAE/g)	Antioxidant	DPPH, ABTS, FRAP, β-carotene	In vitro	Synergistic compound interaction	DPPH: 38.83 µg/mL, ABTS: 23.76 µg/mL, FRAP: 2628.87 µmol Fe^{2+} Eq/mL, and β-carotene: 58.91%	[156]
ns	AqE	QUE	Anticarcinogenic	WB, RT-PCR	In vitro	Downregulation of ROS, PKC, PI3K, and COX-2; upregulation of p53 and BAX	QUE modulated OS and apoptotic pathways in HepG2 cells	[157]
Anti-inflammatory activity								
Tylophora ovata L.	Natural and synthetic PAs	*O*-methyltylophorinidine (1, 1s)	Antitumor against TNBC	NFκB inhibition, 3D co-culture	In vitro	Stabilizes IκBα, blocks NFκB	Inhibited spheroid growth, surpassed paclitaxel	[158]
Mangifera indica L.	ns	Polyphenols	Anti-inflammatory, anticancer	Real-time PCR analysis and protein expression	In vitro	Modulates PI3K/AKT/mTOR, NF-κB, PARP-1, and Bcl-2	Reduced cancer cell growth by 90%	[159]
Helicteres isora L.	DCM-E/HeE	Rosmarinic acid	Anti-inflammatory, antioxidant	ELISAs	In vitro	Differentiation in cancer cells, shows no cytotoxic effect at high levels	Reduced TNF-α, PGE-2, and NO levels; highest COX-2 inhibition	[160]

Table 2. *Cont.*

Plant Species	Extract	Phytochemicals	Activity	Analysis Method	Assay	Mechanism of Action	Results	Ref.
Waltheria indica L.	Roots and aerial parts/CH_2Cl_2 extract	Flavonoids	Anti-inflammatory, chemopreventive	NF-κB inhibition, luciferase reporter assay, QR-inducing assay	In vitro	Induces phase 2 enzyme activity via QR induction assay	A total of 7/29 compounds showed inhibitory activity on the NF-κB pathway	[161]
Commiphora leptophloeos L.	Hydroalcoholic leaf extract	Phenolic acids and flavonoids	Anti-inflammatory	NO radical inhibition analysis, qPCR, physicochemical tests	In vitro/in vivo	Downregulates NF-κB and COX-2, reduces cytokines	Reduced inflammatory markers, promising for inflammatory bowel disease	[162]
Matricaria chamomilla L.	ns	β-Amyrin, β-eudesmol, β-sitosterol, apigenin, lupeol, quercetin, myricetin	Anti-inflammatory, anticancer	Proteome analysis, WB, qRT-PCR, thermophoresis	In silicoin vitro	Inhibition of NF-κB, reduces IL-1β and IL6 mRNA expression and G2/M cell cycle arrest	Cancer prevention, reduced pro-inflammatory cytokine expression	[163]
Asparagus densiflorus meyeri L	Root and aerial parts/DCM-E	Saponins, glycosides, sterols, triterpenes	Cytotoxic, anti-inflammatory	MTT assay, MCF-7 cell stimulation using TNF-α, RT-PRC	In vitro	Reduces NO release and NF-κB gene expression	Significant cytotoxicity (IC_{50} 26.13 μg/mL)	[164]
Capparis cartilaginea L.	Ethanolic leaf extract	Alkaloids, flavonoids, phenols, fatty acids, carotenes	Antioxidant, cytotoxic, anti-inflammatory	FBRC, FRAP, MTT assay, COX-1 inhibition	In vitro	Dose-dependent inhibition of thermally induced protein denaturation	Anti-inflammatory effects (IC_{50} 60.23 and 17.67 μg/mL) were better than standards	[165]
Corchorus olitorius L., *Amaranthus hybridus* L.	Hydroethanolic leaf extract	Tannins, flavonoids, phenolics, terpenoids, cardiac glycosides, coumarins	Pro-estrogenic, anti-inflammatory	Phytochemical and ELISA analyses	In vivo	Lowers IL-6 and inhibits proliferation by binding phytoestrogens to ER-β	Antioxidant because of its high tannin content/reduction in tumor size and incidence	[166]
Euphorbia hirta L.	Whole extract	Phytol, fatty acids, 5-HMF	Anti-inflammatory	NO production	In vitro	Suppression of PG generation	Inhibition of iNOS directly involved in inflammation	[167]
Antidiabetic activity								
Tradescantia pallida L.	Leaf extract	Syringic acid, p-coumaric acid, morin, catechin	Glycosylation and hemoglobin activity	α-Amylase assay	In vitro	Glycosylation inhibition non-enzymatically	Boosted insulin production, revitalized β-cells, inhibited AGEs, stimulated glucose transporters and AMPK	[168]
Cissampelos capensis L.	Leaf, stem, and rhizome	Glaziovine, pronuciferine, cissamanine	Antihyperglycemic	α-Amylase assay	In vitro	Enzyme inhibition pathway	Reduced glucose levels	[169]
Phyllanthus emblica L.	ns	Flavonoids	Antihyperglycemic	Molecular docking assay	In silico	Hypoglycemic action, reduces relative risk of T2D, PPAR inhibition of T2D	High binding affinity and selectivity for T2D therapeutic targets	[170]
Ocinum sanctum L.	Leaves	Eugenol	Antihyperglycemic	ELISA, RIA, and Neutral Red assay	In vitro	Physiological pathway	Decreased plasma glucose levels in T2D, increased islet insulin secretion, perfused pancreas	[171]
Ocinum basilicum L.	Leaves	TPC and FC	Antihyperglycemic	Enzyme inhibitory activity assay	In vitro	Enzyme inhibition pathway (α-glucosidase, α-amylase, DPP-IV, PTP1B, and SGLT2)	Inhibition of intestinal sucrase, maltase, and porcine pancreatic *α-amylase*	[172]
Derris elliptica L.	Leaves	QUE and ceramide	Antihyperglycemic	Biochemical analysis and histopathology study	In vivo	Enzyme inhibition pathway	Increased insulin secretion, protected pancreatic β-cells from oxidative stress	[173]
Carica papaya L.	Seeds	Hexadecanoic acid methyl ester, 11-ODA oleic acid	Antihyperglycemic	α-Amylase and α-glucosidase inhibition assay	In vitro	Enzyme inhibition pathway	Reduced glucose levels	[174]

Table 2. *Cont.*

Plant Species	Extract	Phytochemicals	Activity	Analysis Method	Assay	Mechanism of Action	Results	Ref.
Rhazya stricta L.	Roots	Hexadecanoic acid, methyl ester	Antihyperlipidemic, hepatoprotective	DPP-IV, α-amylase, α-secretase inhibition assay, GLP-1 measurement	In vitro/in vivo	Enzyme inhibition pathway	Reduced blood glucose and HbA1c, reduced cholesterol and triglyceride levels, reduced liver enzyme activity	[175]
Halooxylon stocksii L.	Root and aerial parts	8-ODA methyl ester	Antidiabetic	α-Amylase and α-glucosidase assay	In vitro	Enzyme inhibition pathway	Reduced glucose levels	[176]
Anti-obesity activity								
Rosa centifolia L.	Petals	Ellagic acid (polyphenols)	Lipid metabolism improvement	PCR	In vivo	Suppression of lipid synthesis, inhibition of intestinal absorption, downregulation of *Scd1* and *Hmcgr* mRNAs in the liver	Reduced body weight and adipose tissue, increased fecal triglycerides, lipid, and cholesterol metabolism	[177]
Rheum rhabarbarum L.	ns	Emodin, rhein (anthraquinones)	Lipid-lowering	ELISA and histological evaluation	In vitro/in vivo	FAS-ACC production prevention through decreased PPARγ and C/EBPα expression, reduced lipid accumulation	Body weight and adipose tissue reduction	[178]
Brassica juncea L.	ns	Sinigrin (glucosinolate)	Anti-obesity	Cell culture and XTT assay, WB, histological analysis	In vitro/in vivo	Reduces expression of adipogenic and lipid synthesis proteins	Inhibited lipid accumulation in 3T3-L1 and decreased eWAT mass in obese mice fed a high-fat diet	[179]
Anthophycus longifolius L.	ns	Rhodomycinone, salsolinol, 5-HCO, 2-COS, demethylalangiside	Anti-obesity, anti-hyperglycemia	α-Amylase, α-glucosidase, pancreatic lipase assay	In vitro	Enzyme inhibition pathway	Delayed lipid, CH digestion, and absorption	[180]
Solanum xanthocarpum L.	Fresh and dry leaves	Solasodine, carpesterol, β-sitosterol, diosgenin	Hypoglycemic, hepatoprotective, hypotensive	Pancreatic lipase inhibition assay, MTT	In vitro	ns	At 62.5 µg/mL, the fresh leaf extract reduced cancer cell viability by 50%	[181]
Rumex roth-schildianus L.	Acetone fraction	Flavonoids, phenolics	Anti-α-amylase, anti-α-glucosidase, anti-lipase	Lipase inhibition activity	In vitro	Inhibits OS, α-amylase, α-glucosidase, and lipase	Strong lipase inhibition (acetone fraction IC_{50} 26.3 µg/mL), close to orlistat (IC_{50} 12.3 µg/mL)	[182]
Neuroprotective activity								
Paeonia ostii	Stamen	(+)-3′′-methoxy-oxylactiflorin	Anti-inflammatory	Molecular docking, NO inhibition assay	In vitro/in silico	Inhibition of NO production by binding with protein COX-2	Reduced NO production to values of EC_{50} 3.02 µM	[183]
Phyllanthus emblica	Fruit extract	ns	Anti-inflammatory	NO inhibition assay	In vivo	Reduces IL-1β and TNF-α/increases expression of 5-HT1D, 5-HT2A, and D2 receptors	Amelioration of social interaction, social affiliation, anxiety, and motor coordination	[184]
Tabebuia impetiginosa	Leaves	Iridoids and organic acids	Anti-inflammatory	AChE inhibitory activity, LA detection, Y-maze test, PA assay	In vitro/in vivo	CP attenuates cognitive impairment, effects rat performance in Y-maze and PA tests	Reduced CP-induced chemo-brain, restored hippocampal function	[185]

Abbreviations: CCK-8: Cell Counting Kit-8; STAT3: Signal Transducer and Activator of Transcription 3; WB: Western Blotting; FC: Flow Cytometry; SHP-1: Src Homology 2 Domain-Containing Phosphatase-1; EMT: Epithelial–Mesenchymal Transition; Treg: Regulatory T Cell; TAM: Tumor-Associated Macrophage; LYM: Lymphocyte; IL-10: Interleukin-10; TGF-β: Transforming Growth Factor-Beta; ME: Methanol Extract; OSCC: Oral Squamous Cell Carcinoma; AP: apoptosis; DC: Dicoumarol; BALB/c: Bagg Albino Laboratory Strain/c; Vero: African Green Monkey Kidney Epithelial Cells; MCF-7: Human Breast Cancer Cells; siRNA: Small Interfering RNA; MTT assay: 3-(4,5-Dimethylthiazol-2-yl)-2,5-Diphenyltetrazolium Bromide Assay; p53: tumor protein p53; FOXO3a: Forkhead Box O3; QUE: quercetin; ROS: reactive oxygen species; OS: oxidative stress; PKC: Protein Kinase C; PI3K: Phosphatidylinositol 3-Kinase; COX-2: Cyclooxygenase-2; RT-PCR: Reverse Transcription Polymerase Chain Reaction; HepG2 cells: Human Liver Cancer Cell Line; AMPK: Adenosine Monophosphate-Activated Protein Kinase;

AGES: advanced glycation end products; T2D: Type 2 Diabetes; PPAR: Peroxisome Proliferator-Activated Receptor; NFκB: Nuclear Factor Kappa B; IκBα: Inhibitor of Nuclear Factor Kappa B Alpha; HT-29 cells: Human Colorectal Adenocarcinoma Cell Line; DPPH: 2,2-Diphenyl-1-picrylhydrazyl; FRAP: Ferric-Reducing Antioxidant Power; NCI-H23 cells: Human Lung Adenocarcinoma Cell Line; ORAC: Oxygen Radical Absorbance Capacity; ER: Endoplasmic Reticulum; HeE: Hexane Extract; DCM-E: Dichloromethane Extract; MetE: Methanolic Extract; EAE: Ethyl Acetate Extract; ChE: Chloroform Extract; BolE: Butanol Extract; AqE: aqueous extract; PEE: Petroleum Ether Extract; CVT: Cyclic Voltammetry Technique; PI3K/AKT/mTOR: Phosphoinositide 3-Kinase/Protein Kinase B/Mechanistic Target of Rapamycin; PARP-1: Poly (ADP-ribose) Polymerase 1; Bcl-2: B-cell Lymphoma 2; TNF-α: tumor necrosis factor-alpha; THP-1 cells: Human Monocyte Cell Line; ELISA: Enzyme-Linked Immunosorbent Assay; QR: Quinone Reductase; NO: nitric oxide; 5-HMF: Hydroxymethyl-2-furancarboxaldehyde; iNOS: Inducible Nitric Oxide Synthase; PG: prostaglandin; RIA: Radioimmune Assay; TPCs: Total Polyphenol Compounds; FC: flavonoid content; 11-ODA: 11-Octadecanoic Acid; HbA1c: Glycosylated Hemoglobin; PCR: Polymerase Chain Reaction; FAS: Fatty Acid Synthase; ACC: Acetyl-CoA Carboxylase; 5-HCA: 5-Hydroxyconiferyl; 2-COS: 2-Caffeoylisocitrate; DPP-IV: Dipeptidyl Peptidase-IV; PTP1B: Protein Tyrosine Phosphatase 1B; SGLT2: Sodium–Glucose Co-Transporter 2; qPCR: Quantitative Polymerase Chain Reaction; FBRC: Ferric-Bipyridine-Reducing Capacity of Total Antioxidant; ER-β: Estrogen Receptor Beta; eWAT: epididymal white adipose tissue; IL-1β: Interleukin 1 Beta; TNF-α: tumor necrosis factor-alpha; 5-HT1D: Serotonin 1D Receptor; 5-HT2A: Serotonin 2A Receptor; D2: Dopamine 2 Receptor; AChE: Acetylcholinesterase; LA: Locomotor Activity; PA: Passive Avoidance; ns: Not Specified.

3.1. Antioxidant and Anticancer Properties

Antioxidants and phytochemicals act through two main mechanisms: chain breaking by donating electrons to free radicals and quenching ROS and other free radicals. As mentioned, there is a preeminent intertwining between oxidative stress and various diseases, including being considered a major contributor to the onset of cancer [186]. As such, various antioxidants from common and uncommon sources have been studied, considering both properties, by direct and indirect pathways (Table 2). The authors of a recent study examined the antioxidant properties of allicin, a major and fragrant sulfur organic compound from garlic, in in vitro and in vivo models of cholangiocarcinoma (CCA) [146]. Allicin was found to reduce oxidative stress and inhibit the STAT3 signaling pathway, which promotes tumor growth. The study demonstrated that allicin suppressed CCA cell proliferation, migration, and invasion while inducing apoptosis. Its dual action as an antioxidant and inhibitor of tumor-promoting signals suggests allicin could be a promising therapeutic option for CCA [146].

An in vivo study evaluated the phytochemicals in green tea (*Camellia sinensis*), including catechins and curcuma (*Curcuma longa*) curcumin, for chemopreventive effects in a hamster buccal pouch carcinoma model [144]. Green tea PCs, which mainly comprise flavan-3-ols and, mainly, epicatechin gallate, inhibited angiogenesis, while curcumin induced apoptosis through cytochrome c release and caspase activation. Both phytochemicals reduced cell proliferative and angiogenic activity, and when combined, they were more effective at inducing apoptosis and cell cycle arrest [144]. The PC-rich aqueous extract of virgin's rose (*Fagonia cretica*) was tested for its effects on breast cancer cell lines. The results proved that the extract induced cell cycle arrest and apoptosis through both p53-dependent and -independent mechanisms. Specifically, it led to a significant increase in tumor protein p53 and FOXO3a expression [151]. On the other hand, plant-derived quercetin exhibited chemopreventive effects in HepG2 (human liver cancer) cells by reducing ROS, downregulating PI3K, PKC, and COX-2, and increasing pro-apoptotic p53 and BAX expression, highlighting its antioxidant and anticarcinogenic action [157]. One in silico and in vitro study attributed Nile tulip's (*Markhamia lutea*) strong antioxidant properties primarily to its rich PC content, mainly including flavonoids and phenolic acids, which effectively scavenged free radicals and chelated metal ions [155].

A study on tobacco weed (*Elephantopus mollis*) highlighted 3,4-di-*O*-caffeoyl quinic acid for its significant antioxidant activity and chemopreventive properties, including apoptosis induction in NCI-H23 (human lung adenocarcinoma) cells and β-glucosidase inhibition [153]. Additionally, the leaf extract of *Onobrychis argyrea* has demonstrated significant antioxidant, antidiabetic, anti-Alzheimer's disease, and anticancer in vitro properties. It induced apoptosis in human colorectal adenocarcinoma HT-29 cells by disrupting mitochondrial membranes and activating caspases, and it showed high antioxidant activity, enzyme inhibitory activities, and strong anticancer capacity [152].

3.2. Anti-Inflammatory Properties

Anti-inflammatory phytochemicals may help prevent cancer by targeting inflammation-related mechanisms. They induce cell cycle arrest by inhibiting cell cycle regulators, leading to halted cancer cell proliferation [187]. Moreover, they promote apoptosis by upregulating pro-apoptotic proteins and caspases while downregulating anti-apoptotic factors, thus facilitating cancer cell death [188]. They also promote autophagy by upregulating autophagy-related proteins and inhibit mTOR signaling, leading to the elimination of damaged cells. These combined effects decrease inflammation and suppress tumor progression [189].

It has been reported that mango (*Mangifera indica*) polyphenols (mainly gallic acid, hydroxybenzoic acid hexoside, and hydrolysable tannins, such as monogalloyl-glucoside) reduced NF-κB expression and phosphorylation in tumor necrosis factor-alpha TNF-α-treated MCF-12A cells, indicating anti-inflammatory effects. Phytochemicals identified in mango modulated the phosphoinositide 3-kinase/protein kinase B/mechanistic target of the rapamycin (PI3K/AKT/mTOR) signaling pathway [159]. *Commiphora leptophloeos* extract, rich in phenolic acids and flavonoids, showed anti-inflammatory effects by down-regulating NF-κB/COX-2 and reducing pro-inflammatory cytokines (TNF-α, IL-1β, IL-6). This suggests its potential in the treatment of inflammatory bowel disease (IBD) [162]. Studies on the anti-inflammatory activity of *Helicteres isora* fruit extracts have been focused on its effects on pro-inflammatory mediators such as prostaglandin E2, COX-2, and TNF-α. Compared to celecoxib, a known COX-2 inhibitor, a hexane extract showed the strongest inhibition of PGE-2 (69.68%), followed by an ethanol extract (57.17%). All extracts were less effective than dexamethasone, a standard anti-inflammatory drug, but the hexane extract showed 51.61% inhibition of TNF-α [160]. The extract of *Euphorbia hirta* has been shown to have anti-inflammatory activity in vitro, which is relevant to its potential chemoprotective effects. By reducing nitric oxide (NO) production through the inhibition of inducible NO synthase (iNOS), *E. hirta* can help to lower inflammation, a known contributor to cancer advancement. Chronic inflammation can facilitate tumor growth and spread, so the extract's ability to suppress NO and inflammatory cytokines may play a role in its anticancer properties [167]. Chamomile phytochemicals, including quercetin and the terpenoids β-amyrin and lupeol, act as NF-κB inhibitors and induce G2/M cell cycle arrest. These compounds lower pro-inflammatory cytokines IL-1β and IL-6 and interact with microtubules, which supports their potential as anticancer modulators of inflammatory pathways [163].

3.3. *Antidiabetic Properties*

Diabetes mellitus (DM) is a metabolic syndrome that increases the levels of glucose in the blood [173,190]. The main chronic complications of this disease are linked to the level of exposure of patients with hyperglycemia [190]. Despite the availability of chemical drugs that act as therapeutic agents, there is a need to find natural alternatives to control the complications derived from the disease and improve glycemic control and glucose intolerance [173]. In this sense, traditional medicine revealed several molecules present in plants with antidiabetic activity, as well as their mechanisms of action [190].

For instance, leaf extracts have been widely studied to determine their antidiabetic potential. An in vivo and in vitro study investigated the volatile and phenolic compounds present in basil (*Ocimun basilicum*) essential oil to determine its potential antidiabetic activity. Streptozotocin (50 mg/kg of body weight) was used as a model to induce DM in Wistar Albino rats [191]. After 28 days of feeding basil essential oil (70 mg/kg of body weight) rich in PCs, the rats showed a protective effect in pancreatic β-cells and a decrease in blood glucose levels. Regarding in vitro assays, the inhibitory effects on the digestive enzymes α-amylase and α-glucosidase inhibitory activity were studied, showing a strong inhibition potential. The authors linked the antidiabetic effect of basil essential oil to the phenolic and flavonoid compound content. However, the mechanism of action was not elucidated [191]. In another in vivo study, *Derris elliptica* methanolic leaf extract's phytochemical profile and antidiabetic potential were determined [173]. Streptozotocin was used to induce DM in Sprague Dawley rats. After 14 days of feeding with the plant's extract at two different concentrations (200 and 400 mg/kg body weight), the results showed a hypoglycemic effect and reduced cholesterol levels in a dose-dependent manner. Moreover, the highest dose increased serum insulin levels, and an antihyperglycemic effect was shown. Although the authors did not explain the mechanism of action, quercetin was the compound linked to these antidiabetic properties [173]. *Eugenia sonderiana* leaf extracts were also studied to determine their antidiabetic potential and to develop a structure–activity correlation [190]. Both in vivo and in vitro studies were assessed, which showed great α-amylase and α-glucosidase inhibitory activity and reduced glucose, pancreatic enzymes, and triglycerides, while HDL cholesterol levels were maintained within normal ranges. To explain the antidiabetic ability of *E. sonderiana* leaf extracts, the authors suggested a synergistic effect between PCs by increasing glycogen synthesis, α-amylase, and saponins by a complex formation with cholesterol [190].

3.4. Anti-Obesity Properties

Obesity contributes to the increase in the prevalence of chronic diseases, including type II DM, hypertension, dyslipidemia, cardiovascular diseases, and some types of cancer [177,192]. Common strategies to decrease the incidence of obesity are physical activity, surgery, drugs, and diet modifications [192]. In this regard, the natural compounds present in plants have been suggested as potential chemicals to be used for both the prevention and treatment of obesity [192]. Several studies have assessed the anti-obesity potential of phytochemicals. For instance, green tea has been traditionally used as a medical plant, and scientific studies have revealed its potential for obesity prevention, linked to its high content of PCs. In vitro results regarding green tea leaves suggest the anti-obesity potential of this plant by inhibiting adipocyte differentiation and proliferation [193].

In a study, C57BL/6J mice fed a high-fat diet supplemented with 0.25% Provence rose (*Rosa centifolia*) polyphenol extract for 35 days showed decreased body weight and adipose tissue, as well as reduced serum cholesterol and hepatic triglyceride levels. The treatment increased fecal triglycerides and increased lipolysis-related proteins while suppressing lipid synthesis enzymes. Ellagic acid, a key component of the Providence rose, is likely responsible for these anti-obesity effects, emphasizing its capability for treating obesity [177].

Mustard (*Brassica juncea*) attenuated lipid accumulation in 3T3-L1 adipocytes and lowered epididymal white adipose tissue mass in obese mice fed a high-fat diet [179]. Finally, acetone extract fractions from *Rumex rothschildianus*, which is rich in flavonoids and phenolics, strongly inhibited lipase activity, comparable to orlistat, and also affected α-amylase and α-glucosidase [182].

3.5. Neuroprotective Properties

In the past decades, there has been a rise in the incidence of neurodegenerative disorders, which has been linked to the increase in longevity that has occurred. Neurodegenerative diseases encompass a wide range of conditions that affect the central nervous system, including alterations in the neuronal structure and cellular dysfunction that lead to progressive degeneration [194]. The most prevalent are Alzheimer's disease, Parkinson's disease, and lateral amyotrophic sclerosis. Although there are some drugs available to ameliorate these conditions, their efficacy is limited. In the past years, natural compounds have been suggested as potential alternatives with the ability to delay disease onset, reduce disease progression, and regenerate damage via their anti-amyloid, antioxidant, and anti-inflammatory properties [195].

Paeonia ostia has traditionally been used as a medical and ornamental plant in China. The neuroprotective effect of the stamens of this plant has been studied by the NO inhibition assay in lipopolysaccharide-induced BV-2 cells. The (+)-3′′-methoxy-oxylactiflorin present in the plant has excellent inhibitory properties in inhibiting NO production. Moreover, molecular docking results showed that the anti-inflammatory mechanism of this molecule was a binding action with COX-2 [183]. In another study, Indian gooseberry (*Phyllantys emblica*) fruit extract, which is rich in PCs, was tested in rats with sodium valproate-induced postnatal autism. After treatment with the fruit's extract at 100 mg/kg of body weight, the amelioration of social interactions, social affiliation, anxiety, and motor coordination was achieved. Moreover, the results showed the restoring effect of the extract in glutathione-S-transferase and glutathione reductase, which are oxidative enzymes, and a reduction in malondialdehyde (MDA) and NO levels [184]. These activities showcase its ability to induce neuroprotection by anti-inflammatory and antioxidant effects, as well.

4. Development of Phytochemicals as Nutraceuticals

4.1. Extraction, Purification, and Encapsulation

The direct ingestion of phytochemicals within their native plant matrices presents several challenges [196]. The bioavailability of these compounds is often restricted by factors such as low concentrations, poor solubility, instability under gastrointestinal conditions, and rapid metabolic degradation and excretion [197]. These obstacles highlight the need for developing methods to concentrate, purify, and encapsulate phytochemicals, thereby enhancing their effectiveness as nutraceuticals (Figure 4) [198].

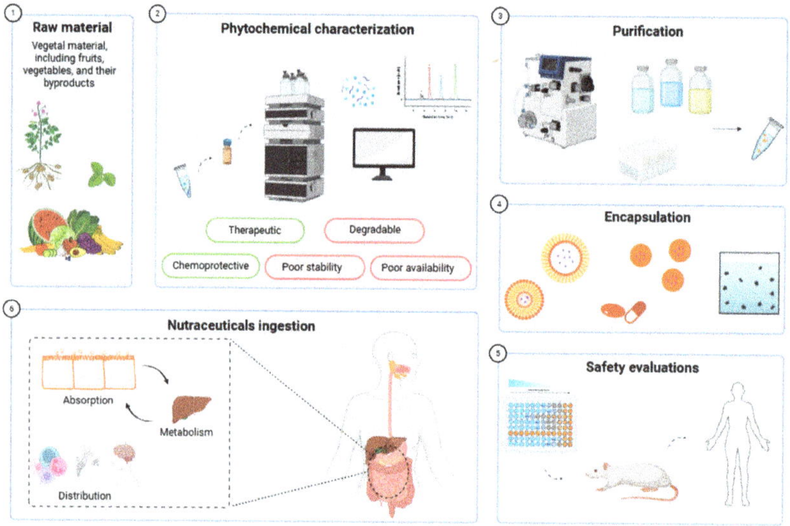

Figure 4. Process of development of phytochemicals as nutraceuticals.

The initial phase in the utilization of phytochemicals for nutraceutical purposes involves their extraction from plant sources [199]. This extraction process is complex and must be carefully optimized to preserve the structure and bioactivity of the target compounds [200]. Traditional extraction techniques, such as solvent extraction, have been improved by emerging methods, like microwave-assisted extraction and ultrasound-assisted extraction, since they offer increased efficiency and selectivity [201]. A major challenge in phytochemical extraction and recovery is to incorporate environmental sustainability into the design and optimization of production processes, with the aim of developing more eco-friendly methodologies [202] and enhancing economic outcomes within the food industry [203]. The choice of appropriate extraction techniques is crucial, as it directly impacts the quality and characteristics of the final product [204]. The key factors influencing the efficacy of both conventional and emerging extraction techniques include the solid-to-solvent ratio, solvent concentration, particle size, and the mode of extraction (flow or batch) [205]. These parameters critically affect the efficiency of extraction and the phytochemical composition of the resultant extract [206].

Following extraction, the subsequent step involves the purification of the bioactive compounds from the crude extract. This is necessary because the initial phytochemical mixture typically contains a combination of active and inactive substances, along with impurities that could affect the safety and efficacy of the final product [207]. The degree of purification achieved is linked to the pharmacokinetic properties of the compounds, which influence their bioavailability, distribution, metabolism, and excretion [208]. Recent advancements in the purification and isolation techniques for bioactive compounds from plant sources have addressed some of the challenges posed by the complexity and diversity of phytochemicals [209,210]. These advancements focus on optimizing methods to concentrate the desired compounds, thereby enhancing bioactivity—such as antioxidant, antibacterial, or cytotoxic effects—while maintaining simplicity, specificity, and efficiency in processing [211,212]. Techniques such as high-performance liquid chromatography (HPLC), solid-phase extraction (SPE), and preparative thin-layer chromatography (TLC) are commonly used for the isolation and purification of bioactive molecules because of their practicality, cost-effectiveness, and the availability of various stationary phases [204,211]. Notably, silica, alumina, cellulose, and polyamide have proven effective in separating phytochemicals [211]. To achieve the separation of complex mixtures, the use of multiple mobile phases with varying polarities is often required [204,211].

Despite the successful purification of phytochemicals, the formulation of these compounds into nutraceuticals remains challenged by issues of stability, solubility, and bioavailability. Encapsulation technologies have emerged as vital strategies to overcome these challenges [213,214]. Additionally, encapsulation can improve the quality of food products by masking undesirable odors and flavors and extending shelf life [203]. This process involves enclosing phytochemicals within a protective matrix or carrier system, which shields them from environmental degradation, enhances their solubility in biological fluids, and facilitates controlled release in the gastrointestinal tract [215,216]. Various encapsulation methods have been developed, each offering unique advantages depending on the physicochemical properties of the phytochemical and its intended application [217]. Microencapsulation, for instance, is widely used to protect sensitive compounds and improve their stability during processing and storage [218,219]. Conversely, nanoencapsulation provides enhanced bioavailability and targeted delivery due to small particle sizes and large surface areas [220,221], which improve interactions with biological membranes and allow for more precise control over release kinetics [222,223].

In conclusion, the effective application of phytochemicals as nutraceuticals is closely linked to advances in extraction, purification, and encapsulation techniques. These processes not only enhance the pharmacological potential of phytochemicals but also ensure their stability, bioavailability, and functional integration into various nutraceutical products.

4.2. *Considerations of Bioavailability*

Phytochemicals offer a wide range of health benefits, yet their effectiveness is often compromised by challenges related to bioavailability, including issues with water dispersibility, chemical stability, and gastrointestinal absorption [224]. For phytochemicals to be effective, they must be released from their food matrices, solubilized in gastrointestinal fluids, and absorbed by enterocytes in the gastrointestinal tract. However, these compounds may precipitate, interact with other dietary molecules, or be chemically altered by digestive enzymes, metabolic pathways, or gut microbiota, which can significantly reduce their bioactive forms by the time they enter systemic circulation [225].

Therefore, the bioavailability of phytochemicals is affected by several factors, including the complexity of the food matrix, the chemical form of the compound, and individual physiological differences, such as gut microbiota composition, mucosal health, and metabolic activity. These factors contribute to considerable inter- and intra-individual variability in bioavailability, complicating the prediction of their efficacy [226]. Additionally, the impact of food processing on phytochemical stability and bioavailability is critical, as processing techniques can either enhance or diminish their bioactive potential [227].

To address these issues, in vitro digestion models and cellular uptake assays are employed to assess factors affecting both bioaccessibility and bioavailability. These models are essential for screening and optimizing formulations prior to more expensive in vivo studies. Despite the availability of various assessment methods, a universally accepted standard for bioaccessibility remains elusive, and the current methods must account for the diverse structural characteristics of phytochemicals [208]. For lipophilic compounds, such as carotenoids and fat-soluble vitamins, formulation techniques like encapsulation and emulsification have shown potential in enhancing bioaccessibility and protecting these compounds from degradation [208].

In silico modeling, commonly used in pharmaceutical research, offers an interesting approach as it allows for predicting phytochemical bioavailability based on physicochemical properties. However, these models are still being developed for dietary phytochemicals, indicating a need for further research to establish effective predictive tools [228]. Moreover, technological processes, such as thermal processing and biotechnological interventions, including enzymatic hydrolysis and fermentation, have demonstrated their influence in improving the bioavailability of phenolic compounds. However, while these strategies have shown promise in animal studies, further human trials are necessary to validate their effectiveness [229]. Thus, both research and innovation in processing and formulation techniques are essential for advancing nutraceuticals and optimizing the bioavailability of phytochemicals.

5. Current Applications of Nutraceuticals

Nutraceuticals, which combine nutritional and medicinal properties, are increasingly being incorporated into food products to promote health and prevent disease. These bioactive compounds, derived from a variety of matrices, offer a wide range of benefits, such as improving cardiovascular and immune system function and managing chronic diseases. In addition, advances in encapsulation and nanotechnology have further improved their efficacy in food applications, expanding the potential of health-promoting foods [204].

5.1. Ingredients in Functional Foods

The incorporation of antioxidants into functional foods has been widely researched because of their health benefits and potential to enhance food quality and shelf life. This incorporation can be performed in several ways, including through the addition of crude extracts, fibers, color additives, and other food enrichment (Figure 5) [230].

Crude extracts are raw, unrefined substances obtained from plants, animals, and other natural sources that contain bioactive compounds. The most widely utilized are plant polyphenol extracts, such as green tea extract, grape seed extract, and rosemary extract, but also mushroom, spice, and herbal extracts (garlic extract, ginger extract, turmeric extract), vanilla extract, and, more recently, microalgae extract (Spirulina, *Chlorella* extracts) [231]. These extracts also exhibit antimicrobial activity, with green tea extract having major antimicrobial activity against major foodborne pathogens, such as *Listeria monocytogenes*, *Salmonella typhimurium*, *Escherichia coli* O157:H7, and *Campylobacter jejuni*, contributing to food safety [232,233]. Some crude extracts also contribute to flavoring, such as mushroom extracts, garlic extracts, rosemary extracts, and spice and herbal extracts, in general. These can also provide distinctive aromas and colors, enhancing their sensory attributes. For example, Spirulina extracts that are naturally greenish blue can modify the color of a food [234].

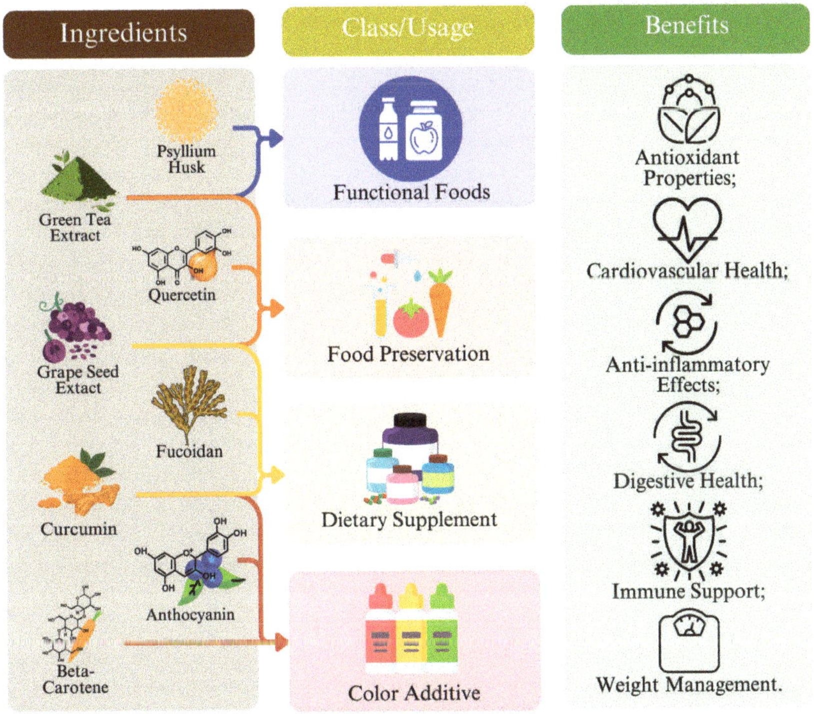

Figure 5. Ingredient types, uses, and benefits in food applications.

Fiber is added to foods because of its functional and health benefits. Fiber can be roughly divided into soluble and insoluble fibers. Soluble fiber includes pectin and inulin, while insoluble fiber includes cellulose, psyllium, and wheat bran. Functionally, adding fiber to foods can impact their sensory properties, such as appearance, texture, taste, and flavor [235]. Pectin can be found in fruit jams and jellies; it acts as a gelling agent, giving jams and jellies their thick, spreadable consistency. Additionally, it allows for a lower amount of sugar to be used while improving product shelf life and stability [236]. Inulin, usually extracted from chicory root, is used as a fat replacer and provides a creamy texture in low-fat or reduced-calorie foods [237]. It also serves as a prebiotic, promoting the growth of beneficial gut bacteria [238]. Cellulose and its derivatives, such as microcrystalline cellulose and bacterial cellulose, are commonly used as stabilizers and thickeners in food products, including meat products, emulsions, beverages, dairy products, and bakery items [239]. Furthermore, microcrystalline cellulose and other cellulose derivatives are used in functional and nutraceutical foods for their positive effects on gastrointestinal health and lipid metabolism [239]. Cellulose gels are also widely used in the food industry for their high-water absorption capacity and biocompatibility. They are applied in food packaging, functional foods, and food safety because of their structural flexibility and stimuli-responsive properties [240]. Other more unrefined fibers, such as psyllium husk and wheat bran, are also often used in the food industry. Unmodified wheat bran and psyllium husk are used as nutritional enhancements as they are a source of dietary fiber, minerals, vitamins, and bioactive compounds, such as phenolic acids, which contribute to improved bowel health, the prevention of diseases, like colon cancer, and cardiovascular health [241]. These fibers, when modified through processes such as mechanical milling, enzymatic hydrolysis, and thermal treatments, can be utilized as functional ingredients. For example, modified wheat bran can be incorporated into cereal foods, baked products, and fried snacks to reduce their oil content and increase their fiber content [241].

To specifically modify the color of a food, color additives can be used. The main natural color additives used in the food industry are anthocyanins, beta-carotene, betanin, and curcumin. Unlike synthetic dyes, these are usually less resistant to degradation via sunlight or heat exposure, but they do not carry risks, like causing hyperactivity in children [242]. Anthocyanins are widely used as natural food colorants because of their vibrant colors, which range from red to blue depending on pH levels [243,244]. Often, these anthocyanins are chemically modified, including acylation and glycosylation, which offers enhanced color stability, making them more suitable for industrial applications as natural colorants [245]. These also provide additional health benefits, improving carbohydrate metabolism and decreasing the risk factors of metabolic disorders [245]. Anthocyanins are used in beverages, such as fruit juices, smoothies, teas, and energy drinks, dairy products, such as flavored milk and ice creams, candies, jams and jellies, bakery products, and many others [246]. β-carotene is a pro-vitamin A carotenoid that contributes to the recommended intake of this essential nutrient in foods where it is added [60]. β-carotene is used to provide an orange hue to various food products, such as margarine and butter substitutes, fruit juices and smoothies, dairy products, such as cheese and yogurt, snacks, baked goods, and infant formula, among many others [247]. Betanin, a red-violet pigment found in beetroots, is also widely used in the food industry. Betanin exhibits high antioxidant activity, which helps to scavenge reactive oxygen species and prevent lipid oxidation in foods, thereby preserving food quality [248,249]. Betanin shows significant stability at low temperatures, which is beneficial for its use in frozen and refrigerated foods [249]. Encapsulation techniques, such as liposomal nanocarriers and microencapsulation, improve the stability and bioavailability of betanin [250,251]. Furthermore, betanin can delay the retrogradation of starches, which is beneficial for maintaining the quality and texture of starchy foods, like bread and pastries [252].

Curcumin, whether as a turmeric extract or an unprocessed form, is commonly used to impart a yellowish-orange hue to foods [253]. It has also demonstrated potential in the prevention and management of various health conditions, including cardiovascular diseases, diabetes, metabolic syndrome, arthritis, and mental disorders [254,255].

In conclusion, the incorporation of antioxidants, fibers, crude extracts, and natural color additives into functional foods has been shown to enhance not only the nutritional value but also the sensory quality, shelf life, and safety of various products. Both soluble and insoluble fibers, such as pectin, inulin, cellulose, psyllium husk, and wheat bran, play crucial roles in improving texture and stability while providing health benefits, like enhanced digestive health and disease prevention. Some examples of the mentioned nutraceuticals include Curcumin Elite™ or Mega Green Tea Extract™ from Life Extension® Ltd. [256], Inülin from Töufood® Ltd. [257], and Blue Spirulina from Bluetec Colorants® Ltd. [258]

5.2. Isolated Phytochemicals as Nutraceuticals

Isolated phytochemicals are bioactive compounds derived from plants that have gained significant attention as nutraceuticals because of their potent health-promoting properties. These compounds, which include flavonoids, alkaloids, terpenoids, and polyphenols, are being increasingly isolated and studied for their potential to prevent and manage various chronic diseases, such as cancer, cardiovascular disorders, and neurodegenerative conditions (Table 3). Phytochemicals are often extracted from agricultural and food waste streams, promoting a circular economy by converting waste into value-added functional ingredients [259]. By concentrating the active ingredients, isolated phytochemicals offer more targeted therapeutic effects compared to whole plant extracts, making them valuable in the development of functional foods and dietary supplements [200].

Table 3. Nutraceutical applications and functionality.

Nutraceutical	Source	Applications	Functionality	Health Benefits	Ref.
Pectin	Fruits (apple, citrus…)	Jams, jellies, dairy products	Gelling agent, thickener	Anticancer, immunomodulatory, anti-inflammatory, cholesterol-lowering	[260]
Inulin	Chicory root	Low-fat foods, fiber supplements	Prebiotic, fat replacer	Gut microbiota regulation, lipid metabolism regulation, mineral absorption enhancement, anti-inflammatory	[261]
Cellulose	Plants	Low-fat foods, plant-based meats, bakery products	Stabilizer, thickener	Gut microbiota regulation, cholesterol reduction, blood glucose level regulation, anti-inflammatory	[239]
Wheat bran	Wheat	Cereals, bread, bakery products	Texture enhancer, fiber source	Gut microbiota regulation, cancer risk reduction, cardioprotective	[262]
Psyllium husk	*Plantago ovata* seeds	Fiber supplement, cereals	Fiber source, thickener	Antidiabetic, reduces cholesterol levels, aids in gastrointestinal health	[263,264]
Catechins	Green tea leaves	Beverages, supplements, snacks	Antioxidant, antimicrobial	Antioxidant, anti-inflammatory, antiviral, anti-obesity	[265,266]
Grape seed extract	Grape seeds	Beverages, supplements	Antioxidant, antimicrobial	Anti-inflammatory, antioxidant, cardioprotective, antimicrobial, anticancer	[267]
β-carotene	Carrots, sweet potatoes	Supplements, snacks, beverages, candies	Colorant	Antioxidant, supports immune function	[268]
Anthocyanins	Berries, red cabbage	Supplements, snacks, beverages, candies	Colorant	Antioxidant, anti-inflammatory, antidiabetic, anti-obesity	[269]
Betanins	Beetroot	Supplements, snacks, beverages, candies	Colorant	Antioxidative, anti-inflammatory, antidiabetic, potential anticancer benefits	[248]
Curcumin	Turmeric root	Supplements, snacks, beverages, candies	Colorant	Antioxidant, anti-inflammatory, anticancer, and immune-regulatory properties	[270]
Resveratrol	Grapes	Beverages, supplements	Antioxidant, antimicrobial	Antioxidant, anti-inflammatory, anticancer, cardioprotective	[271]
Fucoidans	Brown algae	Supplements, fortified foods	Gelling agent, thickener	Antioxidant, anti-inflammatory, anticoagulant, antitumor, antiviral	[272]
Agar	Red algae	Jellies, jams, candy, plant-based gelatin	Gelling agent, texture enhancer	Antioxidant, antiviral, antibacterial, prebiotic, antitumor	[273]
Carrageenan	Red algae	Jellies, jams, candy, plant-based gelatin	Thickener, gelling agent	Cardioprotective, anticancer, antiviral, anticoagulant, antioxidant	[274]
Quercetin	Onions, apple peels	Supplements, fortified foods	Antioxidant, preservative	Anti-inflammatory, antimicrobial, anticancer, cardioprotective	[275,276]

Polyphenol extracts, which are concentrated from fruits, vegetables, herbs, and seeds, are well known for their antioxidant properties. Green tea catechins, particularly epigallocatechin-3-gallate, are effective in delaying lipid oxidation and extending the shelf life of lipid-containing foods [231,254]. These polyphenols are also incorporated into convenient products, such as chewing gum and gelatin gummies, enhancing consumer accessibility and palatability [277]. Furthermore, green tea extracts demonstrate antimicrobial properties against common foodborne pathogens, enhancing both food safety and quality [232]. Similarly, resveratrol, often sourced from grape industry waste, exhibits strong antioxidant and antibacterial activity, making it suitable for use in oil-in-water emulsions, bulk oils, and ground meat and as a potential natural preservative in food packaging and processing [278,279].

Omega-3 FAs, particularly eicosatetraenoic acid (EPA) and docosahexaenoic acid (DHA), are widely recognized for their roles in supporting brain development and cardiovascular health [280]. When sourced from marine species, such as some microalgae like *Schizochytrium* and *Nannochloropsis*, these oils are extracted through solvent methods [281]. Because of the growing consumer demand for plant-based omega-3s, microalgae supplements have become popular as they contain lower levels of heavy metals compared to traditional fish oils [282]. However, the high cost and instability of algae oils limit their use in food products, although nano-emulsification techniques have shown promise in stabilizing omega-3s for functional food applications [283]. Additionally, algae-derived omega-3s are used in livestock feed, which enriches meat, milk, and eggs with bioactive compounds that benefit human health [284], including increasing omega-3 blood indices and reducing diastolic blood pressure [285].

Fucoidans are polysaccharides extracted from brown algae that have gained recognition for their multifunctional bioactive properties. These include anticoagulant, antiviral, antitumor, antibacterial, and immunomodulatory effects [286]. Fucoidans are also valuable as gelling and thickening agents in the food industry, offering both functional and health benefits [287]. Their solubility and ease of incorporation into food formulations make them suitable for various applications, including the development of UVB-protective cosmetics and functional foods [288,289].

Agar and carrageenan, two polysaccharides derived from red algae, are widely used as gelling, thickening, and stabilizing agents in food products. Agar's ability to form gels at low concentrations makes it ideal for candies, soups, snacks, and plant-based formulations [273,290]. Agar is also often used in plant-based and halal formulations of products that otherwise require animal gelatin [291]. Additionally, agar and its derivatives exhibit multiple bioactive functions, such as antioxidant, antiviral, antibacterial, prebiotic, antitumor, anticoagulant, and antidiabetic activities. These bioactive properties make agar a valuable ingredient in functional foods and nutraceuticals [273,290]. Carrageenan is similarly utilized as a thickening, gelling, and protein-suspending agent, enhancing texture and water retention in food products while also providing various health benefits [292]. In low-fat meat formulations, the incorporation of carrageenan promotes gel formation and water retention, thereby improving textural properties [292]. Similar to agar, carrageenan exhibits anti-inflammatory, anticoagulant, antitumor, and antithrombotic effects [293].

Quercetin, a flavonoid present in a variety of plants, is predominantly extracted from onions and apple peels because of their commercial availability [294]. In food products like bread, quercetin demonstrates antioxidant and antiglycation effects in a dose-dependent manner. Furthermore, it inhibits the formation of advanced glycation end products (AGEs), thereby enhancing the shelf life of these products [295]. Quercetin has also been shown to offer protective benefits against cardiovascular and neurodegenerative diseases, underscoring its potential as a valuable bioactive compound for the development of functional foods aimed at promoting health [296].

In summary, phytochemicals, such as polyphenols, omega-3-rich oils, fucoidans, agar, carrageenan, and quercetin, are increasingly recognized for their dual roles in improving both the quality and health benefits of food products. These compounds, sourced from plants and algae, contribute to enhanced shelf life, nutritional value, and functional properties while offering a wide range of health benefits, including antioxidant, antimicrobial, and cardiovascular protective effects. Some commercialized available examples of these ingredients include Ceamsea ™ from Ceamsa Ltd. [297], Fucoidan Extract from Swanson Ltd. [298], Agar-Agar from Algamar Ltd. [299], and Omega 3 formula from Kohbo Labs Ltd. [300].

6. Conclusions

This review uncovers the versatile functions of antioxidants in plant organisms and their involvement in protecting against OS. Several antioxidant compounds, such as flavonoids, carotenoids, and phenolic acids, promote plant resistance and are also beneficial for human health. A better understanding of these mechanisms may allow the optimization of agricultural practices and food supplies. Further research should be directed towards the more efficient use of plant antioxidants to improve their productive capacity and nutraceutical value. In this way, the integration of advanced omics technologies, such as metabolomics and transcriptomics, may improve the understanding of the biosynthetic pathways and regulatory networks of plant antioxidants. Furthermore, the convergence of plant biotechnology with sustainable agricultural practices offers promising avenues not only for increasing crop yields but also for enriching the nutritional and functional qualities of food products.

Author Contributions: Conceptualization: A.G.P., J.E., A.O.S.J., R.N.-M., E.N.Y., P.B., A.P.-V., F.C., M.B.P.P.O., M.C. and M.A.P.; methodology: A.G.P., J.E., A.O.S.J., R.N.-M., E.N.Y., P.B., A.P.-V., F.C. and M.C.; validation, A.G.P., M.B.P.P.O., M.C. and M.A.P.; formal analysis: A.G.P., J.E., A.O.S.J., R.N.-M., E.N.Y., P.B., A.P.-V., F.C. and M.C., investigation: A.G.P., J.E., A.O.S.J., R.N.-M., E.N.Y., P.B., A.P.-V., F.C. and M.C.; resources: A.G.P., J.E., A.O.S.J., R.N.-M., E.N.Y., P.B., A.P.-V., F.C. and M.C.; data curation: A.G.P., A.O.S.J., P.B., A.P.-V. and M.C.; writing—original draft preparation: A.G.P., J.E., A.O.S.J., R.N.-M., E.N.Y., P.B., A.P.-V., F.C. and M.C.; writing—review and editing: A.G.P., A.O.S.J., M.C., M.B.P.P.O. and M.A.P.; visualization: A.G.P., A.O.S.J., P.B., A.P.-V. and F.C.; supervision: A.G.P., M.C., M.B.P.P.O. and M.A.P.; project administration: M.B.P.P.O. and M.A.P.; funding acquisition: M.B.P.P.O. and M.A.P. All authors have read and agreed to the published version of the manuscript.

Funding: The research leading to these results was supported by MICIU/AEI/10.13039/501100011033, in support of the predoctoral industrial grant for A. Perez-Vazquez (DIN2024-013416) in collaboration with Mercantia Desarrollos Alimentarios S.L.; Xunta de Galicia in support of the post-doctoral grant of A.G. Pereira (IN606B-2024/011); and the pre-doctoral grant of P. Barciela (ED481A-2024-230). The authors thank the EU-FORA Fellowship Program (EUBA-EFSA-2023-ENREL-01) that supports the work of F. Chamorro (INNOV2SAFETY-GA13). The authors are grateful for the national funding from FCT, Foundation for Science and Technology, through the individual research grants of J. Echave (2023.04987.BD) and A.O.S. Jorge (2023.00981.BD).

Institutional Review Board Statement: Not applicable.

Informed Consent Statement: Not applicable.

Data Availability Statement: No new data were created or analyzed in this study. Data sharing is not applicable to this article.

Conflicts of Interest: The authors declare no conflicts of interest.

References

1. Carpena, M.; Garcia-Oliveira, P.; Pereira, A.G.; Soria-Lopez, A.; Chamorro, F.; Collazo, N.; Jarboui, A.; Simal-Gandara, J.; Prieto, M.A. Plant Antioxidants from Agricultural Waste: Synergistic Potential with Other Biological Properties and Possible Applications. In *Plant Antioxidants and Health*; Ekiert, H.M., Ramawat, K.G., Arora, J., Eds.; Springer: Cham, Switzerland, 2021; pp. 1–38. ISBN 9783030452995.
2. Shahidi, F.; Ambigaipalan, P. Phenolics and Polyphenolics in Foods, Beverages and Spices: Antioxidant Activity and Health Effects–A Review. *J. Funct. Foods* **2015**, *18*, 820–897. [CrossRef]
3. Jimenez-Lopez, C.; Fraga-Corral, M.; Carpena, M.; García-Oliveira, P.; Echave, J.; Pereira, A.G.; Lourenço-Lopes, C.; Prieto, M.A.; Simal-Gandara, J. Agriculture Waste Valorisation as a Source of Antioxidant Phenolic Compounds within a Circular and Sustainable Bioeconomy. *Food Funct.* **2020**, *11*, 4853–4877. [CrossRef]
4. Reuter, S.; Gupta, S.C.; Chaturvedi, M.M.; Aggarwal, B.B. Oxidative Stress, Inflammation, and Cancer: How Are They Linked? *Free Radic. Biol. Med.* **2010**, *49*, 1603–1616. [CrossRef]
5. Pistollato, F.; Battino, M. Role of Plant-Based Diets in the Prevention and Regression of Metabolic Syndrome and Neurodegenerative Diseases. *Trends Food Sci. Technol.* **2014**, *40*, 62–81. [CrossRef]
6. Zhang, Y.-J.; Gan, R.-Y.; Li, S.; Zhou, Y.; Li, A.-N.; Xu, D.-P.; Li, H.-B. Antioxidant Phytochemicals for the Prevention and Treatment of Chronic Diseases. *Molecules* **2015**, *20*, 21138–21156. [CrossRef]
7. Allaqaband, S.; Dar, A.H.; Patel, U.; Kumar, N.; Nayik, G.A.; Khan, S.A.; Ansari, M.J.; Alabdallah, N.M.; Kumar, P.; Pandey, V.K.; et al. Utilization of Fruit Seed-Based Bioactive Compounds for Formulating the Nutraceuticals and Functional Food: A Review. *Front. Nutr.* **2022**, *9*, 902554. [CrossRef]
8. Jabri, A.; Kumar, A.; Verghese, E.; Alameh, A.; Kumar, A.; Khan, M.S.; Khan, S.U.; Michos, E.D.; Kapadia, S.R.; Reed, G.W.; et al. Meta-Analysis of Effect of Vegetarian Diet on Ischemic Heart Disease and All-Cause Mortality. *Am. J. Prev. Cardiol.* **2021**, *7*, 100182. [CrossRef]
9. Selinger, E.; Neuenschwander, M.; Koller, A.; Gojda, J.; Kühn, T.; Schwingshackl, L.; Barbaresko, J.; Schlesinger, S. Evidence of a Vegan Diet for Health Benefits and Risks–an Umbrella Review of Meta-Analyses of Observational and Clinical Studies. *Crit. Rev. Food Sci. Nutr.* **2023**, *63*, 9926–9936. [CrossRef]
10. Jablonský, M.; Škulcová, A.; Malvis, A.; Šima, J. Extraction of Value-Added Components from Food Industry Based and Agro-Forest Biowastes by Deep Eutectic Solvents. *J. Biotechnol.* **2018**, *282*, 46–66. [CrossRef]
11. Oswell, N.J.; Gunstone, F.D.; Pegg, R.B. Vegetable Oils. In *Bailey's Industrial Oil and Fat Products*; Wiley: Hoboken, NJ, USA, 2020; pp. 1–30. ISBN 9781439816851.
12. Chouaibi, M.; Rezig, L.; Hamdi, S.; Ferrari, G. Chemical Characteristics and Compositions of Red Pepper Seed Oils Extracted by Different Methods. *Ind. Crops Prod.* **2019**, *128*, 363–370. [CrossRef]
13. García-Lomillo, J.; González-SanJosé, M.L. Applications of Wine Pomace in the Food Industry: Approaches and Functions. *Compr. Rev. Food Sci. Food Saf.* **2017**, *16*, 3–22. [CrossRef]
14. Lu, Q.; Li, H.; Xiao, Y.; Liu, H. A State-of-the-Art Review on the Synthetic Mechanisms, Production Technologies, and Practical Application of Polyunsaturated Fatty Acids from Microalgae. *Algal Res.* **2021**, *55*, 102281. [CrossRef]
15. Russo, G.L.; Moccia, S.; Russo, M.; Spagnuolo, C. Redox Regulation by Carotenoids: Evidence and Conflicts for Their Application in Cancer. *Biochem. Pharmacol.* **2021**, *194*, 114838. [CrossRef]
16. Bohn, T. Provitamin a Carotenoids: Occurrence, Intake and Bioavailability. *Food Nutr. Components Focus* **2012**, *1*, 142–161.
17. Poojary, M.M.; Barba, F.J.; Aliakbarian, B.; Donsì, F.; Pataro, G.; Dias, D.A.; Juliano, P. Innovative Alternative Technologies to Extract Carotenoids from Microalgae and Seaweeds. *Mar. Drugs* **2016**, *14*, 214. [CrossRef]
18. Rodriguez-Concepcion, M.; Avalos, J.; Bonet, M.L.; Boronat, A.; Gomez-Gomez, L.; Hornero-Mendez, D.; Limon, M.C.; Meléndez-Martínez, A.J.; Olmedilla-Alonso, B.; Palou, A.; et al. A Global Perspective on Carotenoids: Metabolism, Biotechnology, and Benefits for Nutrition and Health. *Prog. Lipid Res.* **2018**, *70*, 62–93. [CrossRef]
19. Nie, Q.; Hu, J.; Gao, H.; Li, M.; Sun, Y.; Chen, H.; Zuo, S.; Fang, Q.; Huang, X.; Yin, J.; et al. Bioactive Dietary Fibers Selectively Promote Gut Microbiota to Exert Antidiabetic Effects. *J. Agric. Food Chem.* **2021**, *69*, 7000–7015. [CrossRef]
20. Baiano, A. Recovery of Biomolecules from Food Wastes—A Review. *Molecules* **2014**, *19*, 14821–14842. [CrossRef]
21. Saeed, F.; Pasha, I.; Arshad, M.U.; Anjum, F.M.; Hussain, S.; Rasheed, R.; Nasir, M.A.; Shafique, B. Physiological and Nutraceutical Perspectives of Fructan. *Int. J. Food Prop.* **2015**, *18*, 1895–1904. [CrossRef]
22. Pereira, A.G.; Garcia-Perez, P.; Cassani, L.; Chamorro, F.; Cao, H.; Barba, F.J.; Simal-Gandara, J.; Prieto, M.A. Camellia Japonica: A Phytochemical Perspective and Current Applications Facing Its Industrial Exploitation. *Food Chem. X* **2022**, *13*, 100258. [CrossRef]
23. Salehi, B.; Martorell, M.; Arbiser, J.L.; Sureda, A.; Martins, N.; Maurya, P.K.; Sharifi-Rad, M.; Kumar, P.; Sharifi-Rad, J. Antioxidants: Positive or Negative Actors? *Biomolecules* **2018**, *8*, 124. [CrossRef]

24. Li, S.; Fasipe, B.; Laher, I. Potential Harms of Supplementation with High Doses of Antioxidants in Athletes. *J. Exerc. Sci. Fit.* **2022**, *20*, 269–275. [CrossRef]
25. García-Cortés, M.; Robles-Díaz, M.; Ortega-Alonso, A.; Medina-Caliz, I.; Andrade, R. Hepatotoxicity by Dietary Supplements: A Tabular Listing and Clinical Characteristics. *Int. J. Mol. Sci.* **2016**, *17*, 537. [CrossRef]
26. Stickel, F.; Kessebohm, K.; Weimann, R.; Seitz, H.K. Review of Liver Injury Associated with Dietary Supplements. *Liver Int.* **2011**, *31*, 595–605. [CrossRef]
27. Li, S.; Li, Y.; Smolke, C.D. Strategies for Microbial Synthesis of High-Value Phytochemicals. *Nat. Chem.* **2018**, *10*, 395–404. [CrossRef]
28. Reguengo, L.M.; Salgaço, M.K.; Sivieri, K.; Júnior, M.R.M. Agro-Industrial by-Products: Valuable Sources of Bioactive Compounds. *Food Res. Int.* **2022**, *152*, 110871. [CrossRef]
29. Watkins, P.A. Fatty Acids: Metabolism. In *Encyclopedia of Human Nutrition*; Elsevier: Amsterdam, The Netherlands, 2013; Volume 2–4, pp. 220–230. ISBN 9780123848857.
30. Ristic-Medic, D.; Vucic, V.; Takic, M.; Karadžic, I.; Glibetic, M. Polyunsaturated Fatty Acids in Health and Disease. *J. Serbian Chem. Soc.* **2013**, *78*, 1269–1289. [CrossRef]
31. Moghadasian, M.H.; Shahidi, F. Fatty Acids. In *International Encyclopedia of Public Health*; Elsevier: Amsterdam, The Netherlands, 2017; Volume 3, pp. 114–122. ISBN 9780128037089.
32. He, M.; Ding, N.Z. Plant Unsaturated Fatty Acids: Multiple Roles in Stress Response. *Front. Plant Sci.* **2020**, *11*, 562785. [CrossRef]
33. Tsao, R. Chemistry and Biochemistry of Dietary Polyphenols. *Nutrients* **2010**, *2*, 1231–1246. [CrossRef]
34. Jimenez-Lopez, C.; Pereira, A.G.; Lourenço-Lopes, C.; Garcia-Oliveira, P.; Cassani, L.; Fraga-Corral, M.; Prieto, M.A.; Simal-Gandara, J. Main Bioactive Phenolic Compounds in Marine Algae and Their Mechanisms of Action Supporting Potential Health Benefits. *Food Chem.* **2021**, *341*, 128262. [CrossRef]
35. Di Lorenzo, C.; Colombo, F.; Biella, S.; Stockley, C.; Restani, P. Polyphenols and Human Health: The Role of Bioavailability. *Nutrients* **2021**, *13*, 273. [CrossRef]
36. Han, X.; Shen, T.; Lou, H. Dietary Polyphenols and Their Biological Significance. *Int. J. Mol. Sci.* **2007**, *8*, 950–988. [CrossRef]
37. Zhang, Z.; Li, X.; Sang, S.; McClements, D.J.; Chen, L.; Long, J.; Jiao, A.; Jin, Z.; Qiu, C. Polyphenols as Plant-Based Nutraceuticals: Health Effects, Encapsulation, Nano-Delivery, and Application. *Foods* **2022**, *11*, 2189. [CrossRef]
38. Haminiuk, C.W.I.; Maciel, G.M.; Plata-Oviedo, M.S.V.; Peralta, R.M. Phenolic Compounds in Fruits-an Overview. *Int. J. Food Sci. Technol.* **2012**, *47*, 2023–2044. [CrossRef]
39. Smeriglio, A.; Barreca, D.; Bellocco, E.; Trombetta, D. Proanthocyanidins and Hydrolysable Tannins: Occurrence, Dietary Intake and Pharmacological Effects. *Br. J. Pharmacol.* **2017**, *174*, 1244–1262. [CrossRef]
40. Rashmi, H.B.; Negi, P.S. Phenolic Acids from Vegetables: A Review on Processing Stability and Health Benefits. *Food Res. Int.* **2020**, *136*, 109298. [CrossRef]
41. Caleja, C.; Ribeiro, A.; Barreiro, M.F.; Ferreira, I.C.F.R. Phenolic Compounds as Nutraceuticals or Functional Food Ingredients. *Curr. Pharm. Des.* **2017**, *23*, 2787–2806. [CrossRef]
42. Vuolo, M.M.; Lima, V.S.; Junior, M.R.M. Phenolic Compounds. In *Bioactive Compounds: Health Benefits and Potential Applications*; Campos, M.R.S., Ed.; Elsevier: Amsterdam, The Netherlands, 2019; pp. 33–50. ISBN 9780128147740.
43. Sorrenti, V.; Burò, I.; Consoli, V.; Vanella, L. Recent Advances in Health Benefits of Bioactive Compounds from Food Wastes and By-Products: Biochemical Aspects. *Int. J. Mol. Sci.* **2023**, *24*, 2019. [CrossRef]
44. Lin, J.K.; Weng, M.S. Flavonoids as Nutraceuticals. *Sci. Flavonoids* **2006**, *7*, 213–238. [CrossRef]
45. Dias, M.; Romaní-Pérez, M.; Romaní, A.; de la Cruz, A.; Pastrana, L.; Fuciños, P.; Amado, I.R. Recent Technological Advances in Phenolic Compounds Recovery and Applications: Source of Nutraceuticals for the Management of Diabetes. *Appl. Sci.* **2022**, *12*, 9271. [CrossRef]
46. Rahman, M.M.; Rahaman, M.S.; Islam, M.R.; Rahman, F.; Mithi, F.M.; Alqahtani, T.; Almikhlafi, M.A.; Alghamdi, S.Q.; Alruwaili, A.S.; Hossain, M.S.; et al. Role of Phenolic Compounds in Human Disease: Current Knowledge and Future Prospects. *Molecules* **2022**, *27*, 233. [CrossRef]
47. Vignesh, A.; Amal, T.C.; Sarvalingam, A.; Vasanth, K. A Review on the Influence of Nutraceuticals and Functional Foods on Health. *Food Chem. Adv.* **2024**, *5*, 100749. [CrossRef]
48. Maoka, T. Carotenoids as Natural Functional Pigments. *J. Nat. Med.* **2020**, *74*, 1–16. [CrossRef]
49. Crupi, P.; Faienza, M.F.; Naeem, M.Y.; Corbo, F.; Clodoveo, M.L.; Muraglia, M. Overview of the Potential Beneficial Effects of Carotenoids on Consumer Health and Well-Being. *Antioxidants* **2023**, *12*, 1069. [CrossRef]
50. Lakshminarayana, R.; Paul, B. Free Radical Chemistry of Carotenoids and Oxidative Stress Physiology of Cancer. In *Handbook of Oxidative Stress in Cancer: Therapeutic Aspects: Volume 1*; Chakraborti, S., Ed.; Springer: Singapore, 2022; Volume 1, pp. 3077–3097. ISBN 9789811654220.

51. Srivastava, R. Physicochemical, Antioxidant Properties of Carotenoids and Its Optoelectronic and Interaction Studies with Chlorophyll Pigments. *Sci. Rep.* **2021**, *11*, 18365. [CrossRef]
52. Saini, R.K.; Prasad, P.; Lokesh, V.; Shang, X.; Shin, J.; Keum, Y.S.; Lee, J.H. Carotenoids: Dietary Sources, Extraction, Encapsulation, Bioavailability, and Health Benefits—A Review of Recent Advancements. *Antioxidants* **2022**, *11*, 795. [CrossRef]
53. Pereira, A.G.; Otero, P.; Echave, J.; Carreira-Casais, A.; Chamorro, F.; Collazo, N.; Jaboui, A.; Lourenço-Lopes, C.; Simal-Gandara, J.; Prieto, M.A. Xanthophylls from the Sea: Algae as Source of Bioactive Carotenoids. *Mar. Drugs* **2021**, *19*, 188. [CrossRef]
54. Widomska, J.; Zareba, M.; Subczynski, W.K. Can Xanthophyll-Membrane Interactions Explain Their Selective Presence in the Retina and Brain? *Foods* **2016**, *5*, 7. [CrossRef]
55. Ojulari, O.V.; Lee, S.G.; Nam, J.O. Therapeutic Effect of Seaweed Derived Xanthophyl Carotenoid on Obesity Management; Overview of the Last Decade. *Int. J. Mol. Sci.* **2020**, *21*, 2502. [CrossRef]
56. Gürbüz, M.; Aktaç, Ş. Understanding the Role of Vitamin A and Its Precursors in the Immune System. *Nutr. Clin. Metab.* **2022**, *36*, 89–98. [CrossRef]
57. Ebadi, M.; Mohammadi, M.; Pezeshki, A.; Jafari, S.M. Health Benefits of Beta-Carotene. In *Handbook of Food Bioactive Ingredients*; Jafari, S.M., Rashidinejad, A., Simal-Gandara, J., Eds.; Springer International Publishing: Cham, Switzerland, 2023; pp. 1–26. ISBN 978-3-030-81404-5.
58. Aziz, E.; Batool, R.; Akhtar, W.; Rehman, S.; Shahzad, T.; Malik, A.; Shariati, M.A.; Laishevtcev, A.; Plygun, S.; Heydari, M.; et al. Xanthophyll: Health Benefits and Therapeutic Insights. *Life Sci.* **2020**, *240*, 117104. [CrossRef] [PubMed]
59. Singh, A.; Omer, K. Recent Advancement in Therapeutic Activity of Carotenoids. In *Dietary Carotenoids*; Rao, A.V., Rao, L., Eds.; IntechOpen: Rijeka, Croatia, 2024.
60. Grune, T.; Lietz, G.; Palou, A.; Ross, A.C.; Stahl, W.; Tang, G.; Thurnham, D.; Yin, S.A.; Biesalski, H.K. β-Carotene Is an Important Vitamin A Source for Humans. *J. Nutr.* **2010**, *140*, 2268S–2285S. [CrossRef]
61. Murphy, E.J.; Fehrenbach, G.W.; Abidin, I.Z.; Buckley, C.; Montgomery, T.; Pogue, R.; Murray, P.; Major, I.; Rezoagli, E. Polysaccharides—Naturally Occurring Immune Modulators. *Polymers* **2023**, *15*, 2373. [CrossRef]
62. Pedrosa, L.d.F.; de Vos, P.; Fabi, J.P. Nature's Soothing Solution: Harnessing the Potential of Food-Derived Polysaccharides to Control Inflammation. *Curr. Res. Struct. Biol.* **2023**, *6*, 100112. [CrossRef]
63. Baghel, R.S.; Choudhary, B.; Pandey, S.; Pathak, P.K.; Patel, M.K.; Mishra, A. Rehashing Our Insight of Seaweeds as a Potential Source of Foods, Nutraceuticals, and Pharmaceuticals. *Foods* **2023**, *12*, 3642. [CrossRef]
64. Lin, G.p.; Wu, D.s.; Xiao, X.w.; Huang, Q.y.; Chen, H.b.; Liu, D.; Fu, H.q.; Chen, X.h.; Zhao, C. Structural Characterization and Antioxidant Effect of Green Alga Enteromorpha Prolifera Polysaccharide in Caenorhabditis Elegans via Modulation of MicroRNAs. *Int. J. Biol. Macromol.* **2020**, *150*, 1084–1092. [CrossRef]
65. Zhang, Y.; Xu, M.; Hu, C.; Liu, A.; Chen, J.; Gu, C.; Zhang, X.; You, C.; Tong, H.; Wu, M.; et al. Sargassum Fusiforme Fucoidan SP2 Extends the Lifespan of Drosophila Melanogaster by Upregulating the Nrf2-Mediated Antioxidant Signaling Pathway. *Oxidative Med. Cell. Longev.* **2019**, *2019*, 8918914. [CrossRef]
66. Cui, H.; Wu, S.; Shang, Y.; Li, Z.; Chen, M.; Li, F.; Wang, C. Pleurotus Nebrodensis Polysaccharide(PN50G) Evokes A549 Cell Apoptosis by the ROS/AMPK/PI3K/AKT/MTOR Pathway to Suppress Tumor Growth. *Food Funct.* **2016**, *7*, 1616–1627. [CrossRef]
67. Zeng, F.S.; Yao, Y.F.; Wang, L.F.; Li, W.J. Polysaccharides as Antioxidants and Prooxidants in Managing the Double-Edged Sword of Reactive Oxygen Species. *Biomed. Pharmacother.* **2023**, *159*, 114221. [CrossRef]
68. Wang, J.; Hu, S.; Nie, S.; Yu, Q.; Xie, M. Reviews on Mechanisms of in Vitro Antioxidant Activity of Polysaccharides. *Oxidative Med. Cell. Longev.* **2016**, *2016*, 5692852. [CrossRef]
69. Devi, N.; Sarmah, M.; Khatun, B.; Maji, T.K. Encapsulation of Active Ingredients in Polysaccharide–Protein Complex Coacervates. *Adv. Colloid Interface Sci.* **2017**, *239*, 136–145. [CrossRef] [PubMed]
70. Liu, Z.; Sun, X. A Critical Review of the Abilities, Determinants, and Possible Molecular Mechanisms of Seaweed Polysaccharides Antioxidants. *Int. J. Mol. Sci.* **2020**, *21*, 7774. [CrossRef] [PubMed]
71. Chen, N.; Jiang, T.; Xu, J.; Xi, W.; Shang, E.; Xiao, P.; Duan, J.a. The Relationship between Polysaccharide Structure and Its Antioxidant Activity Needs to Be Systematically Elucidated. *Int. J. Biol. Macromol.* **2024**, *270*, 132391. [CrossRef] [PubMed]
72. Fernandes, P.A.R.; Coimbra, M.A. The Antioxidant Activity of Polysaccharides: A Structure-Function Relationship Overview. *Carbohydr. Polym.* **2023**, *314*, 120965. [CrossRef]
73. Bai, L.; Xu, D.; Zhou, Y.M.; Zhang, Y.B.; Zhang, H.; Chen, Y.B.; Cui, Y.L. Antioxidant Activities of Natural Polysaccharides and Their Derivatives for Biomedical and Medicinal Applications. *Antioxidants* **2022**, *11*, 2491. [CrossRef]
74. Niyigaba, T.; Liu, D.; Habimana, J.d.D. The Extraction, Functionalities and Applications of Plant Polysaccharides in Fermented Foods: A Review. *Foods* **2021**, *10*, 3004. [CrossRef]

75. Sheibani, E.; Hosseini, A.; Nasab, A.S.; Adib, K.; Ganjali, M.R.; Pourmortazavi, S.M.; Ahmadi, F.; Khosrowshahi, E.M.; Mirsadeghi, S.; Rahimi-Nasrabadi, M.; et al. Application of Polysaccharide Biopolymers as Natural Adsorbent in Sample Preparation. *Crit. Rev. Food Sci. Nutr.* **2023**, *63*, 2626–2653. [CrossRef]
76. Teng, C.; Qin, P.; Shi, Z.; Zhang, W.; Yang, X.; Yao, Y.; Ren, G. Structural Characterization and Antioxidant Activity of Alkali-Extracted Polysaccharides from Quinoa. *Food Hydrocoll.* **2021**, *113*, 106392. [CrossRef]
77. Bhuyan, P.P.; Nayak, R.; Patra, S.; Abdulabbas, H.S.; Jena, M.; Pradhan, B. Seaweed-Derived Sulfated Polysaccharides; The New Age Chemopreventives: A Comprehensive Review. *Cancers* **2023**, *15*, 715. [CrossRef]
78. Livingstone, K.M.; Ramos-Lopez, O.; Pérusse, L.; Kato, H.; Ordovas, J.M.; Martínez, J.A. Precision Nutrition: A Review of Current Approaches and Future Endeavors. *Trends Food Sci. Technol.* **2022**, *128*, 253–264. [CrossRef]
79. Rhazi, L.; Depeint, F.; Gotor, A.A. Loss in the Intrinsic Quality and the Antioxidant Activity of Sunflower (*Helianthus annuus* L.) Oil during an Industrial Refining Process. *Molecules* **2022**, *27*, 916. [CrossRef] [PubMed]
80. Ambra, R.; Natella, F.; Lucchetti, S.; Forte, V.; Pastore, G. α-Tocopherol, β-Carotene, Lutein, Squalene and Secoiridoids in Seven Monocultivar Italian Extra-Virgin Olive Oils. *Int. J. Food Sci. Nutr.* **2017**, *68*, 538–545. [CrossRef] [PubMed]
81. Fagali, N.; Catalá, A. Antioxidant Activity of Conjugated Linoleic Acid Isomers, Linoleic Acid and Its Methyl Ester Determined by Photoemission and DPPH Techniques. *Biophys. Chem.* **2008**, *137*, 56–62. [CrossRef] [PubMed]
82. Lewinska, A.; Zebrowski, J.; Duda, M.; Gorka, A.; Wnuk, M. Fatty Acid Profile and Biological Activities of Linseed and Rapeseed Oils. *Molecules* **2015**, *20*, 22872–22880. [CrossRef]
83. Hulatt, C.J.; Wijffels, R.H.; Bolla, S.; Kiron, V. Production of Fatty Acids and Protein by Nannochloropsis in Flat-Plate Photobioreactors. *PLoS ONE* **2017**, *12*, e0170440. [CrossRef]
84. Xiao, B.; Li, Y.; Lin, Y.; Lin, J.; Zhang, L.; Wu, D.; Zeng, J.; Li, J.; Liu, J.w.; Li, G. Eicosapentaenoic Acid (EPA) Exhibits Antioxidant Activity via Mitochondrial Modulation. *Food Chem.* **2022**, *373*, 131389. [CrossRef]
85. Li, G.; Li, Y.; Xiao, B.; Cui, D.; Lin, Y.; Zeng, J.; Li, J.; Cao, M.-J.; Liu, J. Antioxidant Activity of Docosahexaenoic Acid (DHA) and Its Regulatory Roles in Mitochondria. *J. Agric. Food Chem.* **2021**, *69*, 1647–1655. [CrossRef]
86. Li, Z.; Chen, X.; Li, J.; Meng, T.; Wang, L.; Chen, Z.; Shi, Y.; Ling, X.; Luo, W.; Liang, D.; et al. Functions of PKS Genes in Lipid Synthesis of Schizochytrium Sp. by Gene Disruption and Metabolomics Analysis. *Mar. Biotechnol.* **2018**, *20*, 792–802. [CrossRef]
87. Rodrigues, N.; Casal, S.; Pinho, T.; Cruz, R.; Peres, A.M.; Baptista, P.; Pereira, J.A. Fatty Acid Composition from Olive Oils of Portuguese Centenarian Trees Is Highly Dependent on Olive Cultivar and Crop Year. *Foods* **2021**, *10*, 496. [CrossRef]
88. Petraru, A.; Ursachi, F.; Amariei, S. Nutritional Characteristics Assessment of Sunflower Seeds, Oil and Cake. Perspective of Using Sunflower Oilcakes as a Functional Ingredient. *Plants* **2021**, *10*, 2487. [CrossRef]
89. Ali, Y.M.; Kadir, A.A.; Ahmad, Z.; Yaakub, H.; Zakaria, Z.A.; Abdullah, M.N.H. Free Radical Scavenging Activity of Conjugated Linoleic Acid as Single or Mixed Isomers. *Pharm. Biol.* **2012**, *50*, 712–719. [CrossRef] [PubMed]
90. Zhou, W.; Wang, H.; Chen, L.; Cheng, W.; Liu, T. Bioresource Technology Heterotrophy of Filamentous Oleaginous Microalgae Tribonema Minus for Potential Production of Lipid and Palmitoleic Acid. *Bioresour. Technol.* **2017**, *239*, 250–257. [CrossRef]
91. Borges, T.H.; Pereira, J.A.; Cabrera-Vique, C.; Lara, L.; Oliveira, A.F.; Seiquer, I. Characterization of Arbequina Virgin Olive Oils Produced in Different Regions of Brazil and Spain: Physicochemical Properties, Oxidative Stability and Fatty Acid Profile. *Food Chem.* **2017**, *215*, 454–462. [CrossRef]
92. Hu, W.; Fitzgerald, M.; Topp, B.; Alam, M.; O'Hare, T.J. A Review of Biological Functions, Health Benefits, and Possible de Novo Biosynthetic Pathway of Palmitoleic Acid in Macadamia Nuts. *J. Funct. Foods* **2019**, *62*, 103520. [CrossRef]
93. Mohammed, D.; Freije, A.; Abdulhussain, H.; Khonji, A.; Hasan, M.; Ferraris, C.; Gasparri, C.; Aljar, M.A.A.; Redha, A.A.; Giacosa, A.; et al. Analysis of the Antioxidant Activity, Lipid Profile, and Minerals of the Skin and Seed of Hazelnuts (*Corylus avellana* L.), Pistachios (*Pistacia vera*) and Almonds (*Prunus dulcis*)—A Comparative Analysis. *AppliedChem* **2023**, *3*, 110–118. [CrossRef]
94. Shuai, X.; Dai, T.; Chen, M.; Liang, R.; Du, L.; Chen, J.; Liu, C. Comparative Study of Chemical Compositions and Antioxidant Capacities of Oils Obtained from 15 Macadamia (*Macadamia integrifolia*) Cultivars in China. *Foods* **2021**, *10*, 1031. [CrossRef]
95. Górnaś, P.; Rudzińska, M.; Grygier, A.; Lācis, G. Diversity of Oil Yield, Fatty Acids, Tocopherols, Tocotrienols, and Sterols in the Seeds of 19 Interspecific Grapes Crosses. *J. Sci. Food Agric.* **2019**, *99*, 2078–2087. [CrossRef]
96. Henry, G.E.; Momin, R.A.; Nair, M.G.; Dewitt, D.L. Antioxidant and Cyclooxygenase Activities of Fatty Acids Found in Food. *J. Agric. Food Chem.* **2002**, *50*, 2231–2234. [CrossRef]
97. Wang, J.; Zhang, Y.; Fang, Z.; Sun, L.; Wang, Y.; Liu, Y.; Xu, D.; Nie, F.; Gooneratne, R. Oleic Acid Alleviates Cadmium-Induced Oxidative Damage in Rat by Its Radicals Scavenging Activity. *Biol. Trace Elem. Res.* **2019**, *190*, 95–100. [CrossRef]
98. Xavier, A.A.O.; Pérez-Gálvez, A. Carotenoids as a Source of Antioxidants in the Diet. In *Sub-Cellular Biochemistry*; Stange, C., Ed.; Springer International Publishing: Cham, Switzerland, 2016; Volume 79, pp. 359–375. ISBN 978-3-319-39126-7.

99. Bozalan, N.K.; Karadeniz, F. Carotenoid Profile, Total Phenolic Content, and Antioxidant Activity of Carrots. *Int. J. Food Prop.* **2011**, *14*, 1060–1068. [CrossRef]
100. Lourenço-Lopes, C.; Fraga-Corral, M.; Jimenez-Lopez, C.; Carpena, M.; Pereira, A.G.; Garcia-Oliveira, P.; Prieto, M.A.; Simal-Gandara, J. Biological Action Mechanisms of Fucoxanthin Extracted from Algae for Application in Food and Cosmetic Industries. *Trends Food Sci. Technol.* **2021**, *117*, 163–181. [CrossRef]
101. Ambati, R.R.; Moi, P.S.; Ravi, S.; Aswathanarayana, R.G. Astaxanthin: Sources, Extraction, Stability, Biological Activities and Its Commercial Applications—A Review. *Mar. Drugs* **2014**, *12*, 128–152. [CrossRef] [PubMed]
102. Iwamoto, H.; Soccol, C.R.; Molina-Aulestia, D.T.; Cardoso, J.; de Melo Pereira, G.V.; de Souza Vandenberghe, L.P.; Manzoki, M.C.; Ambati, R.R.; Ravishankar, G.A.; de Carvalho, J.C. Lutein from Microalgae: An Industrial Perspective of Its Production, Downstream Processing, and Market. *Fermentation* **2024**, *10*, 106. [CrossRef]
103. Tudor, C.; Pintea, A. A Brief Overview of Dietary Zeaxanthin Occurrence and Bioaccessibility. *Molecules* **2020**, *25*, 4067. [CrossRef]
104. Burri, B.J.; La Frano, M.R.; Zhu, C. Absorption, Metabolism, and Functions of β-Cryptoxanthin. *Nutr. Rev.* **2016**, *74*, 69–82. [CrossRef]
105. Brahma, D.; Dutta, D. Evaluating β-Cryptoxanthin Antioxidant Properties against ROS-Induced Macromolecular Damages and Determining Its Photo-Stability and in-Vitro SPF. *World J. Microbiol. Biotechnol.* **2023**, *39*, 310. [CrossRef]
106. Galla, R.; Ruga, S.; Aprile, S.; Ferrari, S.; Brovero, A.; Grosa, G.; Molinari, C.; Uberti, F. New Hyaluronic Acid from Plant Origin to Improve Joint Protection—An In Vitro Study. *Int. J. Mol. Sci.* **2022**, *23*, 8114. [CrossRef]
107. Mohammed, A.A.; Niamah, A.K. Identification and Antioxidant Activity of Hyaluronic Acid Extracted from Local Isolates of Streptococcus Thermophilus. *Mater. Today Proc.* **2022**, *60*, 1523–1529. [CrossRef]
108. Schiraldi, C.; Cimini, D.; De Rosa, M. Production of Chondroitin Sulfate and Chondroitin. *Appl. Microbiol. Biotechnol.* **2010**, *87*, 1209–1220. [CrossRef]
109. Campo, G.M.; Avenoso, A.; Campo, S.; Ferlazzo, A.M.; Calatroni, A. Antioxidant Activity of Chondroitin Sulfate. In *Advances in Pharmacology*; Academic Press: Cambridge, MA, USA, 2006; Volume 53, pp. 417–431. ISBN 0120329557.
110. Pavão, M.S.G.; Mourão, P.A.S. Challenges for Heparin Production: Artificial Synthesis or Alternative Natural Sources? *Glycobiol. Insights* **2012**, *3*, 1–6. [CrossRef]
111. Bijak, M.; Saluk, J.; Szelenberger, R.; Nowak, P. Popular Naturally Occurring Antioxidants as Potential Anticoagulant Drugs. *Chem. Biol. Interact.* **2016**, *257*, 35–45. [CrossRef] [PubMed]
112. Medlej, M.K.; Batoul, C.; Olleik, H.; Li, S.; Hijazi, A.; Nasser, G.; Maresca, M.; Pochat-Bohatier, C. Antioxidant Activity and Biocompatibility of Fructo-Polysaccharides Extracted from a Wild Species of Ornithogalum from Lebanon. *Antioxidants* **2021**, *10*, 68. [CrossRef] [PubMed]
113. Delattre, C.; Fenoradosoa, T.A.; Michaud, P. Galactans: An Overview of Their Most Important Sourcing and Applications as Natural Polysaccharides. *Brazilian Arch. Biol. Technol.* **2011**, *54*, 1075–1092. [CrossRef]
114. Srivastava, P.; Malviya, R. Sources of Pectin, Extraction and Its Applications in Pharmaceutical Industry—an Overview. *Indian J. Nat. Prod. Resour.* **2011**, *2*, 10–18.
115. Wathoni, N.; Shan, C.Y.; Shan, W.Y.; Rostinawati, T.; Indradi, R.B.; Pratiwi, R.; Muchtaridi, M. Characterization and Antioxidant Activity of Pectin from Indonesian Mangosteen (*Garcinia mangostana* L.) Rind. *Heliyon* **2019**, *5*, e02299. [CrossRef]
116. Indrianingsih, A.W.; Rosyida, V.T.; Apriyana, W.; Hayati, S.N.; Darsih, C.; Nisa, K.; Ratih, D. Antioxidant and Antibacterial Properties of Bacterial Cellulose-Indonesian Plant Extract Composites for Mask Sheet. *J. Appl. Pharm. Sci.* **2020**, *10*, 37–42. [CrossRef]
117. Cabañas-Romero, L.V.; Valls, C.; Valenzuela, S.V.; Roncero, M.B.; Pastor, F.I.J.; Diaz, P.; Martínez, J. Bacterial Cellulose-Chitosan Paper with Antimicrobial and Antioxidant Activities. *Biomacromolecules* **2020**, *21*, 1568–1577. [CrossRef]
118. Hashem, A.H.; Al Abboud, M.A.; Alawlaqi, M.M.; Abdelghany, T.M.; Hasanin, M. Synthesis of Nanocapsules Based on Biosynthesized Nickel Nanoparticles and Potato Starch: Antimicrobial, Antioxidant, and Anticancer Activity. *Starch/Staerke* **2022**, *74*, 2100165. [CrossRef]
119. Mikulíková, D.; Kraic, J.Á.N. Natural Sources of Health-Promoting Starch Natural Sources of Health-Promoting Starch. *J. Food Nutr. Res.* **2018**, *45*, 69–76.
120. Zhang, R.; Edgar, K.J. Properties, Chemistry, and Applications of the Bioactive Polysaccharide Curdlan. *Biomacromolecules* **2014**, *15*, 1079–1096. [CrossRef]
121. Li, H.; Xu, S.; Xie, Y.; Zhang, Q.; Ding, S.; Wang, R.; Fu, F.; Zhan, X. Curdlan-Polyphenol Complexes Prepared by PH-Driven Effectively Enhanced Their Physicochemical Stability, Antioxidant and Prebiotic Activities. *Int. J. Biol. Macromol.* **2024**, *267*, 131579. [CrossRef] [PubMed]
122. Kareem, A.J.; Salman, J.A.S. Production of Dextran from Locally Lactobacillus Spp. Isolates. *Reports Biochem. Mol. Biol.* **2019**, *8*, 278–286.

123. Rosca, I.; Turin-Moleavin, I.A.; Sarghi, A.; Lungoci, A.L.; Varganici, C.D.; Petrovici, A.R.; Fifere, A.; Pinteala, M. Dextran Coated Iron Oxide Nanoparticles Loaded with Protocatechuic Acid as Multifunctional Therapeutic Agents. *Int. J. Biol. Macromol.* **2024**, *256*, 128314. [CrossRef] [PubMed]
124. Jonas, R.; Farah, L.F. Production and Application of Microbial Cellulose. *Polym. Degrad. Stab.* **1998**, *59*, 101–106. [CrossRef]
125. Rop, O.; Mlcek, J.; Jurikova, T. Beta-Glucans in Higher Fungi and Their Health Effects. *Nutr. Rev.* **2009**, *67*, 624–631. [CrossRef]
126. Mebrek, S.; Djeghim, H.; Mehdi, Y.; Meghezzi, A.; Anwar, S.; Awadh, N.A.A.; Benali, M. Antioxidant, Anti-Cholinesterase, Anti-α-Glucosidase and Prebiotic Properties of Beta-Glucan Extracted from Algerian Barley. *Int. J. Phytomedicine* **2018**, *10*, 58. [CrossRef]
127. Huq, T.; Khan, A.; Brown, D.; Dhayagude, N.; He, Z.; Ni, Y. Sources, Production and Commercial Applications of Fungal Chitosan: A Review. *J. Bioresour. Bioprod.* **2022**, *7*, 85–98. [CrossRef]
128. Xing, R.; Yu, H.; Liu, S.; Zhang, W.; Zhang, Q.; Li, Z.; Li, P. Antioxidant Activity of Differently Regioselective Chitosan Sulfates in Vitro. *Bioorganic Med. Chem.* **2005**, *13*, 1387–1392. [CrossRef]
129. De Souza, M.C.R.; Marques, C.T.; Dore, C.M.G.; Da Silva, F.R.F.; Rocha, H.A.O.; Leite, E.L. Antioxidant Activities of Sulfated Polysaccharides from Brown and Red Seaweeds. *J. Appl. Phycol.* **2007**, *19*, 153–160. [CrossRef]
130. Rhein-Knudsen, N.; Reyes-Weiss, D.; Horn, S.J. Extraction of High Purity Fucoidans from Brown Seaweeds Using Cellulases and Alginate Lyases. *Int. J. Biol. Macromol.* **2023**, *229*, 199–209. [CrossRef]
131. Pereira, L.; COTAS, J. *Alginates-Recent Uses of This Natural Polymer*; IntechOpen: Rijeka, Croatia, 2019; ISBN 978-1-78985-642-2.
132. El-Sheekh, M.; Kassem, W.M.A.; Alwaleed, E.A.; Saber, H. Optimization and Characterization of Brown Seaweed Alginate for Antioxidant, Anticancer, Antimicrobial, and Antiviral Properties. *Int. J. Biol. Macromol.* **2024**, *278*, 134715. [CrossRef] [PubMed]
133. Machado, B.; Costa, S.M.; Costa, I.; Fangueiro, R.; Ferreira, D.P. The Potential of Algae as a Source of Cellulose and Its Derivatives for Biomedical Applications. *Cellulose* **2024**, *31*, 3353–3376. [CrossRef]
134. Petchsomrit, A.; Chanthathamrongsiri, N.; Jiangseubchatveera, N.; Manmuan, S.; Leelakanok, N.; Plianwong, S.; Siranonthana, N.; Sirirak, T. Extraction, Antioxidant Activity, and Hydrogel Formulation of Marine Cladophora Glomerata. *Algal Res.* **2023**, *71*, 103011. [CrossRef]
135. Al Harthi, S.S.; Mavazhe, A.; Al Mahroqi, H.; Khan, S.A. Quantification of Phenolic Compounds, Evaluation of Physicochemical Properties and Antioxidant Activity of Four Date (*Phoenix dactylifera* L.) Varieties of Oman. *J. Taibah Univ. Med. Sci.* **2015**, *10*, 346–352. [CrossRef]
136. Palmieri, M.G.S.; Cruz, L.T.; Bertges, F.S.; Húngaro, H.M.; Batista, L.R.; da Silva, S.S.; Fonseca, M.J.V.; Rodarte, M.P.; Vilela, F.M.P.; Amaral, M.d.P.H.d. Enhancement of Antioxidant Properties from Green Coffee as Promising Ingredient for Food and Cosmetic Industries. *Biocatal. Agric. Biotechnol.* **2018**, *16*, 43–48. [CrossRef]
137. Chiou, S.Y.; Sung, J.M.; Huang, P.W.; Lin, S.D. Antioxidant, Antidiabetic, and Antihypertensive Properties of Echinacea Purpurea Flower Extract and Caffeic Acid Derivatives Using in Vitro Models. *J. Med. Food* **2017**, *20*, 171–179. [CrossRef]
138. Gharat, N.N.; Rathod, V.K. Extraction of Ferulic Acid from Rice Bran Using NADES-Ultrasound-Assisted Extraction: Kinetics and Optimization. *J. Food Process Eng.* **2023**, *46*, e14158. [CrossRef]
139. Rivero-Cruz, J.F.; Granados-Pineda, J.; Pedraza-Chaverri, J.; Pérez-Rojas, J.M.; Kumar-Passari, A.; Diaz-Ruiz, G.; Rivero-Cruz, B.E. Phytochemical Constituents, Antioxidant, Cytotoxic, and Antimicrobial Activities of the Ethanolic Extract of Mexican Brown Propolis. *Antioxidants* **2020**, *9*, 70. [CrossRef]
140. Priyandoko, D.; Widowati, W.; Kusuma, H.S.W.; Afifah, E.; Wijayanti, C.R.; Wahyuni, C.D.; Idris, A.M.; Putdayani, R.A.; Rizal, R. Antioxidant Activity of Green Tea Extract and Myricetin. In Proceedings of the InHeNce 2021—2021 IEEE International Conference on Health, Instrumentation and Measurement, and Natural Sciences, Medan, Indonesia, 14–16 July 2021; pp. 1–5.
141. Dou, X.; Zhou, Z.; Ren, R.; Xu, M. Apigenin, Flavonoid Component Isolated from Gentiana Veitchiorum Flower Suppresses the Oxidative Stress through LDLR-LCAT Signaling Pathway. *Biomed. Pharmacother.* **2020**, *128*, 110298. [CrossRef]
142. Akullo, J.O.; Kiage-Mokua, B.N.; Nakimbugwe, D.; Ng'ang'a, J.; Kinyuru, J. Phytochemical Profile and Antioxidant Activity of Various Solvent Extracts of Two Varieties of Ginger and Garlic. *Heliyon* **2023**, *9*, e18806. [CrossRef]
143. Antasionasti, I.; Datu, O.S.; Lestari, U.S.; Abdullah, S.S.; Jayanto, I. Correlation Analysis of Antioxidant Activities with Tannin, Total Flavonoid, and Total Phenolic Contents of Nutmeg (Myristica Fragrans Houtt) Fruit Precipitated by Egg White. *Borneo J. Pharm.* **2021**, *4*, 301–310. [CrossRef]
144. Saleh, M.M.; Darwish, Z.E.; El Nouaem, M.I.; Fayed, N.A.; Mourad, G.M.; Ramadan, O.R. The Potential Preventive Effect of Dietary Phytochemicals In Vivo. *BDJ Open* **2023**, *9*, 30. [CrossRef] [PubMed]
145. Molina-Montes, E.; Salamanca-Fernández, E.; Garcia-Villanova, B.; Sánchez, M.J. The Impact of Plant-based Dietary Patterns on Cancer-related Outcomes: A Rapid Review and Meta-analysis. *Nutrients* **2020**, *12*, 2010. [CrossRef]

146. Chen, H.; Zhu, B.; Zhao, L.; Liu, Y.; Zhao, F.; Feng, J.; Jin, Y.; Sun, J.; Geng, R.; Wei, Y. Allicin Inhibits Proliferation and Invasion in Vitro and in Vivo via SHP-1-Mediated STAT3 Signaling in Cholangiocarcinoma. *Cell. Physiol. Biochem.* **2018**, *47*, 641–653. [CrossRef]
147. Kim, J.S.; Jeong, S.K.; Oh, S.J.; Lee, C.G.; Kang, Y.R.; Jo, W.S.; Jeong, M.H. The Resveratrol Analogue, HS-1793, Enhances the Effects of Radiation Therapy through the Induction of Anti-Tumor Immunity in Mammary Tumor Growth. *Int. J. Oncol.* **2020**, *56*, 1405–1416. [CrossRef]
148. Bone, S.M.K. (Ed.) Principles of Herbal Pharmacology. In *Principles and Practice of Phytotherapy*; Elsevier: Amsterdam, The Netherlands, 2013; pp. 17–82. ISBN 9780443069925.
149. Aras, D.; Cinar, O.; Cakar, Z.; Ozkavukcu, S.; Can, A. Can Dicoumarol Be Used as a Gonad-Safe Anticancer Agent: An in Vitro and in Vivo Experimental Study. *Mol. Hum. Reprod.* **2015**, *22*, 57–67. [CrossRef]
150. Chen, Y.C.; Chia, Y.C.; Huang, B.M. Phytochemicals from Polyalthia Species: Potential and Implication on Anti-Oxidant, Anti-Inflammatory, Anti-Cancer, and Chemoprevention Activities. *Molecules* **2021**, *26*, 5369. [CrossRef]
151. Lam, M.; Carmichael, A.R.; Griffiths, H.R. An Aqueous Extract of Fagonia Cretica Induces DNA Damage, Cell Cycle Arrest and Apoptosis in Breast Cancer Cells via FOXO3a and P53 Expression. *PLoS ONE* **2012**, *7*, e102655. [CrossRef]
152. Yeniçeri, E.; Altay, A.; Koksal, E.; Altın, S.; Taslimi, P.; Yılmaz, M.A.; Cakir, O.; Tarhan, A.; Kandemir, A. Phytochemical Profile by LC-MS/MS Analysis and Evaluation of Antioxidant, Antidiabetic, Anti-Alzheimer, and Anticancer Activity of *Onobrychis Argyrea* Leaf Extracts. *Eur. J. Integr. Med.* **2024**, *66*, 102337. [CrossRef]
153. Ooi, K.L.; Muhammad, T.S.T.; Tan, M.L.; Sulaiman, S.F. Cytotoxic, Apoptotic and Anti-α-Glucosidase Activities of 3,4-Di-O-Caffeoyl Quinic Acid, an Antioxidant Isolated from the Polyphenolic-Rich Extract of Elephantopus Mollis Kunth. *J. Ethnopharmacol.* **2011**, *135*, 685–695. [CrossRef]
154. Amamra, S.; Cartea, M.E.; Belhaddad, O.E.; Soengas, P.; Baghiani, A.; Kaabi, I.; Arrar, L. Determination of Total Phenolics Contents, Antioxidant Capacity of *Thymus Vulgaris* Extracts Using Electrochemical and Spectrophotometric Methods. *Int. J. Electrochem. Sci.* **2018**, *13*, 7882–7893. [CrossRef]
155. Magdy, M.; Elosaily, A.H.; Mohsen, E.; EL Hefnawy, H.M. Chemical Profile, Antioxidant and Anti-Alzheimer Activity of Leaves and Flowers of Markhamia Lutea Cultivated in Egypt: In Vitro and in Silico Studies. *Futur. J. Pharm. Sci.* **2024**, *10*, 103. [CrossRef]
156. Abdelouhab, K.; Guemmaz, T.; Karamać, M.; Kati, D.E.; Amarowicz, R.; Arrar, L. Phenolic Composition and Correlation with Antioxidant Properties of Various Organic Fractions from *Hertia Cheirifolia* Extracts. *J. Pharm. Biomed. Anal.* **2023**, *235*, 115673. [CrossRef] [PubMed]
157. Maurya, A.K.; Vinayak, M. Anticarcinogenic Action of Quercetin by Downregulation of Phosphatidylinositol 3-Kinase (PI3K) and Protein Kinase C (PKC) via Induction of P53 in Hepatocellular Carcinoma (HepG2) Cell Line. *Mol. Biol. Rep.* **2015**, *42*, 1419–1429. [CrossRef]
158. Reimche, I.; Yu, H.; Ariantari, N.P.; Liu, Z.; Merkens, K.; Rotfuß, S.; Peter, K.; Jungwirth, U.; Bauer, N.; Kiefer, F.; et al. Phenanthroindolizidine Alkaloids Isolated from *Tylophora Ovata* as Potent Inhibitors of Inflammation, Spheroid Growth, and Invasion of Triple-Negative Breast Cancer. *Int. J. Mol. Sci.* **2022**, *23*, 10319. [CrossRef]
159. Arbizu-Berrocal, S.H.; Kim, H.; Fang, C.; Krenek, K.A.; Talcott, S.T.; Mertens-Talcott, S.U. Polyphenols from Mango (*Mangifera indica* L.) Modulate PI3K/AKT/MTOR-Associated Micro-RNAs and Reduce Inflammation in Non-Cancer and Induce Cell Death in Breast Cancer Cells. *J. Funct. Foods* **2019**, *55*, 9–16. [CrossRef]
160. Rattanamaneerusmee, A.; Thirapanmethee, K.; Nakamura, Y.; Bongcheewin, B.; Chomnawang, M.T. Chemopreventive and Biological Activities of Helicteres Isora L. Fruit Extracts. *Res. Pharm. Sci.* **2018**, *13*, 476–483. [CrossRef]
161. Monteillier, A.; Cretton, S.; Ciclet, O.; Marcourt, L.; Ebrahimi, S.N.; Christen, P.; Cuendet, M. Cancer Chemopreventive Activity of Compounds Isolated from Waltheria Indica. *J. Ethnopharmacol.* **2017**, *203*, 214–225. [CrossRef]
162. Da Silva, V.C.; Guerra, G.C.B.; Araújo, D.F.D.S.; De Araújo, E.R.; De Araújo, A.A.; Dantas-Medeiros, R.; Zanatta, A.C.; Da Silva, I.L.G.; De Araújo Júnior, R.F.; Esposito, D.; et al. Chemopreventive and Immunomodulatory Effects of Phenolic-Rich Extract of Commiphora Leptophloeos against Inflammatory Bowel Disease: Preclinical Evidence. *J. Ethnopharmacol.* **2024**, *328*, 118025. [CrossRef]
163. Drif, A.I.; Yücer, R.; Damiescu, R.; Ali, N.T.; Abu Hagar, T.H.; Avula, B.; Khan, I.A.; Efferth, T. Anti-Inflammatory and Cancer-Preventive Potential of Chamomile (*Matricaria chamomilla* L.): A Comprehensive In Silico and In Vitro Study. *Biomedicines* **2024**, *12*, 1484. [CrossRef]
164. Sobhy, Y.S.; Abo-zeid, Y.S.; Mahgoub, S.S.; Mina, S.A.; Mady, M.S. In-vitro Cytotoxic and Anti-inflammatory Potential of Asparagus Densiflorus Meyeri and Its Phytochemical Investigation. *Chem. Biodivers.* **2024**, *21*, e202400959. [CrossRef] [PubMed]
165. Thamer, F.H.; Al-opari, A.M.; Al-Gani, A.M.S.; Al-jaberi, E.A.; Allugam, F.A.; Almahboub, H.H.; Mosik, H.M.; Khalil, H.H.; Abduljalil, M.M.; Alpogosh, M.Y.; et al. Capparis Cartilaginea Decne. as a Natural Source of Antioxidant, Anti-Inflammatory, and Anti-Cancer Herbal Drug. *Phytomedicine Plus* **2024**, *4*, 100502. [CrossRef]

166. Dania, O.E.; Dokunmu, T.M.; Adegboye, B.E.; Adeyemi, A.O.; Chibuzor, F.C.; Iweala, E.E.J. Pro-Estrogenic and Anti-Inflammatory Effects of Corchorus Olitorius and Amaranthus Hybridus Leaves in DMBA-Induced Breast Cancer. *Phytomedicine Plus* **2024**, *4*, 100567. [CrossRef]
167. Sharma, N.; Samarakoon, K.W.; Gyawali, R.; Park, Y.H.; Lee, S.J.; Oh, S.J.; Lee, T.H.; Jeong, D.K. Evaluation of the Antioxidant, Anti-Inflammatory, and Anticancer Activities of *Euphorbia Hirta* Ethanolic Extract. *Molecules* **2014**, *19*, 14567–14581. [CrossRef]
168. Imtiaz, F.; Islam, M.; Saeed, H.; Ishaq, M.; Shareef, U.; Qaisar, M.N.; Ullah, K.; Mansoor Rana, S.; Yasmeen, A.; Saleem, A. HPLC Profiling for the Simultaneous Estimation of Antidiabetic Compounds from Tradescantia Pallida. *Arab. J. Chem.* **2024**, *17*, 105703. [CrossRef]
169. Latolla, N.; Reddy, S.; van de Venter, M.; Hlangothi, B. Phytochemical Composition and Antidiabetic Potential of the Leaf, Stem, and Rhizome Extracts of Cissampelos Capensis L.F. *S. Afr. J. Bot.* **2023**, *163*, 468–477. [CrossRef]
170. Sharma, P.; Joshi, T.; Joshi, T.; Chandra, S.; Tamta, S. In Silico Screening of Potential Antidiabetic Phytochemicals from Phyllanthus Emblica against Therapeutic Targets of Type 2 Diabetes. *J. Ethnopharmacol.* **2020**, *248*, 112268. [CrossRef]
171. Hannan, J.M.A.; Marenah, L.; Ali, L.; Rokeya, B.; Flatt, P.R.; Abdel-Wahab, Y.H.A. Ocimum Sanctum Leaf Extracts Stimulate Insulin Secretion from Perfused Pancreas, Isolated Islets and Clonal Pancreatic β-Cells. *J. Endocrinol.* **2006**, *189*, 127–136. [CrossRef]
172. El-Beshbishy, H.A.; Bahashwan, S.A. Hypoglycemic Effect of Basil (*Ocimum basilicum*) Aqueous Extract Is Mediated through Inhibition of α-Glucosidase and α-Amylase Activities: An in Vitro Study. *Toxicol. Ind. Health* **2012**, *28*, 42–50. [CrossRef]
173. Rahman, R.A.; Chellammal, H.S.J.; Shah, S.A.A.; Zohdi, R.M.; Ramachandran, D.; Mohsin, H.F. Exploring the Therapeutic Potential of *Derris elliptica* (Wall.) Benth in Streptozotocin-Induced Diabetic Rats: Phytochemical Characterization and Antidiabetic Evaluation: Exploring the Therapeutic Potential of *Derris elliptica* (Wall.). *Saudi Pharm. J.* **2024**, *32*, 102016. [CrossRef]
174. Agada, R.; Thagriki, D.; Lydia, D.E.; Khusro, A.; Alkahtani, J.; Al Shaqha, M.M.; Alwahibi, M.S.; Elshikh, M.S. Antioxidant and Anti-Diabetic Activities of Bioactive Fractions of Carica Papaya Seeds Extract. *J. King Saud Univ.-Sci.* **2021**, *33*, 101342. [CrossRef]
175. Mahmood, R.; Kayani, W.K.; Ahmed, T.; Malik, F.; Hussain, S.; Ashfaq, M.; Ali, H.; Rubnawaz, S.; Green, B.D.; Calderwood, D.; et al. Assessment of Antidiabetic Potential and Phytochemical Profiling of Rhazya Stricta Root Extracts. *BMC Complement. Med. Ther.* **2020**, *20*, 293. [CrossRef] [PubMed]
176. Rizvi, S.N.R.; Afzal, S.; Khan, K.U.R.; Aati, H.Y.; Rao, H.; Ghalloo, B.A.; Shahzad, M.N.; Khan, D.A.; Esatbeyoglu, T.; Korma, S.A. Chemical Characterisation, Antidiabetic, Antibacterial, and In Silico Studies for Different Extracts of *Haloxylon stocksii* (Boiss.) Benth: A Promising Halophyte. *Molecules* **2023**, *28*, 3847. [CrossRef]
177. Ye, Y.; Kawaguchi, Y.; Takeuchi, A.; Zhang, N.; Mori, R.; Mijiti, M.; Banno, A.; Okada, T.; Hiramatsu, N.; Nagaoka, S. Rose Polyphenols Exert Antiobesity Effect in High-Fat–Induced Obese Mice by Regulating Lipogenic Gene Expression. *Nutr. Res.* **2023**, *119*, 76–89. [CrossRef]
178. Fang, J.Y.; Huang, T.H.; Chen, W.J.; Aljuffali, I.A.; Hsu, C.Y. Rhubarb Hydroxyanthraquinones Act as Antiobesity Agents to Inhibit Adipogenesis and Enhance Lipolysis. *Biomed. Pharmacother.* **2022**, *146*, 112497. [CrossRef]
179. Im, J.H.; Lim, J.s.; Han, X.; Men, X.; Oh, G.; Fu, X.; Cho, G.h.; Hwang, W.s.; Choi, S.I.; Lee, O.H. Anti-Obesogenic Effect of Standardized Brassica Juncea Extract on Bisphenol A-Induced 3T3-L1 Preadipocytes and C57BL/6J Obese Mice. *Food Sci. Biotechnol.* **2024**, *16*, 13. [CrossRef]
180. Magwaza, S.N.; Erukainure, O.L.; Olofinsan, K.; Meriga, B.; Islam, M.S. Evaluation of the Antidiabetic, Antiobesity and Antioxidant Potential of Anthophycus Longifolius ((Turner) Kützing). *Sci. African* **2024**, *23*, e02051. [CrossRef]
181. Brattiya, K.; Jagdish, R.; Ahamed, A.; Anuradha, C.R.; Jeyasudha, J. Anti-Obesity and Anticancer Activity of Solanum Xanthocarpum Leaf Extract: An in Vitro Study. *Natl. J. Physiol. Pharm. Pharmacol.* **2023**, *13*, 1. [CrossRef]
182. Jaradat, N.; Hawash, M. Anti-Obesity Activities of *Rumex Rothschildianus* Aarons. Extracts. *BMC Complement. Med. Ther.* **2021**, *21*, 107. [CrossRef]
183. Peng, X.N.; Zhou, Y.; Liu, Y.X.; Huo, Y.L.; Ren, J.Y.; Bai, Z.Z.; Zhang, Y.L.; Tang, J.J. Neuroprotective Potential of Phytochemicals Isolated from *Paeonia Ostii* 'Feng Dan' Stamen. *Ind. Crops Prod.* **2023**, *200*, 116808. [CrossRef]
184. Gouda, B.; Sinha, S.N.; Sangaraju, R.; Huynh, T.; Patangay, S.; Mullapudi, S.V.; Mungamuri, S.K.; Patil, P.B.; Periketi, M.C. Extraction, Phytochemical Profile, and Neuroprotective Activity of Phyllanthus Emblica Fruit Extract against Sodium Valproate-Induced Postnatal Autism in BALB/c Mice. *Heliyon* **2024**, *10*, e34992. [CrossRef] [PubMed]
185. Khaled, N.; Ibrahim, N.; Ali, A.E.; Youssef, F.S.; El-Ahmady, S.H. LC-QTOF-MS/MS Phytochemical Profiling of Tabebuia Impetiginosa (Mart. Ex DC.) Standl. Leaf and Assessment of Its Neuroprotective Potential in Rats. *J. Ethnopharmacol.* **2024**, *331*, 118292. [CrossRef]
186. Hayes, J.D.; Dinkova-Kostova, A.T.; Tew, K.D. Oxidative Stress in Cancer. *Cancer Cell* **2020**, *38*, 167–197. [CrossRef]
187. Almilaibary, A. Phyto-Therapeutics as Anti-Cancer Agents in Breast Cancer: Pathway Targeting and Mechanistic Elucidation. *Saudi J. Biol. Sci.* **2024**, *31*, 103935. [CrossRef]

188. Yun, C.W.; Jeon, J.; Go, G.; Lee, J.H.; Lee, S.H. The Dual Role of Autophagy in Cancer Development and a Therapeutic Strategy for Cancer by Targeting Autophagy. *Int. J. Mol. Sci.* **2021**, *22*, 179. [CrossRef]
189. Koh, Y.-C.; Ho, C.-T.; Pan, M.-H. Recent Advances in Cancer Chemoprevention with Phytochemicals. *J. Food Drug Anal.* **2020**, *28*, 14–37. [CrossRef] [PubMed]
190. Bastos, R.G.; de, O. Rodrigues, S.; Marques, L.A.; Oliveira, C.M. d.; Salles, B.C.C.; Zanatta, A.C.; Rocha, F.D.; Vilegas, W.; Pagnossa, J.P.; Fernanda, F.B.; et al. Eugenia Sonderiana O. Berg Leaves: Phytochemical Characterization, Evaluation of in Vitro and in Vivo Antidiabetic Effects, and Structure-Activity Correlation. *Biomed. Pharmacother.* **2023**, *165*, 115126. [CrossRef]
191. Kanmaz, H.; Gokce, Y.; Hayaloglu, A.A. Volatiles, Phenolic Compounds and Bioactive Properties of Essential Oil and Aqueous Extract of Purple Basil (*Ocimum basilicum* L.) and Antidiabetic Activity in Streptozotocin-Induced Diabetic Wistar Rats. *Food Chem. Adv.* **2023**, *3*, 100429. [CrossRef]
192. Tung, Y.C.; Hsieh, P.H.; Pan, M.H.; Ho, C.T. Cellular Models for the Evaluation of the Antiobesity Effect of Selected Phytochemicals from Food and Herbs. *J. Food Drug Anal.* **2017**, *25*, 100–110. [CrossRef]
193. Wolfram, S.; Wang, Y.; Thielecke, F. Anti-Obesity Effects of Green Tea: From Bedside to Bench. *Mol. Nutr. Food Res.* **2006**, *50*, 176–187. [CrossRef]
194. Suescun, J.; Chandra, S.; Schiess, M.C. The Role of Neuroinflammation in Neurodegenerative Disorders. In *Translational Cell Biology*; Elsevier: Amsterdam, The Netherlands, 2019; pp. 241–267. [CrossRef]
195. Ayaz, M.; Mosa, O.F.; Nawaz, A.; Hamdoon, A.A.E.; Elkhalifa, M.E.M.; Sadiq, A.; Ullah, F.; Ahmed, A.; Kabra, A.; Khan, H.; et al. Neuroprotective Potentials of Lead Phytochemicals against Alzheimer's Disease with Focus on Oxidative Stress-Mediated Signaling Pathways: Pharmacokinetic Challenges, Target Specificity, Clinical Trials and Future Perspectives. *Phytomedicine* **2024**, *124*, 155272. [CrossRef] [PubMed]
196. Gil-Martín, E.; Forbes-Hernández, T.; Romero, A.; Cianciosi, D.; Giampieri, F.; Battino, M. Influence of the Extraction Method on the Recovery of Bioactive Phenolic Compounds from Food Industry By-Products. *Food Chem.* **2022**, *378*, 131918. [CrossRef] [PubMed]
197. Aqil, F.; Munagala, R.; Jeyabalan, J.; Vadhanam, M.V. Bioavailability of Phytochemicals and Its Enhancement by Drug Delivery Systems. *Cancer Lett.* **2013**, *334*, 133–141. [CrossRef]
198. Siddiqui, R.A.; Moghadasian, M.H. Nutraceuticals and Nutrition Supplements: Challenges and Opportunities. *Nutrients* **2020**, *12*, 1593. [CrossRef] [PubMed]
199. Xu, J. Harnessing the Power of Plants: A Green Factory for Bioactive Compounds. *Life* **2023**, *13*, 2041. [CrossRef]
200. Kumar, A.; Nirmal, P.; Kumar, M.; Jose, A.; Tomer, V.; Oz, E.; Proestos, C.; Zeng, M.; Elobeid, T.; Sneha, V.; et al. Major Phytochemicals: Recent Advances in Health Benefits and Extraction Method. *Molecules* **2023**, *28*, 887. [CrossRef]
201. El Maaiden, E.; Bouzroud, S.; Nasser, B.; Moustaid, K.; El Mouttaqi, A.; Ibourki, M.; Boukcim, H.; Hirich, A.; Kouisni, L.; El Kharrassi, Y. A Comparative Study between Conventional and Advanced Extraction Techniques: Pharmaceutical and Cosmetic Properties of Plant Extracts. *Molecules* **2022**, *27*, 2074. [CrossRef]
202. Cannavacciuolo, C.; Pagliari, S.; Celano, R.; Campone, L.; Rastrelli, L. Critical Analysis of Green Extraction Techniques Used for Botanicals: Trends, Priorities, and Optimization Strategies-A Review. *TrAC-Trends Anal. Chem.* **2024**, *173*, 117627. [CrossRef]
203. Domínguez, R.; Pateiro, M.; Munekata, P.E.S.; McClements, D.J.; Lorenzo, J.M. Encapsulation of Bioactive Phytochemicals in Plant-Based Matrices and Application as Additives in Meat and Meat Products. *Molecules* **2021**, *26*, 3984. [CrossRef]
204. Bommakanti, V.; Ajikumar, A.P.; Sivi, C.M.; Prakash, G.; Mundanat, A.S.; Ahmad, F.; Haque, S.; Prieto, M.A.; Rana, S.S. An Overview of Herbal Nutraceuticals, Their Extraction, Formulation, Therapeutic Effects and Potential Toxicity. *Separations* **2023**, *10*, 177. [CrossRef]
205. Belwal, T.; Chemat, F.; Venskutonis, P.R.; Cravotto, G.; Jaiswal, D.K.; Bhatt, I.D.; Devkota, H.P.; Luo, Z. Recent Advances in Scaling-up of Non-Conventional Extraction Techniques: Learning from Successes and Failures. *TrAC-Trends Anal. Chem.* **2020**, *127*, 115895. [CrossRef]
206. Pateiro, M.; Barba, F.J.; Domínguez, R.; Sant'Ana, A.S.; Khaneghah, A.M.; Gavahian, M.; Gómez, B.; Lorenzo, J.M. Essential Oils as Natural Additives to Prevent Oxidation Reactions in Meat and Meat Products: A Review. *Food Res. Int.* **2018**, *113*, 156–166. [CrossRef] [PubMed]
207. Bitwell, C.; Indra, S.S.; Luke, C.; Kakoma, M.K. A Review of Modern and Conventional Extraction Techniques and Their Applications for Extracting Phytochemicals from Plants. *Sci. Afr.* **2023**, *19*, e01585. [CrossRef]
208. Nicolescu, A.; Babotă, M.; Barros, L.; Rocchetti, G.; Lucini, L.; Tanase, C.; Mocan, A.; Bunea, C.I.; Crișan, G. Bioaccessibility and Bioactive Potential of Different Phytochemical Classes from Nutraceuticals and Functional Foods. *Front. Nutr.* **2023**, *10*, 1184535. [CrossRef] [PubMed]
209. Altemimi, A.; Lightfoot, D.A.; Kinsel, M.; Watson, D.G. Employing Response Surface Methodology for the Optimization of Ultrasound Assisted Extraction of Lutein and β-Carotene from Spinach. *Molecules* **2015**, *20*, 6611–6625. [CrossRef]

210. Altemimi, A.; Lakhssassi, N.; Baharlouei, A.; Watson, D.G.; Lightfoot, D.A. Phytochemicals: Extraction, Isolation, and Identification of Bioactive Compounds from Plant Extracts. *Plants* **2017**, *6*, 42. [CrossRef]
211. Zhang, Z.; Pang, X.; Xuewu, D.; Ji, Z.; Jiang, Y. Role of Peroxidase in Anthocyanin Degradation in Litchi Fruit Pericarp. *Food Chem.* **2005**, *90*, 47–52. [CrossRef]
212. Mulinacci, N.; Prucher, D.; Peruzzi, M.; Romani, A.; Pinelli, P.; Giaccherini, C.; Vincieri, F.F. Commercial and Laboratory Extracts from Artichoke Leaves: Estimation of Caffeoyl Esters and Flavonoidic Compounds Content. *J. Pharm. Biomed. Anal.* **2004**, *34*, 349–357. [CrossRef]
213. Manocha, S.; Dhiman, S.; Grewal, A.S.; Guarve, K. Nanotechnology: An Approach to Overcome Bioavailability Challenges of Nutraceuticals. *J. Drug Deliv. Sci. Technol.* **2022**, *72*, 103418. [CrossRef]
214. Zheng, B.; McClements, D.J. Formulation of More Efficacious Curcumin Delivery Systems Using Colloid Science: Enhanced Solubility, Stability, and Bioavailability. *Molecules* **2020**, *25*, 2791. [CrossRef]
215. Agriopoulou, S.; Tarapoulouzi, M.; Varzakas, T.; Jafari, S.M. Application of Encapsulation Strategies for Probiotics: From Individual Loading to Co-Encapsulation. *Microorganisms* **2023**, *11*, 2896. [CrossRef] [PubMed]
216. Kato, L.S.; Lelis, C.A.; da Silva, B.D.; Galvan, D.; Conte-Junior, C.A. Micro- and Nanoencapsulation of Natural Phytochemicals: Challenges and Recent Perspectives for the Food and Nutraceuticals Industry Applications. In *Advances in Food and Nutrition Research*; Toldrá, F., Ed.; Academic Press: Cambridge, MA, USA, 2023; Volume 104, pp. 77–137. ISBN 9780443193026.
217. Alu'datt, M.H.; Alrosan, M.; Gammoh, S.; Tranchant, C.C.; Alhamad, M.N.; Rababah, T.; Zghoul, R.; Alzoubi, H.; Ghatasheh, S.; Ghozlan, K.; et al. Encapsulation-Based Technologies for Bioactive Compounds and Their Application in the Food Industry: A Roadmap for Food-Derived Functional and Health-Promoting Ingredients. *Food Biosci.* **2022**, *50*, 101971. [CrossRef]
218. Choudhury, N.; Meghwal, M.; Das, K. Microencapsulation: An Overview on Concepts, Methods, Properties and Applications in Foods. *Food Front.* **2021**, *2*, 426–442. [CrossRef]
219. Ozkan, G.; Franco, P.; De Marco, I.; Xiao, J.; Capanoglu, E. A Review of Microencapsulation Methods for Food Antioxidants: Principles, Advantages, Drawbacks and Applications. *Food Chem.* **2019**, *272*, 494–506. [CrossRef]
220. Pateiro, M.; Munekata, P.E.S.; Barba, F.J.; Putnik, P. Nanoencapsulation of Promising Bioactive Compounds to Improve Their Absorption, Stability, Functionality and the Appearance of the Final Food Products. *Molecules* **2021**, *26*, 1547. [CrossRef]
221. Altemimi, A.B.; Farag, H.A.M.; Salih, T.H.; Awlqadr, F.H.; Al-Manhel, A.J.A.; Vieira, I.R.S.; Conte-Junior, C.A. Application of Nanoparticles in Human Nutrition: A Review. *Nutrients* **2024**, *16*, 636. [CrossRef]
222. Singh, A.K.; Pal, P.; Pandey, B.; Goksen, G.; Sahoo, U.K.; Lorenzo, J.M.; Sarangi, P.K. Development of "Smart Foods" for Health by Nanoencapsulation: Novel Technologies and Challenges. *Food Chem. X* **2023**, *20*, 100910. [CrossRef]
223. Puttasiddaiah, R.; Lakshminarayana, R.; Somashekar, N.L.; Gupta, V.K.; Inbaraj, B.S.; Usmani, Z.; Raghavendra, V.B.; Sridhar, K.; Sharma, M.; Rennes angers, L.I.A.; et al. Advances in Nanofabrication Technology for Nutraceuticals: New Insights and Future Trends. *Bioengineered* **2022**, *9*, 478. [CrossRef]
224. Hu, Y.; Lin, Q.; Zhao, H.; Li, X.; Sang, S.; McClements, D.J.; Long, J.; Jin, Z.; Wang, J.; Qiu, C. Bioaccessibility and Bioavailability of Phytochemicals: Influencing Factors, Improvements, and Evaluations. *Food Hydrocoll.* **2023**, *135*, 108165. [CrossRef]
225. Holst, B.; Williamson, G. Nutrients and Phytochemicals: From Bioavailability to Bioefficacy beyond Antioxidants. *Curr. Opin. Biotechnol.* **2008**, *19*, 73–82. [CrossRef]
226. Epriliati, I.; Ginjom, I.R. Bioavailability of Phytochemicals. In *Phytochemicals-A Global Perspective of Their Role in Nutrition and Health*; Rao, V., Ed.; IntechOpen: Rijeka, Croatia, 2012.
227. Shahidi, F.; Pan, Y. Influence of Food Matrix and Food Processing on the Chemical Interaction and Bioaccessibility of Dietary Phytochemicals: A Review. *Crit. Rev. Food Sci. Nutr.* **2022**, *62*, 6421–6445. [CrossRef] [PubMed]
228. Selby-Pham, S.N.B.; Miller, R.B.; Howell, K.; Dunshea, F.; Bennett, L.E. Physicochemical Properties of Dietary Phytochemicals Can Predict Their Passive Absorption in the Human Small Intestine. *Sci. Rep.* **2017**, *7*, 1931. [CrossRef] [PubMed]
229. Polia, F.; Pastor-Belda, M.; Martínez-Blázquez, A.; Horcajada, M.N.; Tomás-Barberán, F.A.; García-Villalba, R. Technological and Biotechnological Processes To Enhance the Bioavailability of Dietary (Poly)Phenols in Humans. *J. Agric. Food Chem.* **2022**, *70*, 2092–2107. [CrossRef] [PubMed]
230. Helal, N.A.; Eassa, H.A.; Amer, A.M.; Eltokhy, M.A.; Edafiogho, I.; Nounou, M.I. Nutraceuticals' Novel Formulations: The Good, the Bad, the Unknown and Patents Involved. *Recent Pat. Drug Deliv. Formul.* **2019**, *13*, 105–156. [CrossRef]
231. Senanayake, S.P.J.N. Green Tea Extract: Chemistry, Antioxidant Properties and Food Applications—A Review. *J. Funct. Foods* **2013**, *5*, 1529–1541. [CrossRef]
232. Perumalla, A.V.S.; Hettiarachchy, N.S. Green Tea and Grape Seed Extracts—Potential Applications in Food Safety and Quality. *Food Res. Int.* **2011**, *44*, 827–839. [CrossRef]
233. Bora, A.F.M.; Ma, S.; Li, X.; Liu, L. Application of Microencapsulation for the Safe Delivery of Green Tea Polyphenols in Food Systems: Review and Recent Advances. *Food Res. Int.* **2018**, *105*, 241–249. [CrossRef]

234. Nieto, G.; Martínez-Zamora, L.; Peñalver, R.; Marín-Iniesta, F.; Taboada-Rodríguez, A.; López-Gómez, A.; Martínez-Hernández, G.B. Applications of Plant Bioactive Compounds as Replacers of Synthetic Additives in the Food Industry. *Foods* **2023**, *13*, 47. [CrossRef]
235. Duizer, L.M.; West, R.; Campanella, O.H. Fiber Addition to Cereal Based Foods: Effects on Sensory Properties. In *Food Engineering Series*; Springer: Cham, Switzerland, 2020; pp. 419–435.
236. Muñoz-Almagro, N.; Garrido-Galand, S.; Taladrid, D.; Moreno-Arribas, M.V.; Villamiel, M.; Montilla, A. Use of Natural Low-Methoxyl Pectin from Sunflower by-Products for the Formulation of Low-Sucrose Strawberry Jams. *J. Sci. Food Agric.* **2022**, *102*, 5957–5964. [CrossRef]
237. Tiwari, A.; Sharma, H.K.; Kumar, N.; Kaur, M. The Effect of Inulin as a Fat Replacer on the Quality of Low-Fat Ice Cream. *Int. J. Dairy Technol.* **2015**, *68*, 374–380. [CrossRef]
238. Le Bastard, Q.; Chapelet, G.; Javaudin, F.; Lepelletier, D.; Batard, E.; Montassier, E. The Effects of Inulin on Gut Microbial Composition: A Systematic Review of Evidence from Human Studies. *Eur. J. Clin. Microbiol. Infect. Dis.* **2020**, *39*, 403–413. [CrossRef] [PubMed]
239. Nsor-Atindana, J.; Chen, M.; Goff, H.D.; Zhong, F.; Sharif, H.R.; Li, Y. Functionality and Nutritional Aspects of Microcrystalline Cellulose in Food. *Carbohydr. Polym.* **2017**, *172*, 159–174. [CrossRef] [PubMed]
240. Nath, P.C.; Debnath, S.; Sharma, M.; Sridhar, K.; Nayak, P.K.; Inbaraj, B.S. Recent Advances in Cellulose-Based Hydrogels: Food Applications. *Foods* **2023**, *12*, 350. [CrossRef]
241. Onipe, O.O.; Ramashia, S.E.; Jideani, A.I.O. Wheat Bran Modifications for Enhanced Nutrition and Functionality in Selected Food Products. *Molecules* **2021**, *26*, 3918. [CrossRef]
242. Maiani, G.; Castón, M.J.P.; Catasta, G.; Toti, E.; Cambrodón, I.G.; Bysted, A.; Granado-Lorencio, F.; Olmedilla-Alonso, B.; Knuthsen, P.; Valoti, M.; et al. Carotenoids: Actual Knowledge on Food Sources, Intakes, Stability and Bioavailability and Their Protective Role in Humans. *Mol. Nutr. Food Res.* **2009**, *53*, 194–218. [CrossRef]
243. Wang, Y.; McClements, D.J.; Chen, L.; Peng, X.; Xu, Z.; Meng, M.; Ji, H.; Zhi, C.; Ye, L.; Zhao, J.; et al. Progress on Molecular Modification and Functional Applications of Anthocyanins. *Crit. Rev. Food Sci. Nutr.* **2023**, *64*, 11409–11427. [CrossRef]
244. Pereira, A.G.; Fraga, M.; Oliveira, P.G.; Jimenez-Lopez, C.; Lourenço-Lopes, C.; Barros, L.; Ferreira, I.C.F.R.; Prieto, M.A.; Simal-Gandara, J. Identification, Quantification, and Method Validation of Anthocyanins. In *Anthocyanins: Antioxidant Properties, Sources and Health Benefits*; Nova Science Publishers: Hauppauge, NY, USA, 2020; pp. 51–83. ISBN 9781536178173.
245. Jokioja, J.; Yang, B.; Linderborg, K.M. Acylated Anthocyanins: A Review on Their Bioavailability and Effects on Postprandial Carbohydrate Metabolism and Inflammation. *Compr. Rev. Food Sci. Food Saf.* **2021**, *20*, 5570–5615. [CrossRef]
246. Mattioli, R.; Francioso, A.; Mosca, L.; Silva, P. Anthocyanins: A Comprehensive Review of Their Chemical Properties and Health Effects on Cardiovascular and Neurodegenerative Diseases. *Molecules* **2020**, *25*, 3809. [CrossRef]
247. Niu, B.; Shao, P.; Feng, S.; Qiu, D.; Sun, P. Rheological Aspects in Fabricating Pullulan-Whey Protein Isolate Emulsion Suitable for Electrospraying: Application in Improving β-Carotene Stability. *LWT* **2020**, *129*, 109581. [CrossRef]
248. Esatbeyoglu, T.; Wagner, A.E.; Schini-Kerth, V.B.; Rimbach, G. Betanin-A Food Colorant with Biological Activity. *Mol. Nutr. Food Res.* **2015**, *59*, 36–47. [CrossRef]
249. da Silva, D.V.T.; dos Santos Baião, D.; de Oliveira Silva, F.; Alves, G.; Perrone, D.; Del Aguila, E.M.; Paschoalin, V.M.F. Betanin, a Natural Food Additive: Stability, Bioavailability, Antioxidant and Preservative Ability Assessments. *Molecules* **2019**, *24*, 458. [CrossRef] [PubMed]
250. Pagano, A.P.E.; Khalid, N.; Kobayashi, I.; Nakajima, M.; Neves, M.A.; Bastos, E.L. Microencapsulation of Betanin in Monodisperse W/O/W Emulsions. *Food Res. Int.* **2018**, *109*, 489–496. [CrossRef] [PubMed]
251. Amjadi, S.; Ghorbani, M.; Hamishehkar, H.; Roufegarinejad, L. Improvement in the Stability of Betanin by Liposomal Nanocarriers: Its Application in Gummy Candy as a Food Model. *Food Chem.* **2018**, *256*, 156–162. [CrossRef]
252. Dai, T.; He, X.; Xu, J.; Geng, Q.; Li, C.; Sun, J.; Liu, C.; Chen, J.; He, X. Effects of Betanin on Pasting, Rheology and Retrogradation Properties of Different Starches. *Foods* **2022**, *11*, 1600. [CrossRef]
253. Tsuda, T. Curcumin as a Functional Food-Derived Factor: Degradation Products, Metabolites, Bioactivity, and Future Perspectives. *Food Funct.* **2018**, *9*, 705–714. [CrossRef]
254. Munekata, P.E.S.; Pateiro, M.; Zhang, W.; Dominguez, R.; Xing, L.; Fierro, E.M.; Lorenzo, J.M. Health Benefits, Extraction and Development of Functional Foods with Curcuminoids. *J. Funct. Foods* **2021**, *79*, 104392. [CrossRef]
255. Xu, X.Y.; Meng, X.; Li, S.; Gan, R.Y.; Li, Y.; Li, H.B. Bioactivity, Health Benefits, and Related Molecular Mechanisms of Curcumin: Current Progress, Challenges, and Perspectives. *Nutrients* **2018**, *10*, 1553. [CrossRef]
256. Life Extension. Available online: https://www.lifeextensioneurope.com/ (accessed on 6 May 2025).
257. Toüfood. Available online: https://www.toufood.com/ (accessed on 7 May 2025).
258. Bluetec Colorants Ltd. Available online: https://www.bestphycocyanin.com/ (accessed on 7 May 2025).

259. McClements, D.J.; Öztürk, B. Utilization of Nanotechnology to Improve the Application and Bioavailability of Phytochemicals Derived from Waste Streams. *J. Agric. Food Chem.* **2022**, *70*, 6884–6900. [CrossRef]
260. Zhang, S.; Waterhouse, G.I.N.; Xu, F.; He, Z.; Du, Y.; Lian, Y.; Wu, P.; Sun-Waterhouse, D. Recent Advances in Utilization of Pectins in Biomedical Applications: A Review Focusing on Molecular Structure-Directing Health-Promoting Properties. *Crit. Rev. Food Sci. Nutr.* **2023**, *63*, 3386–3419. [CrossRef]
261. Qin, Y.Q.; Wang, L.Y.; Yang, X.Y.; Xu, Y.J.; Fan, G.; Fan, Y.G.; Ren, J.N.; An, Q.; Li, X. Inulin: Properties and Health Benefits. *Food Funct.* **2023**, *14*, 2948–2968. [CrossRef]
262. Deroover, L.; Tie, Y.; Verspreet, J.; Courtin, C.M.; Verbeke, K. Modifying Wheat Bran to Improve Its Health Benefits. *Crit. Rev. Food Sci. Nutr.* **2020**, *60*, 1104–1122. [CrossRef] [PubMed]
263. Ziai, S.A.; Larijani, B.; Akhoondzadeh, S.; Fakhrzadeh, H.; Dastpak, A.; Bandarian, F.; Rezai, A.; Badi, H.N.; Emami, T. Psyllium Decreased Serum Glucose and Glycosylated Hemoglobin Significantly in Diabetic Outpatients. *J. Ethnopharmacol.* **2005**, *102*, 202–207. [CrossRef] [PubMed]
264. Chen, C.; Shang, C.; Xin, L.; Xiang, M.; Wang, Y.; Shen, Z.; Jiao, L.; Ding, F.; Cui, X. Beneficial Effects of Psyllium on the Prevention and Treatment of Cardiometabolic Diseases. *Food Funct.* **2022**, *13*, 7473–7486. [CrossRef] [PubMed]
265. Saeed, M.; Naveed, M.; Arif, M.; Kakar, M.U.; Manzoor, R.; El-Hack, M.E.A.; Alagawany, M.; Tiwari, R.; Khandia, R.; Munjal, A.; et al. Green Tea (*Camellia sinensis*) and L-Theanine: Medicinal Values and Beneficial Applications in Humans—A Comprehensive Review. *Biomed. Pharmacother.* **2017**, *95*, 1260–1275. [CrossRef]
266. Bogdanski, P.; Suliburska, J.; Szulinska, M.; Stepien, M.; Pupek-Musialik, D.; Jablecka, A. Green Tea Extract Reduces Blood Pressure, Inflammatory Biomarkers, and Oxidative Stress and Improves Parameters Associated with Insulin Resistance in Obese, Hypertensive Patients. *Nutr. Res.* **2012**, *32*, 421–427. [CrossRef]
267. Gupta, M.; Dey, S.; Marbaniang, D.; Pal, P.; Ray, S.; Mazumder, B. Grape Seed Extract: Having a Potential Health Benefits. *J. Food Sci. Technol.* **2020**, *57*, 1205–1215. [CrossRef]
268. Bendich, A. From 1989 to 2001: What Have We Learned about the "Biological Actions of Beta-Carotene"? *J. Nutr.* **2004**, *134*, 225S–230S. [CrossRef]
269. Li, D.; Wang, P.; Luo, Y.; Zhao, M.; Chen, F. Health Benefits of Anthocyanins and Molecular Mechanisms: Update from Recent Decade. *Crit. Rev. Food Sci. Nutr.* **2017**, *57*, 1729–1741. [CrossRef]
270. Hewlings, S.J.; Kalman, D.S. Curcumin: A Review of Its Effects on Human Health. *Foods* **2017**, *6*, 92. [CrossRef]
271. Barber, T.M.; Kabisch, S.; Randeva, H.S.; Pfeiffer, A.F.H.; Weickert, M.O. Implications of Resveratrol in Obesity and Insulin Resistance: A State-of-the-Art Review. *Nutrients* **2022**, *14*, 2870. [CrossRef]
272. Wang, Y.; Xing, M.; Cao, Q.; Ji, A.; Liang, H.; Song, S. Biological Activities of Fucoidan and the Factors Mediating Its Therapeutic Effects: A Review of Recent Studies. *Mar. Drugs* **2019**, *17*, 183. [CrossRef] [PubMed]
273. Li, F.; Liu, K. Research Progress in the Preparation, Structural Characterization, Bioactivities, and Potential Applications of Sulfated Agarans from the Genus Gracilaria. *J. Food Biochem.* **2022**, *46*, e14401. [CrossRef] [PubMed]
274. du Preez, R.; Paul, N.; Mouatt, P.; Majzoub, M.E.; Thomas, T.; Panchal, S.K.; Brown, L. Carrageenans from the Red Seaweed Sarconema Filiforme Attenuate Symptoms of Diet-Induced Metabolic Syndrome in Rats. *Mar. Drugs* **2020**, *18*, 97. [CrossRef] [PubMed]
275. Deepika; Maurya, P.K. Health Benefits of Quercetin in Age-Related Diseases. *Molecules* **2022**, *27*, 2498. [CrossRef]
276. Aghababaei, F.; Hadidi, M. Recent Advances in Potential Health Benefits of Quercetin. *Pharmaceuticals* **2023**, *16*, 1020. [CrossRef]
277. Naware, N.; Ambatkar, S.; Kamble, T.; Bangar, S.; Uppar, K.; Shirke, K.; Patil, M.; Jain, A. A Review Focusing on the Benefits of Green Tea Catechins as Nutraceuticals. *Sci. Phytochem.* **2023**, *2*, 138–146. [CrossRef]
278. Ma, D.S.L.; Tan, L.T.H.; Chan, K.G.; Yap, W.H.; Pusparajah, P.; Chuah, L.H.; Ming, L.C.; Khan, T.M.; Lee, L.H.; Goh, B.H. Resveratrol-Potential Antibacterial Agent against Foodborne Pathogens. *Front. Pharmacol.* **2018**, *9*, 102. [CrossRef]
279. Ko, S.K.; Chung, S.J.; Park, Y.H.; Park, A.S.; Kim, H.J.; Park, G.Y.; Oh, Y.H. Determination of Hazardous Metals in Nail Enamel Containing Glitter. *Korean J. Environ. Health Sci.* **2017**, *43*, 103–110. [CrossRef]
280. Cohen, S.; Emery, P.; Greenwald, M.; Yin, D.; Becker, J.C.; Melia, L.A.; Li, R.; Gumbiner, B.; Thomas, D.; Spencer-Green, G.; et al. A Phase I Pharmacokinetics Trial Comparing PF 05280586 (a Potential Biosimilar) and Rituximab in Patients with Active Rheumatoid Arthritis. *Br. J. Clin. Pharmacol.* **2016**, *82*, 129–138. [CrossRef]
281. Khan, M.I.; Shin, J.H.; Kim, J.D. The Promising Future of Microalgae: Current Status, Challenges, and Optimization of a Sustainable and Renewable Industry for Biofuels, Feed, and Other Products. *Microb. Cell Fact.* **2018**, *17*, 36. [CrossRef]
282. Hajipour, H.; Ghorbani, M.; Kahroba, H.; Mahmoodzadeh, F.; Emameh, R.Z.; Taheri, R.A. Arginyl-Glycyl-Aspartic Acid (RGD) Containing Nanostructured Lipid Carrier Co-Loaded with Doxorubicin and Sildenafil Citrate Enhanced Anti-Cancer Effects and Overcomes Drug Resistance. *Process Biochem.* **2019**, *84*, 172–179. [CrossRef]

283. Bush, L.; Stevenson, L.; Lane, K.E. The Oxidative Stability of Omega-3 Oil-in-Water Nanoemulsion Systems Suitable for Functional Food Enrichment: A Systematic Review of the Literature. *Crit. Rev. Food Sci. Nutr.* **2019**, *59*, 1154–1168. [CrossRef]
284. Kusmayadi, A.; Leong, Y.K.; Yen, H.W.; Huang, C.Y.; Chang, J.S. Microalgae as Sustainable Food and Feed Sources for Animals and Humans–Biotechnological and Environmental Aspects. *Chemosphere* **2021**, *271*, 129800. [CrossRef]
285. Stanton, A.V.; James, K.; Brennan, M.M.; O'Donovan, F.; Buskandar, F.; Shortall, K.; El-Sayed, T.; Kennedy, J.; Hayes, H.; Fahey, A.G.; et al. Omega-3 Index and Blood Pressure Responses to Eating Foods Naturally Enriched with Omega-3 Polyunsaturated Fatty Acids: A Randomized Controlled Trial. *Sci. Rep.* **2020**, *10*, 15444. [CrossRef] [PubMed]
286. Vo, T.S.; Kim, S.K. Fucoidans as a Natural Bioactive Ingredient for Functional Foods. *J. Funct. Foods* **2013**, *5*, 16–27. [CrossRef]
287. Vickers, C.; Liu, F.; Abe, K.; Salama-Alber, O.; Jenkins, M.; Springate, C.M.K.; Burke, J.E.; Withers, S.G.; Boraston, A.B. Endo-Fucoidan Hydrolases from Glycoside Hydrolase Family 107 (GH107) Display Structural and Mechanistic Similarities to α-L-Fucosidases from GH29. *J. Biol. Chem.* **2018**, *293*, 18296–18308. [CrossRef]
288. Wang, M.; Veeraperumal, S.; Zhong, S.; Cheong, K.L. Fucoidan-Derived Functional Oligosaccharides: Recent Developments, Preparation, and Potential Applications. *Foods* **2023**, *12*, 878. [CrossRef]
289. Fernando, I.P.S.; Dias, M.K.H.M.; Madusanka, D.M.D.; Han, E.J.; Kim, M.J.; Heo, S.J.; Ahn, G. Fucoidan Fractionated from Sargassum Coreanum via Step-Gradient Ethanol Precipitation Indicate Promising Uvb-Protective Effects in Human Keratinocytes. *Antioxidants* **2021**, *10*, 347. [CrossRef]
290. Chen, X.; Fu, X.; Huang, L.; Xu, J.; Gao, X. Agar Oligosaccharides: A Review of Preparation, Structures, Bioactivities and Application. *Carbohydr. Polym.* **2021**, *265*, 118076. [CrossRef]
291. Karim, A.A.; Bhat, R. Gelatin Alternatives for the Food Industry: Recent Developments, Challenges and Prospects. *Trends Food Sci. Technol.* **2008**, *19*, 644–656. [CrossRef]
292. Trius, A.; Sebranek, J.G. Carrageenans and Their Use in Meat Products. *Crit. Rev. Food Sci. Nutr.* **1996**, *36*, 69–85. [CrossRef] [PubMed]
293. Zhu, B.; Ni, F.; Sun, Y.; Zhu, X.; Yin, H.; Yao, Z.; Du, Y. Insight into Carrageenases: Major Review of Sources, Category, Property, Purification Method, Structure, and Applications. *Crit. Rev. Biotechnol.* **2018**, *38*, 1261–1276. [CrossRef] [PubMed]
294. Dai, Y.; Row, K.H. Application of Natural Deep Eutectic Solvents in the Extraction of Quercetin from Vegetables. *Molecules* **2019**, *24*, 2300. [CrossRef] [PubMed]
295. Lin, J.; Zhou, W. Role of Quercetin in the Physicochemical Properties, Antioxidant and Antiglycation Activities of Bread. *J. Funct. Foods* **2018**, *40*, 299–306. [CrossRef]
296. Bhat, I.U.H.; Bhat, R. Quercetin: A Bioactive Compound Imparting Cardiovascular and Neuroprotective Benefits: Scope for Exploring Fresh Produce, Their Wastes, and by-Products. *Biology* **2021**, *10*, 586. [CrossRef]
297. Ceamsa. Available online: https://www.ceamsa.com/ (accessed on 7 May 2025).
298. Swanson. Available online: https://swansoneurope.com/ (accessed on 7 May 2025).
299. Algamar Ltd. Available online: https://algamar.com/ (accessed on 7 May 2025).
300. Kohbo Labs. Available online: https://kobholabs.com/ (accessed on 7 May 2025).

foods

MDPI

Article

The Anti-Angiogenic Effect of *Cynara cardunculus* L. subsp. *carduncculus* Waste Product

Anna Cacciola [1], Valeria D'Angelo [1], Federica De Gaetano [1], Antonella Fais [2], Maria Paola Germanò [1,*], Valentina Masala [2], Stefania Olla [3], Venerando Pistarà [4], Rosanna Stancanelli [1], Carlo Ignazio Giovanni Tuberoso [2,*] and Cinzia Anna Ventura [1]

1 Department of Chemical, Biological, Pharmaceutical and Environmental Sciences, University of Messina, Viale F. Stagno d'Alcontres 31, 98166 Messina, Italy; anna.cacciola@unime.it (A.C.); valeria.dangelo@unime.it (V.D.); federica.degaetano@unime.it (F.D.G.); rosanna.stancanelli@unime.it (R.S.); cinziaanna.ventura@unime.it (C.A.V.)

2 Department of Life and Environmental Sciences, University of Cagliari, SS 554-bivio per Sestu, 09042 Monserrato, Italy; fais@unica.it (A.F.); valentina.masala2@unica.it (V.M.)

3 Institute for Genetic and Biomedical Research (IRGB), The National Research Council (CNR), 09042 Monserrato, Italy; stefania.olla@irgb.cnr.it

4 Department of Pharmaceutical and Health Sciences, University of Catania, Viale Andrea Doria 6, 95125 Catania, Italy; vpistara@unict.it

* Correspondence: mariapaola.germano@unime.it (M.P.G.); tuberoso@unica.it (C.I.G.T.); Tel.: +39-090-6766544 (M.P.G.); +39-070-6758644 (C.I.G.T.)

Abstract: *Cynara cardunculus* L. subsp. *cardunculus* (*Cynara cardunculus* L. var. *sylvestris* (Lam.) Fiori), the wild cardoon, is known for its culinary applications and potential health benefits. Due to this, and given the growing interest in circular economics, deepening our under-standing of the effects of wild cardoon leaf waste on angiogenesis and collagenase activity represents a valuable opportunity to valorise agricultural byproducts as health-promoting ingredients. In this study, the waste product of wild cardoon leaves was extracted to examine its chemical composition and biological activities. Analytical techniques identified several bioactive compounds, including flavonoids, hydroxycinnamic acids such as dicaffeoyl succinoylquinic acids, and luteolin-7-O-rutinoside. In vivo tests in zebrafish embryos and the chick chorioallantoic membrane demonstrated dose-dependent antiangiogenic effects, particularly enhanced by the complexation with hydroxypropyl-β-cyclodextrin (HP-β-CD). Considering the link between angiogenesis and collagenase, the potential effects of the extract on collagenase activity was investigated. The extract alone inhibited collagenase with an IC_{50} value comparable to that of the standard inhibitor while its complexed form exhibited a 4.5-fold greater inhibitory activity. A molecular docking study examined the interaction between the main compounds and collagenase. In conclusion, wild cardoon leaves can represent a valuable source of bioactive compounds. This study demonstrated that the complexation of the extract with cyclodextrin determines an increase in its biological activity.

Keywords: wild cardoon; collagenase; angiogenesis; chick chorioallantoic membrane; zebrafish; hydroxypropyl-β-cyclodextrin

1. Introduction

The agri-food sector generates a considerable amount of by-products every year, which are often wasted, resulting in the loss of valuable resources. By-products valorisation is fundamental for the effective application of the circular economy and global sustainability [1]. Since numerous by-products from the agricultural or food industry contain bioactive molecules with a high number of properties useful in the prevention of many important diseases, they have recently garnered considerable interest for their reuse and the production of bioactive compounds, as well as for use as natural additives [2,3].

The large amounts of waste produced by packing houses and the food-processing industry can reach ∼60% of the harvested vegetable, as in the case of the industrial manipulation of artichokes and cardoons [4]. For this reason, the artichoke and cardoon supply chain has attracted increasing attention for the valorisation of its by-products. From an agronomic perspective, *C. cardunculus* species—especially in their cultivated and wild forms—generate considerable amounts of leaf biomass, much of which is discarded during industrial processing [5]. The sustainable valorisation of these by-products is gaining increasing attention considering the principles of the circular bioeconomy, which promotes the conversion of agricultural residues into high-value resources. In this context, *Cynara cardunculus*, belonging to the Asteraceae family, comprises three taxa: globe artichoke (var. *scolymus*), cultivated cardoon (var. *altilis*), and the wild cardoon (subsp. *cardunculus*). The wild perennial cardoon, *C. cardunculus* L. subsp. *cardunculus*, previously known as var. *sylvestris*, is a Mediterranean plant that thrives in warm, humid climates, which belongs to a multifunctional taxon recognised for its phytochemical richness, nutraceutical potential, and traditional medicinal uses [6]. Among these, the significance of wild cardoon leaf phenolic composition for human health, plant defence, and possible nutraceutical uses has attracted growing scientific attention. The medicinal potential of this wild taxon is greatly enhanced by phenolic compounds, a varied class of secondary metabolites. These compounds are well-known for their antibacterial, anti-inflammatory, and antioxidant qualities [7,8]. *C. cardunculus* L. subsp. *cardunculus*, a wild relative of the farmed cardoon and globe artichoke, has a distinct and frequently more chemically varied phenolic profile, which is probably affected by its ecological habitat and evolutionary forces [9]. These molecules determine the value of the plant as a source of bioactive compounds in pharmaceutical and functional food formulations, in addition to playing crucial roles in the plant's physiological responses [8]. These metabolites can now be precisely identified and quantified due to recent developments in phytochemical analysis, particularly high-performance liquid chromatography coupled with mass spectrometry (HPLC-MS), making it easier to analyse them in detail [9].

Concerning its biological activity, Acquaviva et al. showed that the wild cardoon leaf extract helps reduce transient or mild hepatic oxidative stress [10]. In addition, wild thistle leaves have been proven to exhibit antibacterial properties [6]. Supporting its nutraceutical potential, dietary supplementation with wild cardoon leaf extract in rats showed a hypolipidemic effect [11]. Several biological activities are attributed to all varieties of *C. cardunculus*, among which are those shown by the methanolic extract of leaves. These have shown both antioxidant activity and the inhibition of proliferation and migration of tumour cells, as well as angiogenesis in vivo [12]. The process of angiogenesis is essential for growth, development, and wound healing and is implicated in several pathological conditions. This mechanism consists of the formation of new vessels from existing ones [13]. Moreover, collagenase is involved in the angiogenic process and plays a role in collagen remodelling [14]. Therefore, inhibiting collagenase activity may improve therapeutic outcomes in managing angiogenesis-related diseases [15]. Velez et al. [12] have highlighted that the methanolic extract of cultivated cardoon exhibits anti-angiogenesis properties. Despite this, to the best of our knowledge, no comprehensive studies have investigated the potential of wild cardoon leaf extracts to modulate collagenase activity and angiogenesis. Another critical limitation in the consumption and therapeutic application of bioactive molecules derived from artichoke and cardoon extracts lies in their unfavourable physical-chemical properties, which significantly influences their ability to cross the viable membranes, resulting in low and erratic bioavailability. For this reason, different delivery systems were studied aimed at improving the biopharmaceutical properties of natural products, and among them, cyclodextrins (CDs) have received great attention. They are cyclic oligosaccharides able to form inclusion and non-inclusion complexes with both synthetic molecules [16,17] or bioactive compounds derived from natural sources [18–20]. Specifically, hydroxypropyl-β-cyclodextrin (HP-β-CD) has shown superior capacity to encapsulate polyphenols from vegetal extracts, resulting in improved oral bioavailability, enhanced antioxidant activity, and sustained biological effects, permitting their application in medicine, cosmetics, and the food industry [21,22]. Lipophilic molecules can accommodate into the cavity of the macrocycle and their physical-chemical properties were improved, such as solubility and stability. The complex quickly dissolves in the physiological environment, improving the availability of the bioactive molecules at the surface of the biological barrier. As a result, the concentration gradient through the membrane was improved, and absorption was enhanced. Furthermore, it is now proven that CDs can act as penetration enhancer agents by interacting with the viable membranes and extracting their components (cholesterol and lipids), thus improving permeability to both lipophilic and hydrophilic drugs [23].

To support this, there are several studies in the literature that testified the effective use of CDs for the encapsulation and delivery of natural compounds [24,25]. Zhong et al. [26] investigated the activity of inclusion complexes between tea polyphenols and HP-β-CD on myofibrillar protein (MP) extracted from lamb tripe and reported a significant enhancement in the antioxidant activity and bioavailability of the inclusion complexes (5–105 μmol/g). This improvement was attributed to a reduction in carbonyl content, hydrophobicity, and the protein aggregation of MP, as well as an increase in sulfhydryl group content. Furthermore, Li et al. [27] demonstrated that the formation of inclusion complexes between curcumin and HP-β-CD doubles the plasma concentration of the molecule, thereby enhancing its bioavailability compared to the free form. In addition, these complexes maintain more stable plasma levels over time, thus prolonging the beneficial effects of curcumin.

Although the health-promoting properties of the wild cardoon are well-documented, its leaf residues remain underexplored for potential use in the functional food sector, with most research focused on pharmaceutical applications. Addressing this gap could support the development of new nutraceutical ingredients and food-based strategies aimed at cardiovascular and metabolic health. In this context, the present study aims to chemically characterise by ^{1}H NMR, LC-MS/MS and LC-PDA the extract obtained from the wild cardoon leaf by-products and to evaluate its collagenase inhibitory activity and antiangiogenic potential. The antiangiogenic effects were investigated through in vivo assays using the chick chorioallantoic membrane (CAM) and zebrafish (*Danio rerio*) embryo models. Additionally, the extract was complexed with hydroxypropyl-β-cyclodextrin (HP-β-CD) to assess the impact of this formulation strategy on biological activity. Finally, molecular docking was conducted to explore the interactions between the target phenolic constituents and collagenase.

2. Materials and Methods

2.1. Plant Material and Extraction Method

Cynara cardunculus L. subsp. *cardunculus* leaf by-products were collected in March 2024 in Serrenti (South Sardinia, Italy). The specimens were identified by Prof. Maria Paola Germanò (University of Messina, Italy), and a voucher sample (number DISVA.ALI.01.2024) was deposited at the Department of Life and Environmental Sciences of the University of Cagliari (Italy). The leaves were carefully cleaned and the hardest parts were removed and prepared for the extraction. From the sample of about 5 kg, extracts were obtained with EtOH:H_2O 80:20% *v/v* according to the following procedure: 200 g of fresh leaves were extracted with 500 mL of solvent mixture, initially using an immersion blender and subsequently completing the extraction with an Ultra Turrax IKA T18 digital (IKA-Werke GmbH & Co. KG, Staufen, Germany). The obtained extract (SyEt80) was coarsely filtered and centrifuged for 15 min at 4000 rpm at 10 °C. The supernatant was concentrated under vacuum to remove the ethanol using a Rotavapor® R-210 (BUCHI, Cornaredo, Italy) with a heated bath set at t < 40 °C. The extract thus obtained was centrifuged for 15 min at 4000 rpm at 10 °C, frozen at −20 °C, and finally freeze-dried using a Lio5P freeze dryer (5Pascal, Trezzano sul Naviglio, Italy).

2.2. ^1H NMR Analysis

Nuclear Magnetic Resonance (NMR) spectra were recorded on a Varian Unity Inova 500 MHz (11.75 T) at 300 K using a 5 mm multinuclear direct detection probe. All spectra were taken in D_2O as solvent. The chemical shifts are indicated as δ ppm values, and the residual water peak (4.79 ppm) was used as the internal reference.

2.3. Identification and Quantification of Polyphenolic Compounds

The SyEt80 was evaluated both qualitatively and quantitatively by (HR) LC-ESI-QToF-MS/MS and LC-PDA using the methodology described by De Luca et al. [28] and Masala et al. [29]. The studies were conducted using an electrospray ionisation (ESI) source that was set up to function in both positive and negative ion modes. In summary, the analytical set up consisted of an advanced ion mobility QToF LC/MS system including a 1290 Infinity II UPLC and a 6560 IM-QToF (Agilent Technologies Inc., Palo Alto, CA, USA). The ESI/QToF MS data were then assessed using the MassHunter Workstation Qualitative Analysis program, version 10.0 (Agilent Technologies). For the tentative identification of the metabolites and to predict fragmentation and molecular formulae, the MassHunter METLIN metabolite PCDLdatabase B.08.00 (Agilent Technologies), comparison between the experimental MS/MS spectra and spectra from a publicly available mass spectral data repository (e.g., Sirius® software v. 5.5.5 [30,31] and MZmine® v. 4.3 [32]), and fragmentation patterns that have been published in the literature and other natural products databases (KNApSAcK®, PubChem®, Coconut®) were used. Phenolic substances were quali-quantitatively analysed using a 1260 Infinity II HPLC system and an Agilent Technologies G4212B photodiode array detector. The separation was obtained with a Kinetex EVO C18 column (150 × 4.60 mm, 2.6 μm, Phenomenex, Castel Maggiore, BO, Italy) using 0.22 M phosphoric acid (solvent A) and acetonitrile (solvent B) as the mobile phase properly mixed in gradient elution at a constant flow rate of 0.8 mL/min. After decreasing solvent A from 100% to 80% in 20 min, 70% in 35 min, and 0% in 45 min, the gradient was returned to 100% solvent A and maintained for 5 min prior to the subsequent injection. The injection volume was 10 μL. The chromatograms and spectra were processed using Agilent Technologies' OpenLab CDS software version 2.5. The calibration curves were built with the external standard method, using the least squares approach to correlate the peak area with the concentration, with $R^2 > 0.999$ for all standards in the 0.2–10.0 mg/L range and data were expressed as mg/g dry extract residue (mg/g dr). The extracts were diluted 1:5 using 0.22 M phosphoric acid for the analysis. Before injection, the solutions were filtered using a 0.22 μm CA syringe filter.

2.4. In Silico Studies

2.4.1. Protein and Ligand Preparation

In the Protein Data Bank (RCSB PDB) [33], the following two X-ray structures were selected: one corresponding to human fibroblast collagenase (PDB code: 1CGL; https:

//www.rcsb.org/structure/1CGL, accessed on 3 February 2025) and the other to collagenase H from *Clostridium histolyticum* (PDB code: 7ZOC; https://www.rcsb.org/structure/7ZOC, accessed on 3 February 2025). The structures were processed using the Protein Preparation Wizard tool (Schrödinger Release 2024-3) [34]. During this step, hydrogen atoms were added, bond orders were corrected, and formal charges were assigned. The appropriate ionisation state was determined at pH 7.4 using the PROPKA tool (https://www.ddl.unimi.it/vegaol/propka.htm, accessed on 3 February 2025) [35]. A subsequent energy minimisation was performed using the OPLS4 force field to resolve molecular overlaps and structural strains. The minimisation process was terminated when the root mean square deviation (RMSD) of the non-hydrogen atoms reached 0.3 Å. Seven compounds (epigallocatechin gallate with co-crystallised compounds used as a control and 1,5-dicaffeoyl-3-succinoylquinic acid, 1,5-dicaffeoyl-4-succinoyl quinic acid, chlorogenic acid, and luteolin-7-*O*-rutinoside) were retrieved from PubChem and prepared with LigPrep (Schrödinger Release 2024-3) using the OPLS4 force field and preserving the specified chirality.

2.4.2. Molecular Docking

Molecular docking was performed using the Glide software (Schrödinger Release 2024-3 Commercial Version) in extra precision (XP) mode, and XP docking scores were used to rank ligand binding affinities [36,37]. Three poses per ligand were retained using a threshold of 0.50 kcal/mol, and the OPLS_2005 [38] force field was employed. To create grids, the Receptor Grid Generator was utilised. The centre of the grid, with dimensions of 46 × 46 × 46 Å, was determined based on the centroid of the residues within the binding site.

2.5. Preparation of the Extract/HP-β-CD Complex

2.5.1. Preparation of the of SyEt80/HP-β-CD Complex

The complex was prepared following the lyophilisation method. Briefly, the extract (30 mg) was dissolved in 10 mL of water containing 0.5% *w*/*v* HP-β-CD. The solution was kept under continuous magnetic stirring at 500 rpm for 30 min and subsequently subjected to lyophilisation for 72 h using a Christ Alpha 2-4 LSCbasic freeze-dryer (Martin Christ Gefriertrocknungsanlagen GmbH, Osterode am Harz, Germany) to obtain a powder formulation. The lyophilised product was stored in a desiccator at room temperature until further analysis.

2.5.2. Thermogravimetric Analysis of the Complex

A Discovery SDT 650 (TA Instruments—Waters LLC, Milford, MA, USA) was used to carry out TGA on the complex, comparative to the free extract and HP-β-CD. The analyses were performed in the temperature range 25–800 °C and a heating rate of 20.0 °C/min. A nitrogen flow rate of 100.0 mL/min was used. Each sample was weighed (5–10 mg), poured onto a ceramic plate, and analysed in triplicate.

2.6. Collagenase Activity Assay

Collagenase derived from *Clostridium histolyticum* (Sigma-Aldrich, Merk Life Science S.r.l., Milan, Italy) at a concentration of 1 U/mL was prepared in a Tricine buffer containing 50 mM tricine, 10 mM calcium chloride, and 400 mM sodium chloride, with a pH adjusted to 7.5. This solution was subsequently incubated both as a free extract and as an HP-β-CD (0.5% *w*/*v*) inclusion complex at various concentrations for a duration of 15 min. Following the incubation, the synthetic substrate N-(3-[2-Furyl]-acryloyl)-Leu-Gly-Pro-Ala (FALGPA) was introduced to achieve a final concentration of 0.8 mM. The absorbance of the resulting mixture was then monitored at a wavelength of 340 nm. Epigallocatechin gallate was utilised as a positive control in this assay.

2.7. In Vivo Studies

2.7.1. Chick Chorioallantoic Membrane (CAM) Assay

The antiangiogenic activity of SyEt80, both as free extract and as a HP-β-CD (0.5% *w/v*) inclusion complex, was evaluated using the chick chorioallantoic membrane (CAM) assay, following the protocol described by Cioni et al. [39], with minor modifications. Fertilised chicken eggs were treated with the extract or the cyclodextrin complex at concentrations of 25, 50, and 100 µg/egg. Retinoic acid (1.0 µg/egg) was used as a positive control. After 48 h post-treatment the CAMs were observed using a stereomicroscope (SMZ-171 Series, Motic, Xiamen, China) equipped with a digital camera (Moticam® 5 plus). Angiogenesis was assessed by quantifying vascular branch points within a standardised area using AngioQuant (software version v1.33, 2005). Results were expressed as mean values ± standard deviation (SD).

2.7.2. Zebrafish Embryo Assay

The in vivo antiangiogenic effects of SyEt80, both in its free form and formulated with HP-β-CD (0.5% *w/v*), were further investigated in zebrafish (*Danio rerio*) embryos following the methodology reported by Cioni et al. [39]. Initially, the dechorionated embryos were treated only with CD to ensure its tolerability. At the chosen dose of 0.5% (*w/v*) it showed no signs of mortality or developmental alterations. Subsequently, dechorionated embryos were treated with the extract, either in the free form or as an inclusion complex with HP-β-CD (0.5% *w/v*) at concentrations of 10, 20, and 40 µg/embryo for 48 h. These concentrations were selected based on preliminary experimental studies aimed at identifying a biologically active range with acceptable safety margins. Concentration above 40 µg/embryo were associated with signs of developmental toxicity, whereas lower concentrations (<10 µg/embryo) did not cause a significant antiangiogenic response, even in the presence of HP-β-CD. Therefore, the selected range (10–40 µg/embryo) represents an optimal balance between efficacy and tolerability. This selection is further supported by similar dose ranges employed in zebrafish models for testing polyphenol-rich extracts and angiogenesis modulators [39]. 2-Methoxyestradiol (2 µM) was used as the positive control. All experiments were conducted in compliance with the European Directive 2010/63/EU and the ethical guidelines described in the "National Institutes of Health Guide for Care and Use of Laboratory Animals". The experiments conducted on zebrafish embryos up to 72 hpf did not require ethical approval.

2.7.3. Evaluation of Endogenous Alkaline Phosphatase (EAP) Activity in Zebrafish Embryos

The evaluation of endogenous alkaline phosphatase (EAP) activity in zebrafish embryos was performed according to the method described by De Leo et al. [40]. At 72 hpf, embryos underwent ethanol dehydration and were incubated with *p*-nitrophenyl phosphate (pNPP). The reaction was stopped with NaOH (2N), and absorbance was measured at 405 nm using a microplate spectrophotometer (AMR-100 T, ALLSHENG).

2.8. Statistical Analysis

Results were expressed as mean ± SD (standard deviation). Statistical analysis was performed using Student's *t*-test using SigmaPlot 12.0 software to compare the means obtained in the different experiments, and results with * $p < 0.05$ or ** $p < 0.01$ were considered statistically significant.

3. Results and Discussion

3.1. *Quali-Quantitative Determination of Bioactive Compounds in Wild Cardoon Leaf Extracts*

To obtain preliminary information on the composition of *C. cardunculus* L. subsp. *cardunculus* extract, the ^{1}H NMR analysis was conducted by dissolving the crude oily sample (2 μL) in 800 μL of D_2O. The spectra showed complex mixtures that did not permit precise NMR characterisation of the extracts. Therefore, the structures of the main components were identified by comparing their ^{1}H NMR data with those reported in the literature [41]. The ^{1}H NMR spectrum (Figure 1) displays a pattern of signals similar to that of 1-caffeoylquinic acid. The caffeic acid moiety was confirmed by the protons of the 1,3,4-trisubstituted aromatic ring; H-6 and H-7 protons resonate at 7.16 and 7.15 ppm, respectively, as two doublets with typical aromatic proton *ortho* coupling constants (J = 8.0 Hz), while H-2 resonates at 7.22 ppm. The remaining two H-7 and H-8 unsaturated cinnamic protons appear at 7.68 and 6.42 ppm, respectively, and resonate as two doublets with a *trans* coupling constant (J = 16.0 Hz). In the region between 1.55–2.40 ppm and 3.8–4.25 ppm, a series of multiplets characteristic of the four methylene protons and the H-3′, H-4′, and H-5′ protons of a cyclohexanecarboxylic unit can be identified, which are associated with the quinic acid moiety.

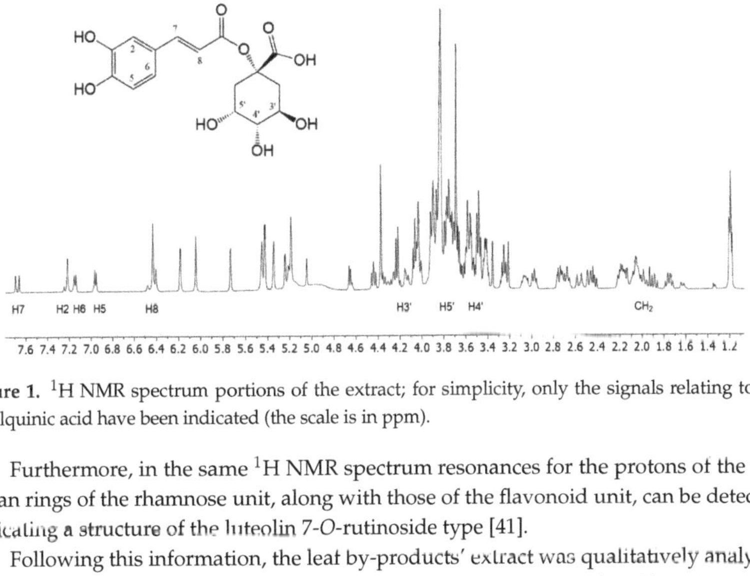

Figure 1. ^{1}H NMR spectrum portions of the extract; for simplicity, only the signals relating to caffeoylquinic acid have been indicated (the scale is in ppm).

Furthermore, in the same ^{1}H NMR spectrum resonances for the protons of the two pyran rings of the rhamnose unit, along with those of the flavonoid unit, can be detected, indicating a structure of the luteolin 7-*O*-rutinoside type [41].

Following this information, the leaf by-products' extract was qualitatively analysed by (HR) LC-ESI-QTOF MS/MS in the negative and positive ion modes. The compounds were identified by comparing the experimental MS/MS spectra with the fragmentation patterns of both pure standard compounds and those reported in the literature and in the public repository of mass spectral data, as well as by comparing the *m/z* values with those described in the literature [29–31]. Additionally, compounds were identified according to their UV-Vis spectra and quantified by measuring their absorption at distinctive wavelengths (flavonols at 360 nm, hydroxycinnamic acids at 313 nm, hydroxybenzoic acids at 280 nm). Figure 2 reports the HPLC-PDA chromatogram at 360 nm of wild cardoon leaf extract (SyEt80), and peaks are identified according to the numeration reported in Table 1.

Table 1 reports the quantification of the target compounds by the LC-PDA method (amount expressed as mg/g dr). The identified phenolic compounds belong to the classes of hydroxycinnamic acids (mono caffeoyl quinic acids (peaks 2, 3, 5, and 6), dicaffeoyl-quinic acids (peaks 8, 9, 10, and 12), dicaffeoyl-succinoyl quinic acids (peaks 13 and 14), flavonoids (luteolin derivatives, peaks 7, 11, and 16), and hydroxybenzoic acids (protocatechuic acid hexoside (peak 1) and syringic acid hexoside (peak 4)). The most abundant class of compounds was represented by hydroxycinnamic acids, showing an average value of 75.5%, followed by flavonoids and hydroxybenzoic acids (24.0% and 0.5%, respectively). Concerning the most concentrated compounds, following the MS and UV-vis data, peak 13 at rt = 24.9 min was attributed to a dicaffeoyl-succinoyl quinic acid due to the specific $[M-H]^-$ at *m/z* 615 and fragments at 515 (loss of a di-caffeoylquinic acid unit), 191 (loss of a quinic acid unit), and 179 (loss of a caffeic acid unit). By comparison with literature data on *C. cardunculus* subsp. *cardunculus* [42] and also on other varieties (var. *altilis* and var. *scolymus*) [43], peak 13 was tentatively attributed to 1,5-dicaffeoyl-3-succinoylquinic acid with an amount of 38.64 ± 0.21 mg/g dr. The second most concentrated compound was identified as luteolin 7-*O*-rutinoside (31.56 ± 0.38 mg/g dr) by comparison with spectral data and pure standard and due to the specific $[M-H]^-$ at *m/z* 593 and a fragment at 285, corresponding to the loss of the luteolin aglycone. Interestingly, other dicaffeoyl-succinoyl quinic acid derivatives were found: one is an isomer of dicaffeoyl-succinoyl quinic acid (peak 14) and was attributed to 1,5-dicaffeoyl-4-succinoyl quinic acid (14.77 ± 0.17 mg/g dr) [42,43] and the other was attributed to a dicaffeoyl-disuccinoylquinic acid (peak 15—5.49 ± 0.10 mg/g dr) [42,43].

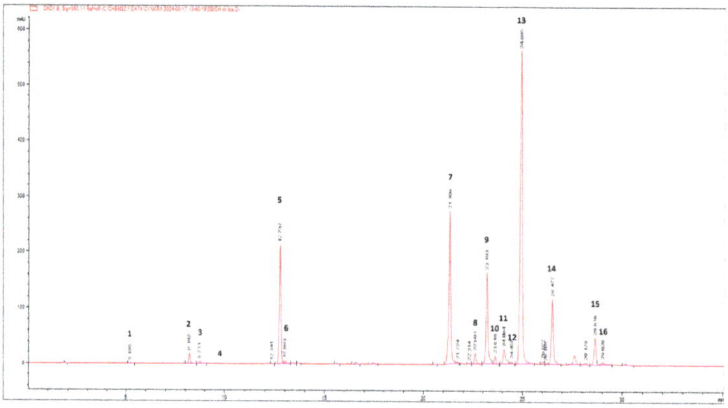

Figure 2. HPLC-PDA chromatogram at 360 nm of wild cardoon leaf extract (SyEt80). Chromatographic conditions are described in the text. Peak identifications are given in Table 1 and text.

Table 1. Concentration of targeted phenolic compounds detected in wild cardoon extract (mg/g dry extract, mean ± SD; *n* = 3).

Compound	Number Figure 2	Wild Cardoon Extract (mg/g dr)
		Mean ± SD
Total hydroxycinnamic acids		**119.33 ± 0.77**
Mono caffeoylquinic acids [a]	2, 3, 5, 6	35.14 ± 0.18
Dicaffeoylquinic acid [b]	8, 9, 10, 12	25.28 ± 0.11
Dicaffeoyl-succinoylquinic acid I [b]	13	38.64 ± 0.21
Dicaffeoyl-succinoylquinic acid II [b]	14	14.77 ± 0.17
Dicaffeoyl-disuccinoylquinic acid	15	5.49 ± 0.10
Total flavonols		**37.87 ± 0.40**
Luteolin 7-*O*-rutinoside	7	31.56 ± 0.38
Other luteolin derivatives [c]	11, 16	6.31 ± 0.02
Total hydroxybenzoic acids		**0.78 ± 0.02**
Protocatechuic acid hexoside [d]	1	0.36 ± 0.01
Syringic acid hexoside [e]	4	0.42 ± 0.01
Total polyphenols		**157.98 ± 1.19**

[a] Dosed as chlorogenic acid equivalents; [b] dosed as 1,3-di-*O*-caffeoylquinic acid equivalents; [c] dosed as luteolin equivalents; [d] dosed as protocatechuic acid equivalents; [e] dosed as syringic acid equivalents.

Several isomers of monocaffeoyl quinic acids and dicaffeoyl quinic acids were found, accounting for a total of 35.14 ± 0.18 and 25.28 ± 0.11 mg/g dr, respectively. Chlorogenic acid (5-caffeoylquinic acid, peak 5) accounted for ca. 90% of monocaffeoyl quinic acids with an amount of 31.33 ± 0.16 mg/g dr. Among flavonoids, other luteolin derivatives stand out (6.31 ± 0.02 mg/g dr). Hydroxybenzoic acids were detected in lower quantity, and the most representative were protocatechuic acid hexoside and syringic acid hexoside (0.36 ± 0.01 and 0.42 ± 0.01 mg/g dr, respectively).

Pandino et al. [44] found in leaves several mono-caffeoyl quinic acids luteolin derivatives such as luteolin rutinoside, confirming them as the most representative compounds. Interestingly, Pandino et al. [44] found monosuccinyl-dicaffeoyl quinic acids, in contrast to the dicaffeoyl-succinoyl quinic acid we found. Regarding dicaffeoyl-succinoyl quinic acids and dicaffeoyl-disuccinoyl quinic acid, they were previously found in wild cardoon leaves [42]. They were also found in other parts of the plant *C. cardunculus* [11], such as var. *altilis* stalks [43] and var. *scolymus* edible sprouts [11,43]. Notably, Pinelli et al. [42] analysed the hydroalcoholic leaf extract of wild cardoon, pointing out that the highest average value among the different classes of compounds belongs to dicaffeoyl-succinoyl quinic acids (32.5%), confirming 1,5-dicaffeoyl-3-succinoyl quinic acid as the most abundant (66.25 ± 8.79 µmol/d wt) followed by 1,5-dicaffeoyl-4-succinoyl quinic acid and 1,5-dicaffeoyl-3,4-disuccinoyl quinic acid (16.39 ± 0.36 and 6.51 ± 3.28 µmol/d wt, respectively), confirming the trend highlighted in our study. Pagliari et al. [45] confirmed hydroxycinnamic acids as the most representative classes of compounds; otherwise, among flavonoids, luteolin 7-*O*-glucoside and luteolin 7-*O*-rutinoside were the most abundant (11.99 ± 0.25 and 5.11 ± 0.16 µg/mg EXT in ethanolic extracts), in contrast with luteolin 7-*O*-rutinoside as the most abundant flavonoid in our study (31.56 ± 0.38 mg/g dr).

3.2. Docking Results

Molecular docking studies were conducted to evaluate the binding affinity between target phenolic compounds detected in the hydroalcoholic extract of wild cardoon and collagenase. The crystallographic structures used were human collagenase [46] with code: 1CGL and *Clostridium histolyticum* collagenase H [47] with code: 7ZOC. The co-crystallised ligands present in both structures were redocked as positive controls. The results showed good agreement between the predicted and experimental poses, with an all-atom RMSD of 1.5 Å for 1CGL and 1.3 Å for 7ZOC, confirming the reliability of the docking protocol used. The RMSD values were calculated considering all atoms of the ligand. Additionally, epigallocatechin gallate (EGCG), a compound known in the literature as a collagenase inhibitor, was also used as a further positive control to support the validity of the method. The selected compounds were the two dicaffeoylsuccinoylquinic acids (1,5-dicaffeoyl-3-succinoylquinic acid and 1,5-dicaffeoyl-4-succinoylquinic acid), luteolin 7-*O*-rutinoside, and chlorogenic acid that accounted for 74% of the phenolic compound dosed in the extract. All results of docking are shown in Table 2.

Table 2. Docking results for the interactions of compounds with the human collagenase (1CGL code) and *Clostridium histolyticum* collagenase (7ZOC code). The results are all expressed in kcal/mol.

Compound	XP-Score 1CGL	XP-Score 7ZOC
Epigallocatechin gallate (EGCG)	−2.9	−6.1
1,5-dicaffeoyl-3-succinoylquinic acid	−6.6	−3.4
1,5-dicaffeoyl-4-succinoyl quinic acid	−5.2	−12.1
Luteolin 7-*O*-rutinoside	−7.1	−11.7

Table 2. *Cont.*

Compound	XP-Score 1CGL	XP-Score 7ZOC
Chlorogenic acid	−8.1	−9.2
Ligand 1CGL	−8.1	
Ligand 7ZOC		−6.1

For human collagenase, the co-crystallised ligand with chlorogenic acid achieved the best XP GScore (−8.1 kcal/mol), followed by luteolin 7-*O*-rutinoside (−7.1 kcal/mol), 1,5-dicaffeoyl-3-succinoylquinic acid (−6.6 kcal/mol), 1,5-dicaffeoyl-4 -succinoylquinic acid (−5.2 kcal/mol), and EGCG (−2.9 kcal/mol). For *C. histolyticum* collagenase, 1,5-dicaffeoyl-4-succinoyl quinic acid showed the best XP GScore (−12.1 kcal/mol), followed by luteolin 7-*O*-rutinoside (−11.7 kcal/mol), and chlorogenic acid (−9.2 kcal/mol). The co-crystallised ligand and EGCG both scored –6.1 kcal/mol, while 1,5-dicaffeoyl-3-succinoylquinic acid had the lowest binding score (–3.4 kcal/mol).

All compounds bind on the same binding site for both collagenases used (Figure 3).

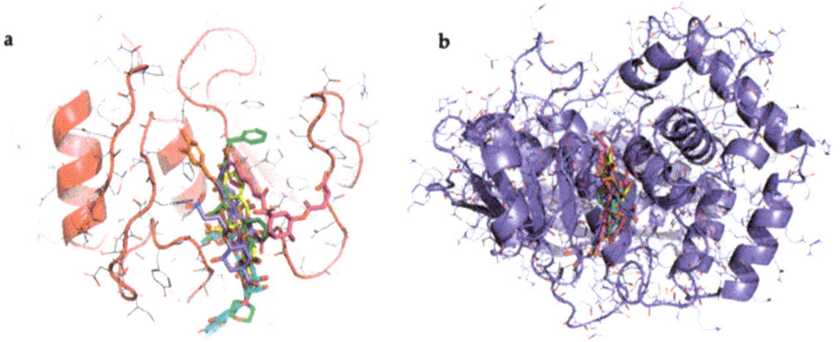

Figure 3. Binding mode of all compounds in collagenase: (**a**) representation of the binding mode of all compounds in human collagenase. In green: ligand 1CGL, in magenta: 1,5-dicaffeoyl-3-succinoylquinic acid, in purple: 1,5-dicaffeoyl-4-succinoylquinic acid, in yellow: epigallocatechin gallate, in cyan: luteolin 7-*O*-rutinoside, in orange: chlorogenic acid. (**b**) Representation of the binding mode of all compounds in collagenase from *Clostridium histolyticum*. In green: ligand 7ZOC, in magenta: 1,5-dicaffeoyl-3-succinoylquinic acid, in purple: 1,5-dicaffeoyl-4-succinoylquinic acid, in yellow: epigallocatechin gallate, in cyan: luteolin 7-*O*-rutinoside, in orange: chlorogenic acid.

Table 2 shows that there is a high binding affinity between three of the compounds identified in the extract and *C. histolyticum* collagenase, the same one used for the experiments. These values are higher than that of the standard inhibitor. The presence of these compounds in the wild cardoon extract may help explain its better IC_{50} value compared to that of the standard ECG inhibitor in experiments with *C. histolyticum* collagenase. The data relating to the interaction between the compounds and human collagenase show that all the compounds exhibit values higher than that of EGCG. The data obtained for luteolin-7-*O*-rutinoside and chlorogenic acid agree with previously published results [48,49]. Taken together, the results help to explain the inhibitory activity of the wild cardoon extract, which contains a high quantity of these compounds.

3.3. Preparation of the Complex

The ability of CDs to act as penetration enhancers for synthetic drugs or natural molecules was now recognised [50]. Starting from this point of view, we prepared the complex between the extract and HP-β-CD to evaluate the influence of complexation on in vitro and in vivo activity of the cardoon extracts. HP-β-CD is a modified CD approved by the FDA for parenteral administration. Currently, it is present in various marketed formulations for oral, intravenous, and ophthalmic administration [51]. The solid complex was prepared by lyophilisation and analysed by thermogravimetric analysis to evaluate the influence of complexation on thermal stability of the extract. Figure 4 reports the obtained scans. The degradation of the free extract proceeds in different steps due to the complexity of its composition and the different thermal stability of the extract's components. The first portion of the thermogram, below 130 °C, showed a very low weight loss (about 4%) due to the dehydration of the sample; then, the successive steps starting from 130 °C and finishing at about 540 °C showed the weight loss attributable to the degradation of the organic compounds and the oligosaccharides present into the extract [52], leaving a residue of about 40%, evidencing the low thermal stability of the components [53].

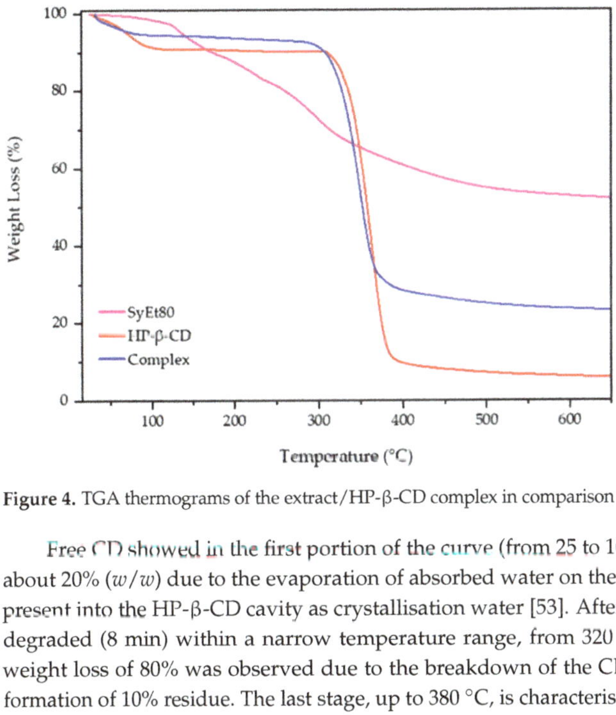

Figure 4. TGA thermograms of the extract/HP-β-CD complex in comparison with the free components.

Free CD showed in the first portion of the curve (from 25 to 100 °C) a weight loss of about 20% (*w*/*w*) due to the evaporation of absorbed water on the macrocycle surface or present into the HP-β-CD cavity as crystallisation water [53]. After that, CD was quickly degraded (8 min) within a narrow temperature range, from 320 to 380 °C, in which a weight loss of 80% was observed due to the breakdown of the CD ring structure and a formation of 10% residue. The last stage, up to 380 °C, is characterised by the slow thermal degradation of the residue. A different trend with respect to the free extract was observed for the complex. First of all, no degradation of different compounds was observed, and the main thermal degradation of the complex proceeded in one step from 300 to 370 °C, leaving a residue of about 30%. This is the result of the complexation and improved thermal stability of the extract. Furthermore, the initial weight loss, at a temperature lower than 75 °C, is of about 10%. This different trend with respect to the free macrocycle could evidence the absence of crystallisation water; thus, the weight loss was exclusively due to the superficial water evaporation. It is conceivable that some components of the extract were included within the cavity, surprising the water molecules allocated here [54]. These results evidenced a positive influence of HP-β-CD on the thermal stability of the extract, confirming its complexation.

3.4. Collagenase Activity

In this study we investigated the potential of wild cardoon (SyEt80) extract to inhibit collagenase, considering the relationship between angiogenesis and collagenase activity. The IC_{50} value obtained from the free extract and in combination with hydroxypropyl-β-cyclodextrin (HP-β-CD, 0.5% *w*/*v*) (Table 3) was compared to those of the standard inhibitor to assess the extract's inhibitory strength.

Table 3. Inhibitory effects (IC_{50}) of SyEt80 extract and SyEt80/HP-β-CD on collagenase enzymatic activity. Epigallocatechin gallate is reported as standard inhibitor. All data represent the mean ± SD of three independent experiments.

Sample	IC_{50} (μg/mL)
SyEt80 extract	148.8 ± 0.5
SyEt80/HP-β-CD	26.8 ± 2.0
Epigallocatechin gallate	120.8 ± 6.2

The IC_{50} value calculated for the SyEt80 extract was found to be 148 μg/mL, comparable with the standard inhibitor EGCG which has an IC_{50} value of 121 μg/mL. A more significant result was achieved with the complexed extract SyEt80/HP-β-CD, which enhanced its inhibitory activity against collagenase by 6-fold compared to the extract alone and 4.5 times more than the single molecule of EGCG. These findings demonstrate that complexation with cyclodextrin significantly enhances the extract's inhibitory activity, most likely due to improved solubility and stability of its key phenolic compounds.

3.5. Chick Chorioallantoic Membrane (CAM) Assay

The antiangiogenic activity of the hydroalcoholic extract from wild cardoon leaves (SyEt80) was investigated using the chick chorioallantoic membrane (CAM) assay, both as a free extract and in combination with HP-β-CD (0.5% *w*/*v*). As shown in Figure 5, treatment with the free extract induced a dose-dependent inhibition of neovascularisation, with inhibition values of 12.20%, 18.91%, and 27.02% at doses of 25, 50, and 100 μg/egg, respectively.

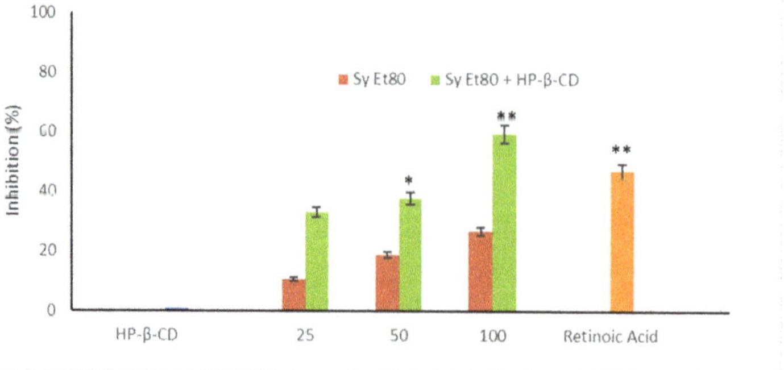

Figure 5. Antiangiogenic activity of SyEt80 and of SyEt80 complexed with HP-β-CD (0.5%, *w*/*v*) in the CAM assay. Results are expressed as % inhibition vs. negative control. Retinoic acid (1 μg/egg) = positive control. Mean ± SD ($n = 10$). * $p < 0.05$; ** $p < 0.01$ vs. control: Student's *t*-test.

The inclusion complex SyEt80/HP-β-CD demonstrated significantly higher antiangiogenic activity, reaching inhibition percentages of 33.33%, 37.89%, and 59.45% at the same concentrations (Figure 5). Retinoic acid was used as the positive control (1 μg/egg, 44.55% inhibition). These results confirm the superior efficacy of the cyclodextrin-complexed formulation, which showed a statistically significant increase in antiangiogenic activity compared to the free extract. The improved efficacy observed in the complexed formulation supports that cyclodextrin-based delivery enhances the solubility, chemical stability, and bioavailability of the active constituents, thereby potentiating their biological effect. Representative photomicrographs of the CAMs are reported in Figure 6. As can be seen, in the control CAM there is a thick vascular network with large vessels (Figure 6A). Conversely, the retinoic acid (Figure 6B) and SyEt80/HP-β-CD (Figure 6D,F,H) treatments produced a less-dense vascular network due to the strong inhibitory effects on capillary formation.

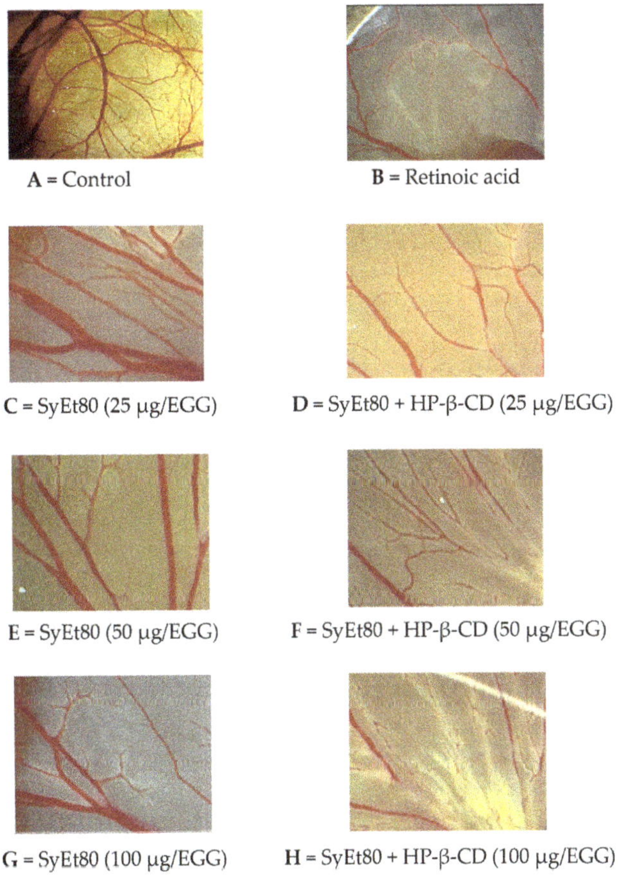

Figure 6. Representative images of the chick chorioallantoic membrane (CAM) assay showing the effects of SyEt80 extract and its complex with HP-β-CD on angiogenesis at different doses. (**A**) Control; (**B**) retinoic acid (positive control); (**C–E**) SyEt80 and (**F–H**) SyEt80 complexed with HP-β-CD at 25, 50, and 100 μg/egg, respectively. The images illustrate a dose-dependent antiangiogenic effect of the extract, which is enhanced by HP-β-CD complexation (scale bar = 1000 μm).

These findings are particularly significant as they demonstrate not only the intrinsic antiangiogenic potential of wild cardoon leaf constituents but also the pivotal role of formulation strategies in modulating biological effects. The dose-dependent inhibition observed with the free extract confirms the presence of active phytochemicals able to interfere with neovascular processes. However, the marked enhancement observed with the HP-β-CD inclusion complex suggests that a considerable fraction of these bioactive compounds is otherwise limited by poor solubility or instability in aqueous environments. This aligns with literature data indicating that polyphenols such as dicaffeoylquinic acids and luteolin derivatives, abundant in the extract, benefit substantially from cyclodextrin encapsulation in terms of bioavailability and biological performance [55]. From a clinical perspective, this formulation approach not only improves the therapeutic applicability but also offers a rational strategy for the valorisation of agricultural by-products through the development of innovative, plant-based antiangiogenic agents.

3.6. *In Vivo Antiangiogenic Potential on Zebrafish Model*

The antiangiogenic potential of the wild cardoon leaf extracts (SyEt80) and the inclusion complex with HP-β-CD (0.5% *w*/*v*) were further assessed in vivo using the zebrafish (*Danio rerio*) embryo model by measuring endogenous alkaline phosphatase (EAP) activity as a biochemical marker of vascular development. Results are reported in Figure 7.

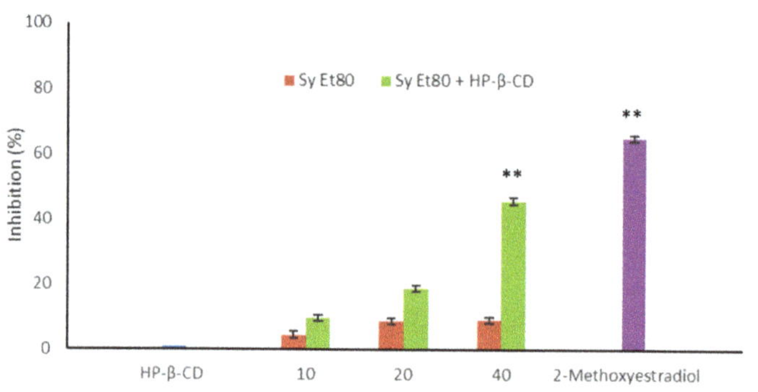

Figure 7. Antiangiogenic activity of SyEt80 (10, 20, and 40 μg/embryo) free and complexed with HP-β-CD (0.5%, *w*/*v*) in zebrafish embryos. Results are expressed as % inhibition vs. negative control. 2-methoxyestradiol (2 μM) = positive control. Mean ± SD (n = 10). ** $p < 0.01$ vs. negative control: Student's *t*-test.

Treatment with the free extract resulted in modest inhibitory activity, with values of 4.54%, 8.77%, and 8.03% at concentrations of 10, 20, and 40 μg/embryo, respectively. In contrast, embryos treated with the SyEt80/HP-β-CD complex exhibited a clear dose-dependent inhibition of EAP activity, with values of 9.80%, 18.15%, and 45.75% at the same concentrations. 2-Methoxyestradiol was used as a positive control (2 μM, 63.00% inhibition).

These results demonstrate that complexation with HP-β-CD markedly enhances the extract's antiangiogenic efficacy in the zebrafish model, likely due to the improved delivery and bioavailability of its bioactive constituents. The significant increase in EAP inhibition observed with the HP-β-CD complex—especially at the highest dose—highlights how formulation strategies can profoundly influence in vivo pharmacodynamics. Considering the modest effect shown by the unformulated extract, it is reasonable to assume that the limited solubility or rapid degradation of polyphenolic compounds in aqueous solutions may compromise their biological activity in vivo. The improved performance of the inclusion complex not only confirms the functional relevance of these phenolic constituents but also underscores the value of delivery systems in addressing the solubility and stability limitations of plant-derived bioactive constituents. This evidence supports a rational, formulation-centred approach to the development of novel antiangiogenic agents from agro-industrial by-products.

Importantly, during zebrafish experiments, no signs of toxicity—such as pericardial oedema, developmental delays, or morphological malformations—were observed with either the free extract (SyEt80) or the HP-β-CD inclusion complex at the tested concentrations (10–40 μg/embryo). Embryos exhibited normal motility and developmental progression up to 72 hpf. This suggests that the antiangiogenic effects of the SyEt80/HP-β-CD complex occur without compromising embryo development. This toxicity profile supports the selection of the current dose range and reinforces the potential of the extract for future applications in the nutraceutical or functional food domain, where safety is a critical parameter.

Taken together, the findings from the CAM and zebrafish in vivo assays represent a clear demonstration of the antiangiogenic potential of wild cardoon leaf extract. The marked inhibition of neovascularisation in the CAM model and the significant suppression of vascular development biomarkers in zebrafish embryos both support a consistent biologically relevant effect. Notably, the improved efficacy of the HP-β-CD inclusion complex across both models highlights the importance of formulation strategies in maximising the therapeutic potential of plant-derived phytocomplexes.

The strength and reproducibility of these in vivo results provide evidence that the bioactive phenolic constituents of wild cardoon—particularly dicaffeoyl-succinoyl quinic acids and luteolin derivatives—are capable of modulating angiogenic mechanisms in whole-organism contexts. These findings support the potential of the extract as a promising candidate for further development in the prevention or co-treatment of angiogenesis-driven disorders. Furthermore, the consistency of results across two independent in vivo models reinforces the therapeutic relevance of the observed effects and sets the groundwork for future clinical validation.

The antiangiogenic effects observed in both the CAM and zebrafish models for SyEt80 and its inclusion complex with HP-β-CD may be mechanistically linked, at least in part, to the significant inhibition of collagenase activity. Matrix metalloproteinases (MMPs), particularly those with collagenolytic activity, are crucial mediators of the extracellular matrix (ECM) remodelling required for endothelial cell migration and vessel formation. This process represents a key step in the angiogenic cascade and is extensively documented in both physiological and pathological settings [14,15]. In our study, collagenase from *C. histolyticum* was employed as a reference enzyme due to its broad collagen-degrading activity and widespread use in preliminary screening assays for collagenase inhibitors, although it differs structurally from human matrix metalloproteinases.

In this context, results clearly demonstrate that SyEt80 and its HP-β-CD complex interfere with collagenase activity. This interpretation is further supported by molecular docking analyses performed on both bacterial and human collagenase structures, which revealed that the major phenolic constituents of the extract bind to the same binding site for both collagenases used. This mechanistic hypothesis is consistent with prior studies showing that angiogenesis is significantly reduced when collagenase-mediated ECM degradation is suppressed [14], and is further supported by comprehensive reviews on the multifunctional role of MMPs in neovascularization.

However, collagenase inhibition is unlikely to be the sole explanation for the observed antiangiogenic effects. Polyphenolic compounds, which are abundantly present in SyEt80, are known to modulate angiogenesis through a range of multiple and interconnected mechanisms—such as the inhibition of VEGF signalling, antioxidant activity, and the downregulation of inflammatory mediators [12,15]. Therefore, it is plausible that the anti-angiogenic activity of the extract is the result of a multi-targeted mechanism, in which collagenase inhibition represents a relevant, but not exclusive, contributor.

4. Conclusions

The in vitro and in vivo findings of this study provide evidence of the anti-angiogenic potential of *Cynara cardunculus* L. subsp. *cardunculus* (wild cardoon) leaf extracts. Notably, the inclusion complex of the extract with hydroxypropyl-β-cyclodextrin markedly enhanced the inhibition of collagenase activity and angiogenic process. The improved activity of the complexed extract is likely due to the increased solubility and stability of key phenolic constituents.

The consistent inhibition observed across both systems highlights the therapeutic relevance of this phytocomplex and the importance of effective formulation strategies. These results suggest that wild cardoon leaf by-products, often considered agricultural wastes, may represent a sustainable and effective source of collagenase inhibitors and antiangiogenic agents. These results could be used for future applications in the management of collagenase and angiogenesis-related disorders. Furthermore, the valorisation of wild cardoon leaf by-products offers promising opportunities beyond pharmaceutical applications. Due to their high content of health-promoting phenolic compounds and favourable profile, these by-products are also suitable for incorporation into functional food products or nutraceutical formulations designed to support vascular health. This dual potential not only broadens their scope of application but also aligns with the principles of circular bioeconomy, providing a sustainable model to transform agro-industrial wastes into high-value bioactive ingredients with significant health benefits.

Author Contributions: Conceptualization, A.F., C.A.V., M.P.G., and C.I.G.T.; methodology, A.F., C.A.V., M.P.G., F.D.G., R.S., and C.I.G.T.; software, S.O., R.S., and V.M.; validation, M.P.G. and C.I.G.T.; formal analysis, A.F., V.P., V.M., C.I.G.T., V.D., F.D.G., R.S., and A.C.; investigation, A.C., V.D., F.D.G., R.S., and S.O.; resources, A.F., M.P.G., and C.I.G.T.; data curation, A.F., M.P.G., A.C., V.D., F.D.G., and V.M.; writing—original draft preparation, A.F., V.P., C.A.V., M.P.G., V.M., C.I.G.T., A.C., and V.D.; writing—review and editing, M.P.G., V.M., and C.I.G.T.; supervision, A.F., C.A.V., M.P.G., and C.I.G.T.; project administration, A.F., V.P., M.P.G., and C.I.G.T.; funding acquisition, V.P., M.P.G., and C.I.G.T. All authors have read and agreed to the published version of the manuscript.

Funding: This research was funded by the European Union—NextGeneration EU Project PRIN 2022 PNRR (P2022SECH4_002) "Exploiting Cynara spp. by-products as a source of molecules with inhibitory activity on skin-related enzymes. CYNINSKINE". This publication was produced while Valentina Masala attended the PhD programme in Life, Environmental and Drug Science at the University of Cagliari, Cycle XXXVIII, with the support of a scholarship financed by the Ministerial Decree no. 351 of 9 April 2022, based on the NRRP—funded by the European Union—NextGenerationEU—Mission 4 "Education and Research", Component 1 "Enhancement of the offer of educational services: from nurseries to universities"—Investment 4.1 "Extension of the number of research doctorates and innovative doctorates for public administration and cultural heritage".

Institutional Review Board Statement: All experiments were conducted in compliance with the European Directive 2010/63/EU and the ethical guidelines described in the "National Institutes of Health Guide for Care and Use of Laboratory Animals". The experiments conducted on zebrafish embryos up to 72 hpf did not require ethical approval.

Informed Consent Statement: Not applicable.

Data Availability Statement: The original contributions presented in the study are included in the article, further inquiries can be directed to the corresponding authors.

Acknowledgments: The authors acknowledge the CeSAR (Centro Servizi d'Ateneo per la Ricerca) core facility of the University of Cagliari (Cagliari, Sardinia, Italy) for the experiments performed with Agilent 6560 IM-QTOF and Giulio Ferino for assistance with the generation of LC–MS data. For contributing to collecting wild cardoon leaves waste, the authors would like to express their gratitude to Sergio Talloru.

Conflicts of Interest: The authors declare no conflicts of interest.

References

1. Fernandes, F.; Delerue-Matos, C.; Grosso, C. Unveiling the potential of agrifood by-products: A comprehensive review of phytochemicals, bioactivities and industrial applications. *Waste Biomass Valoriz.* **2025**, *16*, 2715–2748. [CrossRef]
2. Sorrenti, V.; Burò, I.; Consoli, V.; Vanella, L. Recent advances in health benefits of bioactive compounds from food wastes and by-products: Biochemical aspects. *Int. J. Mol. Sci.* **2023**, *24*, 2019. [CrossRef]
3. Yadav, S.; Malik, K.; Moore, J.M.; Kamboj, B.R.; Malik, S.; Malik, V.K.; Arya, S.; Singh, K.; Mahanta, S.; Bishnoi, D.K. Valorisation of agri-food waste for bioactive compounds: Recent trends and future sustainable challenges. *Molecules* **2024**, *29*, 2055. [CrossRef]
4. Amoriello, T.; Mellara, F.; Ruggeri, S.; Ciorba, R.; Ceccarelli, D.; Ciccoritti, R. Artichoke by-products valorization for phenols-enriched fresh egg pasta: A sustainable food design project. *Sustainability* **2022**, *14*, 14778. [CrossRef]
5. Cravero, V.; Martin, E.; Crippa, I.; López Anido, F.; García, S.M.; Cointry, E. Fresh biomass production and partitioning of aboveground growth in the three botanical varieties of *Cynara cardunculus* L. *Ind. Crops Prod.* **2012**, *37*, 253–258. [CrossRef]
6. Uçar Türker, A.; Başay, Ş.; Çimen, A.; Baba, Y.; Birinci Yıldırım, A. Evaluation of the phenolic content and the nutraceutical potential of ancestor and cultivated artichokes. *Chem. Biodiversity* **2024**, *21*, e202400203. [CrossRef] [PubMed]
7. Pandino, G.; Courts, F.L.; Lombardo, S.; Mauromicale, G.; Williamson, G. Caffeoylquinic acids and flavonoids in the immature Inflorescence of globe artichoke, wild cardoon, and cultivated cardoon. *J. Agric. Food Chem.* **2010**, *58*, 1026–1031. [CrossRef] [PubMed]
8. Bouyahya, A.; Taha, D.; Benali, T.; Zengin, G.; El Omari, N.; El Hachlafi, N.; Khalid, A.; Abdalla, A.N.; Ardianto, C.; Tan, C.S.; et al. Natural sources, biological effects, and pharmacological properties of cynaroside. *Biomed. Pharmacother.* **2023**, *161*, 114337. [CrossRef]
9. De Falco, B.; Incerti, G.; Amato, M.; Lanzotti, V. Artichoke: Botanical, agronomical, phytochemical, and pharmacological overview. *Phytochem. Rev.* **2015**, *14*, 993–1018. [CrossRef]
10. Acquaviva, R.; Malfa, G.A.; Santangelo, R.; Bianchi, S.; Pappalardo, F.; Taviano, M.F.; Miceli, N.; Di Giacomo, C.; Tomasello, B. Wild artichoke (*Cynara cardunculus* subsp. *sylvestris*, Asteraceae) leaf extract: Phenolic profile and oxidative stress inhibitory effects on Hpg2 cells. *Molecules* **2023**, *28*, 2475. [CrossRef] [PubMed]
11. Silva, L.R.; Jacinto, T.A.; Coutinho, P. Bioactive compounds from cardoon as health promoters in metabolic disorders. *Foods* **2022**, *11*, 336. [CrossRef]
12. Velez, Z.; Campinho, M.A.; Guerra, Â.R.; García, L.; Ramos, P.; Guerreiro, O.; Felício, L.; Schmitt, F.; Duarte, M. Biological characterization of *Cynara cardunculus* L. methanolic extracts: Antioxidant, anti-proliferative, anti-migratory and anti-angiogenic activities. *Agriculture* **2012**, *2*, 472–492. [CrossRef]
13. Adair, T.H.; Montani, J.P. *Angiogenesis*; Morgan & Claypool Life Sciences: San Rafael, CA, USA, 2010. [PubMed]
14. Zijlstra, A.; Aimes, R.T.; Quigley, J.P. Collagenolysis-dependent angiogenesis mediated by matrix metalloproteinase-13 (collagenase-3). *J. Biol. Chem.* **2004**, *279*, 27633–27645. [CrossRef] [PubMed]
15. Raza, S.L.; Cornelius, L.A. Matrix metalloproteinases: Pro- and anti-angiogenic activities. *J. Investig. Dermatol. Symp. Proc.* **2000**, *5*, 47–54. [CrossRef] [PubMed]
16. De Gaetano, F.; Mannino, D.; Celesti, C.; Bulzomí, M.; Iraci, N.; Giofrè, S.V.; Esposito, E.; Paterniti, I.; Ventura, C.A. Randomly methylated β-cyclodextrin improves water—Solubility, cellular protection and mucosa permeability of idebenone. *Int. J. Pharm.* **2024**, *665*, 124718. [CrossRef]
17. De Gaetano, F.; Scala, A.; Celesti, C.; Lambertsen Larsen, K.; Genovese, F.; Bongiorno, C.; Leggio, L.; Iraci, N.; Iraci, N.; Mazzaglia, A.; et al. Amphiphilic cyclodextrin nanoparticles as delivery system for idebenone: A preformulation study. *Molecules* **2023**, *28*, 3023. [CrossRef]
18. De Gaetano, F.; Leggio, L.; Celesti, C.; Genovese, F.; Falcone, M.; Giofrè, S.V.; Iraci, N.; Ventura, C.A. Study of host-guest interaction and in vitro neuroprotective potential of cinnamic acid/randomly methylated β-cyclodextrin inclusion complex. *Int. J. Mol. Sci.* **2024**, *25*, 12778. [CrossRef] [PubMed]
19. De Gaetano, F.; Margani, F.; Barbera, V.; D'Angelo, V.; Germanò, M.P.; Pistarà, V.; Ventura, C.A. Characterization and in vivo antiangiogenic activity evaluation of morin-based cyclodextrin inclusion complexes. *Pharmaceutics* **2023**, *15*, 2209. [CrossRef]

20. Spada, G.; Gavini, E.; Cossu, M.; Rassu, G.; Carta, A.; Giunchedi, P. Evaluation of the effect of hydroxypropyl-β-cyclodextrin on topical administration of milk thistle extract. *Carbohydr. Polym.* **2013**, *92*, 40–47. [CrossRef]
21. Kali, G.; Haddadzadegan, S.; Bernkop-Schnürch, A. Cyclodextrins and derivatives in drug delivery: New developments, relevant clinical trials, and advanced products. *Carbohydr. Polym.* **2024**, *324*, 121500. [CrossRef]
22. Christaki, S.; Spanidi, E.; Panagiotidou, E.; Athanasopoulou, S.; Kyriakoudi, A.; Mourtzinos, I.; Gardikis, K. Cyclodextrins for the delivery of bioactive compounds from natural sources: Medicinal, food and cosmetics applications. *Pharmaceuticals* **2023**, *16*, 1274. [CrossRef] [PubMed]
23. Szente, L.; Fenyvesi, É. Cyclodextrin-lipid complexes: Cavity size matters. *Struct. Chem.* **2017**, *28*, 479–492. [CrossRef]
24. Nicolaescu, O.E.; Ionescu, C.; Samide, A.; Tigae, C.; Spînu, C.I.; Oprea, B. Advancements in cyclodextrin complexes with bioactive secondary metabolites and their pharmaceutical applications. *Pharmaceutics* **2025**, *17*, 506. [CrossRef] [PubMed]
25. Pinho, E.; Grootveld, M.; Soares, G.; Henriques, M. Cyclodextrins as encapsulation agents for plant bioactive compounds. *Carbohydr. Polym.* **2014**, *101*, 121–135. [CrossRef]
26. Zhong, Y.; Li, W.; Ran, L.; Hou, R.; Han, P.; Lu, S.; Wang, Q.; Zhao, W.; Zhu, Y.; Dong, J. Inclusion complexes of tea polyphenols with HP-β-cyclodextrin: Preparation, characterization, molecular docking, and antioxidant activity. *J. Food Sci.* **2020**, *85*, 1105–1113. [CrossRef]
27. Li, N.; Wang, N.; Wu, T.; Qiu, C.; Wang, X.; Jiang, S.; Zhang, Z.; Liu, T.; Wei, C.; Wang, T. Preparation of curcumin-hydroxypropyl-β-cyclodextrin inclusion complex by cosolvency-lyophilization procedure to enhance oral bioavailability of the drug. *Drug Dev. Ind. Pharm.* **2018**, *44*, 1966–1974. [CrossRef]
28. De Luca, M.; Tuberoso, C.I.G.; Pons, R.; García, M.T.; Morán, M.D.C.; Ferino, G.; Vassallo, A.; Martelli, G.; Caddeo, C. Phenolic fingerprint, bioactivity and nanoformulation of *Prunus spinosa* L. fruit extract for skin delivery. *Pharmaceutics* **2023**, *15*, 1063. [CrossRef]
29. Masala, V.; Jokić, S.; Aladić, K.; Molnar, M.; Casula, M.; Tuberoso, C.I.G. Chemical profiling and evaluation of antioxidant activity of artichoke (*Cynara cardunculus* var. *scolymus*) leaf by-products' extracts obtained with green extraction techniques. *Molecules* **2024**, *29*, 4816. [CrossRef]
30. Dührkop, K.; Fleischauer, M.; Ludwig, M.; Aksenov, A.A.; Melnik, A.V.; Meusel, M.; Dorrestein, P.C.; Rousu, J.; Böcker, S. SIRIUS4: A rapid tool for turning tandem mass spectra into metabolite structure information. *Nat. Methods* **2019**, *16*, 299–302. [CrossRef]
31. Hoffmann, M.A.; Nothias, L.F.; Ludwig, M.; Fleischauer, M.; Gentry, E.C.; Witting, M.; Dorrestein, P.C.; Dührkop, K.; Böcker, S. High-confidence structural annotation of metabolites absent from spectral libraries. *Nat. Biotechnol.* **2022**, *40*, 411–421. [CrossRef]
32. Schmid, R.; Heuckeroth, S.; Korf, A.; Smirnov, A.; Myers, O.; Dyrlund, T.S.; Bushuiev, R.; Murray, K.J.; Hoffmann, N.; Lu, M.; et al. Integrative analysis of multimodal mass spectrometry data in MZmine 3. *Nat. Biotechnol.* **2023**, *41*, 447–449. [CrossRef] [PubMed]
33. Berman, H.M.; Westbrook, J.; Feng, Z.; Gilliland, G.; Bhat, T.N.; Weissig, H.; Shindyalov, I.N.; Bourne, P.E. The protein data bank. *Nucleic Acids Res.* **2000**, *28*, 235–242. [CrossRef]
34. Madhavi Sastry, G.; Adzhigirey, M.; Day, T.; Annabhimoju, R.; Sherman, W. Protein and ligand preparation: Parameters, protocols, and Influence on virtual screening enrichments. *J. Comput. Aided Mol. Des.* **2013**, *27*, 221–234. [CrossRef] [PubMed]
35. Søndergaard, C.R.; Olsson, M.H.M.; Rostkowski, M.; Jensen, J.H. Improved treatment of ligands and coupling effects in empirical calculation and rationalization of p K a values. *J. Chem. Theory Comput.* **2011**, *7*, 2284–2295. [CrossRef]
36. Friesner, R.A.; Murphy, R.B.; Repasky, M.P.; Frye, L.L.; Greenwood, J.R.; Halgren, T.A.; Sanschagrin, P.C.; Mainz, D.T. Extra precision glide: Docking and scoring incorporating a model of hydrophobic enclosure for protein-ligand complexes. *J. Med. Chem.* **2006**, *49*, 6177–6196. [CrossRef]
37. Friesner, R.A.; Banks, J.L.; Murphy, R.B.; Halgren, T.A.; Klicic, J.J.; Mainz, D.T.; Repasky, M.P.; Knoll, E.H.; Shelley, M.; Perry, J.K.; et al. Glide: A new approach for rapid, accurate docking and scoring. 1. Method and assessment of docking accuracy. *J. Med. Chem.* **2004**, *47*, 1739–1749. [CrossRef]
38. Banks, J.L.; Beard, H.S.; Cao, Y.; Cho, A.E.; Damm, W.; Farid, R.; Felts, A.K.; Halgren, T.A.; Mainz, D.T.; Maple, J.R.; et al. Integrated Modeling Program, Applied Chemical Theory (IMPACT). *J. Comput. Chem.* **2005**, *26*, 1752–1780. [CrossRef]
39. Cioni, E.; Apel, C.; Olivon, F.; Dugo, P.; Fearnley, L.; Sajeva, M.; Tundis, R.; Dugo, G. Exploring the chemical composition, the antioxidant, and the antiangiogenic potential of ancient onion (*Allium cepa* L.) varieties through molecular networks and chemometric tools. *Food Res. Int.* **2025**, *215*, 116692. [CrossRef]
40. De Leo, M.; Iannuzzi, A.M.; Germanò, M.P.; D'Angelo, V.; Camangi, F.; Sevi, F.; Diretto, G.; De Tommasi, N.; Braca, A. Comparative chemical analysis of six ancient Italian sweet cherry (*Prunus avium* L.) varieties showing antiangiogenic activity. *Food Chem.* **2021**, *360*, 129999. [CrossRef]
41. Nguyen, A.N.T.; Vu, T.T.T.; Do, H.T.T.; Nguyen, T.H.; Le, H.V.; Pham, H.K.T.; Truong, P.C.H.; Pham, D.P.; Tran, M.H. Identification of phenolic compounds from vietnamese artichoke (*Cynara scolymus* L.) leaf and their antioxidant activities. *Nat. Prod. Sci.* **2024**, *30*, 39–51. [CrossRef]

42. Pinelli, P.; Agostini, F.; Comino, C.; Lanteri, S.; Portis, E.; Romani, A. Simultaneous quantification of caffeoyl esters and flavonoids in wild and cultivated cardoon leaves. *Food Chem.* **2007**, *105*, 1695–11701. [CrossRef]
43. Ferioli, F.; D'Antuono, L.F. Phenolic compounds in local Italian types of cultivated cardoon (*Cynara cardunculus* L. var. *altilis* DC) stalks and artichoke (*Cynara cardunculus* L. var. *scolymus* L.) edible sprouts. *J. Food Compos. Anal.* **2022**, *106*, 104342. [CrossRef]
44. Pandino, G.; Lombardo, S.; Mauromicale, G.; Williamson, G. Phenolic acids and flavonoids in leaf and floral stem of cultivated and wild *Cynara cardunculus* L. genotypes. *Food Chem.* **2011**, *126*, 417–422. [CrossRef]
45. Pagliari, S.; Cannavacciuolo, C.; Celano, R.; Carabetta, S.; Russo, M.; Labra, M.; Campone, L. Valorisation, green extraction development, and metabolomic analysis of wild cardoon by-product using pressurised liquid extraction UPLC–HRMS and multivariate data analysis. *Molecules* **2022**, *27*, 7157. [CrossRef] [PubMed]
46. Lovejoy, B.; Cleasby, A.; Hassell, A.M.; Longley, K.; Luther, M.A.; Weigl, D.; McGeehan, G.; McElroy, A.B.; Drewry, D.; Lambert, M.H. Structure of the catalytic domain of fibroblast collagenase complexed with an inhibitor. *Science* **1994**, *263*, 375–377. [CrossRef] [PubMed]
47. Voos, K.; Yahiaoui, S.; Konstantinović, J.; Schönauer, E.; Alhayek, A.; Sikandar, A.; Si Chaib, K.; Ramspoth, T.; Rox, K.; Haupenthal, J.; et al. N-Aryl-2-iso-butylmercaptoacetamides: The discovery of highly potent and selective inhibitors of *Pseudomonas aeruginosa* virulence factor LasB and *Clostridium histolyticum* virulence factor ColH. *ChemRxiv* **2022**. [CrossRef]
48. Crascì, L.; Basile, L.; Panico, A.; Puglia, C.; Bonina, F.P.; Basile, P.M.; Rizza, L.; Guccione, S. Correlating in vitro target-oriented screening and docking: Inhibition of matrix metalloproteinases activities by flavonoids. *Planta. Med.* **2017**, *83*, 901–911. [CrossRef]
49. Ajayi, O.O.; Akomolafe, S.F.; Ajayi, O.B.; Oyetayo, F.L.; Bodun, D. Anti-collagenase potentials and ADME Tox analysis of natural phenolic compounds from aqueous extract of *Chrysophyllum albidum* fruit parts: An in silico evaluation. *TJNPR* **2024**, *8*, 8896–8905. [CrossRef]
50. Nicolaescu, O.E.; Belu, I.; Mocanu, A.G.; Manda, V.C.; Rău, G.; Pîrvu, A.S.; Ionescu, C.; Ciulu-Costinescu, F.; Popescu, M.; Ciocîlteu, M.V. Cyclodextrins: Enhancing drug delivery, solubility and bioavailability for modern therapeutics. *Pharmaceutics* **2025**, *17*, 288. [CrossRef]
51. Puskás, I.; Szente, L.; Szőcs, L.; Fenyvesi, É. Recent list of cyclodextrin-containing drug products. *Period. Polytech. Chem. Eng.* **2023**, *67*, 11–17. [CrossRef]
52. Khalfaoui, A.; Khelifi, M.N.; Khelfaoui, A.; Benalia, A.; Derbal, K.; Gisonni, C.; Crispino, G.; Panico, A. The adsorptive removal of bengal rose by artichoke leaves: Optimization by full factorials design. *Water* **2022**, *14*, 2251. [CrossRef]
53. Trotta, F.; Zanetti, M.; Camino, G. Thermal degradation of cyclodextrins. *Polym. Degrad. Stab.* **2000**, *69*, 373–379. [CrossRef]
54. Sandilya, A.A.; Natarajan, U.; Priya, M.H. Molecular view into the cyclodextrin cavity: Structure and hydration. *ACS Omega* **2020**, *5*, 25655–25667. [CrossRef] [PubMed]
55. Baglı, E.; Stefaniotou, M.; Morbidelli, L.; Ziche, M.; Psillas, K.; Murphy, C.; Fotsis, T. Luteolin inhibits vascular endothelial growth factor-induced angiogenesis; inhibition of endothelial cell survival and proliferation by targeting phosphatidylinositol 3′-kinase activity. *Cancer Res.* **2004**, *64*, 7936–7946. [CrossRef] [PubMed]

foods

MDPI

Article

Extraction, Characterization, and Nutraceutical Potential of *Prosthechea karwinskii* Orchid for Insulin Resistance and Oxidative Stress in Wistar Rats

Gabriela Soledad Barragán-Zarate [1], Luicita Lagunez-Rivera [1,*], Alfonso Alexander-Aguilera [2,3,*], Rodolfo Solano [1] and Gerard Vilarem [4]

[1] Laboratorio de Extracción y Análisis de Productos Naturales Vegetales, Centro Interdisciplinario de Investigación para el Desarrollo Integral Regional Unidad Oaxaca, Instituto Politécnico Nacional, Hornos 1003, Santa Cruz Xoxocotlán C.P. 71230, Oaxaca, Mexico; gbarraganz@ipn.mx (G.S.B.-Z.); asolanog@ipn.mx (R.S.)

[2] Facultad de Bioanálisis, Universidad Veracruzana, Carmen Serdán s/n, Col. Flores Magón, Veracruz C.P. 91700, Veracruz, Mexico

[3] Escuela de Medicina, Universidad Cristóbal Colón, Carretera Veracruz-Medellin s/n, Col. Puente Moreno, Boca del Río C.P. 94271, Veracruz, Mexico

[4] Laboratoire de Chimie Agro-Industrielle, ENSIACET, 4 Allée Emile Monso, BP 44362, 31030 Toulouse, Cedex 4, France; gerard.vilarem@ensiacet.fr

* Correspondence: llagunez@ipn.mx (L.L.-R.); aalexander@uv.mx (A.A.-A.)

Abstract: *Prosthechea karwinskii* is an endemic orchid of Mexico with cultural significance for its ornamental, food, religious, and medicinal uses. In traditional medicine, diabetic patients use the leaves of this plant to lower glucose levels. The present study evaluated the effect of *P. karwinskii* leaves extract on the antioxidant enzymes superoxide dismutase (SOD) and catalase (CAT) in a model of obese rats with insulin resistance for its nutraceutical potential to reduce insulin resistance and oxidative stress. Obesity and insulin resistance were induced with 40% sucrose in water for 20 weeks. Four groups (control rats, obese rats, obese rats administered the extract, and obese rats administered metformin) were evaluated. Extract compounds were identified by UHPLC-ESI-qTOF-MS/MS. Glucose, insulin, triglyceride, and insulin resistance indices (HOMA-IR and TyG), as well as the activity of the antioxidant enzymes, increased in rats in the obese group. Administration of *P. karwinskii* extract and metformin reduced glucose, insulin, triglyceride, and insulin resistance indices and antioxidant enzyme activity to values similar to those of the control group. Therefore, this study shows the nutraceutical potential of *P. karwinskii* extract as an ingredient in the formulation of dietary supplements or functional foods to help treat diseases whose pathophysiology is related to oxidative stress and insulin resistance.

Keywords: reactive oxygen species; HOMA-IR; TyG index; SOD; CAT

1. Introduction

Insulin resistance (IR) is a metabolic disorder associated with obesity, where the sensitivity and cellular response to insulin in tissues is decreased, as well as glucose metabolism. IR is involved in the pathophysiology of metabolic disorders such as diabetes, dyslipidemia, non-alcoholic fatty liver disease, and cancer. Insulin resistance is characterized by elevated levels of reactive oxygen species (ROS). The origin of ROS being mainly mitochondrial, hyperglycemia can cause alterations to mitochondrial morphology and stimulate different mitochondrial enzymatic pathways, such as nicotinamide adenine dinucleotide phosphate oxidase (NADPH oxidase), nitric oxide (NO) synthase uncoupling, xanthine oxidase, and lipoxygenases, cyclooxygenases, and peroxidases. In turn, increased levels of oxidative stress with increased ROS production cause insulin resistance. Oxidative stress and insulin resistance are increased in obese individuals [1].

Oxidative stress participates in the pathogenesis of many chronic diseases, including IR and diabetes, and occurs due to an imbalance between elevated ROS generation and/or poor antioxidant defense. Due to their antioxidant activity, medicinal plants are an alternative for the treatment of various diseases [2].

Prosthechea karwinskii (Mart.) J.M.H. Shaw is an endemic orchid of Mexico that is used in the Mixtec region of the state of Oaxaca to decorate church altars during Semana Santa celebrations as well as to prepare various local dishes and in traditional medicine [3]. The extract of its leaves decreased glucose levels [4] and IR [5] in rats with metabolic syndrome and inhibited ROS in an *ex vivo* model with peripheral blood mononuclear cells [6]. Its effect on markers of inflammation and the activity of the enzyme superoxide dismutase (SOD) was evaluated in a model of acute gastric injury, where no clear involvement of SOD in the gastric protection conferred by the extract was observed [7]. However, nothing has been published on the effect of the orchid extract on other parameters related to oxidative stress in an *in vivo* model that would give us more information on the impact of the extract when administered for longer periods. Considering this background, this study aimed to evaluate the effect of *P. karwinskii* leaves extract on the antioxidant enzymes SOD and catalase (CAT) in a model of obese rats with insulin resistance to determine its nutraceutical potential with respect to insulin resistance and oxidative stress and thereby assess its potential as an ingredient in the formulation of food supplements and functional foods to help treat diseases related to insulin resistance and oxidative stress.

2. Materials and Methods

2.1. Materials, Reagents, and Kits

The following commercial kits were obtained: a Catalase assay Kit (707002, Cayman chemical, Ann Arbor, MI, USA), a Superoxide Dismutase assay Kit (706002, Cayman chemical, Ann Arbor, MI, USA), and an Insulin assay Kit (589501, Cayman Chemical, Ann Arbor, MI, USA). Phosphate-buffered saline (PBS), distilled water, formic acid, acetonitrile, and other reagents were purchased from Sigma Aldrich (Toluca, Mexico), and analytical kits for glucose and triglycerides were purchased from Spinreact (Naucalpan, Mexico).

2.2. Plant Material

The leaves of *P. karwinskii* used in this study were collected in 2017 in Zaachila, Oaxaca (16°57′ N latitude, 96°45′ W longitude; 1490 m altitude), dehydrated whole, and stored in paper bags at room temperature away from light and humidity until extraction. Species determination was performed by one of the authors (RS), and a backup specimen was deposited in the OAX herbarium (Solano 4037). In this study, *Prosthechea* is recognized as the accepted generic name of the orchid genus, in agreement with Pridgeon et al. [8], Soto et al. [9], Villaseñor [10], and Solano et al. [11].

2.3. Obtaining the Extract

The extract of the *P. karwinskii* leaves was obtained by an ultrasound-assisted method, following Barragan-Zarate et al. [6] (2020), in a 750 W ultrasound processor (VCX 750, Scientific SENSE). The solvent used was 50% ethanol in water with a sample–solvent ratio of 1:18 (g:mL), and the extraction temperature was maintained at 40 °C with a water bath for 20 min. The extract was stored in an amber glass bottle that was sealed and kept at freezing temperature until analysis to avoid degradation of its compounds.

2.4. Compound Identification with UHPLC-ESI-qTOF-MS/MS

For the identification of plant compounds, the method described by Barragan-Zarate et al. [5] was followed. An ultra-high-performance liquid chromatography (UHPLC) system (Ultimate 3000, Thermo Fisher Scientific, Waltham, MA, USA) combined with an Impact II mass spectrometer (Bruker) with electrospray ionization (ESI) and quadrupole time of flight (qTOF) was used. The column used was the Thermo Scientific Acclaim 120 C18 (2.2 μm, 120 Å, 50 × 2.1 mm), and an elution gradient was used with the following mobile phases: A: 0.1% formic acid in water and B: acetonitrile, with a flow rate of 0.35 mL/min and a temperature of 25 °C. The gradient system was as follows: 0% B (0–2 min), 1% B (2–3 min), 3% B (3–4 min), 32% B (4–5 min), 36% B (5–6 min), 40% B (6–8 min), 45% B (8–9 min), 80% B (9–11 min), 0% B (12–14). The mass spectrometer was operated in negative electrospray mode at 0.4 Bar (5.8 psi), in autoMSMS, with a mass range of 50–700 m/z. The data obtained were processed with Bruker's DataAnalysis software 3.1. Compounds were identified by comparing their exact masses and fragmentation patterns with those collected from libraries and scientific articles.

2.5. Induction of Obesity

A total of 24 Wistar rats, 21 days old, were divided into 2 groups: a control group (CG, $n = 6$) and an obese group (OG, $n = 18$). The rats were fed *ad libitum* with standard food, the CG group drank drinkable water and the OG group drank water with 40% sucrose to induce obesity, and the induction period was 20 weeks, as reported by Barragan-Zarate et al. [5]. The rats were maintained under normal conditions in the biotherium of the Escuela de Medicina of the Universidad Cristóbal Colón, following the official Mexican standard for the care and handling of animals, NOM-062-ZOO-1999 [12], as well as the international norms and standards for the care and use of laboratory animals. The experimental protocol was approved by the Cuerpo académico Genómica y Salud UVE CA-317 of Facultad de Bioanalisis from the Universidad Veracruzana with registration GS-317-1-2010 on August 2022.

2.6. Experimental Design

After 20 weeks of obesity induction, the obese group (OG) was divided into 3 groups: the OG group, which received no treatment; the PK group, which was administered *P. karwinskii* extract at a dose of 300 mg/kg; and the MET group, which was administered metformin at a dose of 200 mg/kg. The selection of the doses was based on a previous study [5], where the effective dose for controlling the evaluated metabolic syndrome parameters was 300 mg/kg. Treatments were administered with an oral feeding cannula. Rats in the OG, PK, and MET groups continued to drink 40% sucrose water during the experimental diet. Comparisons were made with the CG group (the group in which obesity was not induced and which did not receive any treatment). The experimental period was 4 weeks.

2.7. Evaluated Parameters

At the end of the experimental period, the rats were fasted for 12 h and sacrificed. The anesthetic used was sodium pentobarbital. The weight of the rats was recorded, as well as the weight of the total adipose tissue. Blood was collected to determine glucose, insulin, and triglyceride levels and to calculate indicators of insulin resistance. The Homeostatic Model Assessment of Insulin Resistance (HOMA-IR) was calculated as described by Matthews et al. [13] with Equation (1):

$$\text{HOMA-IR} = [\text{fasting insulin } (\mu\text{IU/mL}) \times \text{fasting glucose (mmol/L)}]/22.5. \quad (1)$$

The triglyceride–glucose index (TyG) was calculated according to Simental-Mendía and Guerrero-Romero [14] with Equation (2):

$$\text{TyG} = \text{Ln } [\text{fasting triglycerides (mg/dL)} \times \text{fasting glucose (mg/dL)}]/2. \quad (2)$$

In addition, liver samples were taken and homogenized for the determination of the antioxidant enzymes SOD and CAT, following the respective kit manufacturers' instructions.

2.8. Statistical Analysis

Results are reported as means ± standard deviations ($n = 6$). Analysis of variance (ANOVA) was performed, followed by Neuman's Keul test. Differences were considered statistically significant at $p < 0.05$.

3. Results and Discussion

3.1. Compounds Identified with UHPLC-ESI-qTOF-MS/MS

Table 1 shows the information on the compounds identified in the extract of *P. karwinskii* with UHPLC-ESI-qTOF-MS/MS, which were quinic acid, malic acid, succinic acid, L-(-)-phenylalanine, guanosine, neochlorogenic acid, chlorogenic acid, rutin, kaempferol-3-O-ruthinoside, azelaic acid, pinellic acid, and embelin.

Table 1. Chemical structures and information on compounds identified in leaves extract of *Prosthechea karwinskii* with UPLC-ESI-qTOF-MS/MS.

Peak	RT (min)	*m/z* [M-H]⁻	Error (ppm)	MS/MS Fragments	Compound (Chemical Formula)	Type of Compound	Relative Yield (%)	Chemical Structure
1	0.7	191.0557	1.9	85.0293, 87.0078, 111.0443, 127.6945	Quinic Acid [abd] ($C_7H_{12}O_6$)	Cyclitol, cyclic polyol, and cyclo-hexanecarboxylic acid	22.21	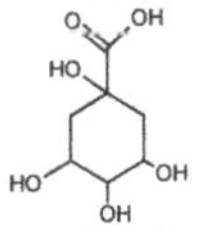
2	0.8	133.0140	1.8	115.0032	Malic acid [d] ($C_4H_6O_5$)	Dicarboxylic organic acid	5.71	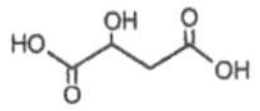
3	1.2	117.0191	2.3	73.0290, 99.0072	Succinic acid [d] ($C_4H_6O_4$)	Dicarboxylic organic acid	1.89	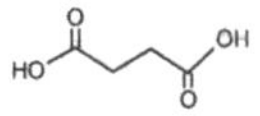
4	2.3	164.0712	3.2	72.0072, 103.0539, 147.0442	L-(-)-Phenylalanine [e] ($C_9H_{11}NO_2$)	Amino acid	0.53	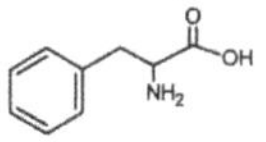
5	2.7	282.0833	3.9	108.5347, 133.0157, 150.0429	Guanosine [e] ($C_{10}H_{13}N_5O_5$)	Nucleoside	0.46	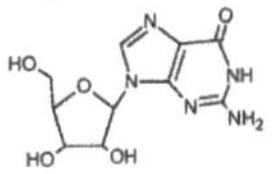
6	6.0	353.0867	3.1	179.0365, 191.0556	Neochlorogenic acid [b] ($C_{16}H_{18}O_9$)	Caffeoylquinic acid, phenolic compound	4.35	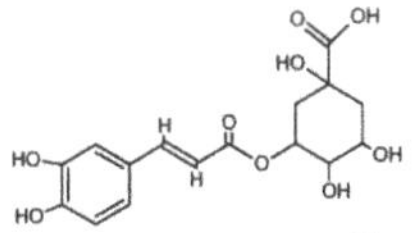
7	6.3	353.0866	3.4	173.0452, 179.0365, 191.0556	Chlorogenic acid [d] ($C_{16}H_{18}O_9$)	Caffeoylquinic acid, phenolic compound	8.44	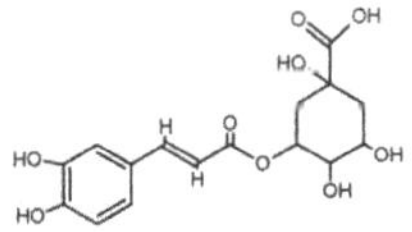
8	6.5	609.1438	3.8	300.0266, 301.0335	Rutin [abe] ($C_{27}H_{30}O_{16}$)	Flavonoid glycoside	15.52	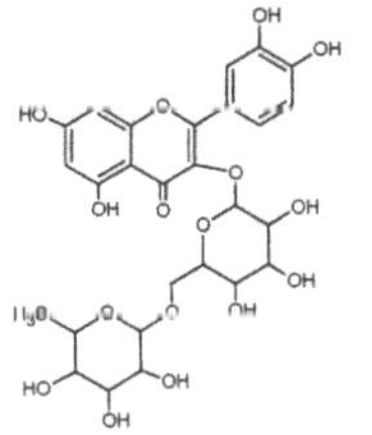
9	6.6	593.1489	3.9	284.0314, 285.0393	Kaempferol-3-O-rutinoside [e] ($C_{27}H_{30}O_{15}$)	Flavonol glycoside	25.30	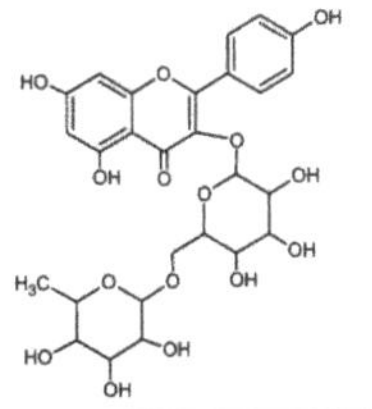

Table 1. *Cont.*

Peak	RT (min)	m/z [M-H]⁻	Error (ppm)	MS/MS Fragments	Compound (Chemical Formula)	Type of Compound	Relative Yield (%)	Chemical Structure
10	6.8	187.0970	3.1	97.0653, 125.0963, 169.0889	Azelaic acid [c] ($C_9H_{16}O_4$)	Dicarboxylic acid	2.10	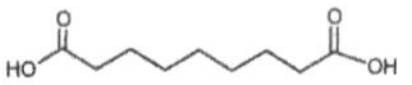
11	8.2	329.2321	3.9	171.1023, 229.1436	Pinellic acid [c] ($C_{18}H_{34}O_5$)	Carboxylic acid	1.09	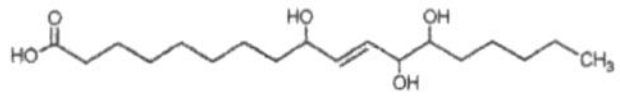
12	11.5	293.2112	3.6	275.2013, 235.1680, 223.1685	Embelin [c] ($C_{17}H_{26}O_4$)	Para-benzoquinone	8.60	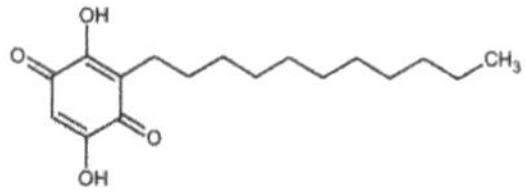

Superscripts correspond to supporting references for compound identification: [a] De Souza et al. [15], [b] Ma et al. [16], [c] Dahibhate et al. [17], [d] MassBank library, [e] Bruker's MetaboBase library. RT: retention time.

The tentative identification of the compounds was performed by comparing their spectral data (m/z values of the precursor ions and fragmentation patterns) with data reported in the literature and in databases (the MassBank and Bruker's MetaboBase libraries). The values reported by De Souza et al. [15] were used for the tentative identification of quinic acid, with a precursor ion m/z value of 191.0575 and fragment 127 corresponding to the neutral loss of formic acid (−46 Da); chlorogenic acid, with a precursor ion m/z value of 353.0890 and fragments 179 and 191 corresponding to caffeic acid, and quinic acid due to the loss of caffeoyl (−162 Da); and rutin, with a precursor ion m/z value of 609.1490 and fragment 301 corresponding to the rutinoside associated with the internal fragmentation of the flavonoid glycoside. Data reported by Ma et al. [16] were employed for the identification of quinic acid, with a precursor ion m/z value of 191.0553 and fragments of 111.0072 and 127.0389; neochlorogenic acid, with a precursor ion m/z value of 353.0886 and fragments of 179.0349 and 191.0562; and rutin, with a precursor ion m/z value of 609.1502 and fragments of 300.0266 and 301.0341. Data reported by Dahibhate et al. [17] were employed for the identification of azelaic acid, with a precursor ion m/z value of 187.0974 and fragments of 169, 125.1, and 97; pinellic acid, with a precursor ion m/z value of 329.2333 and fragments of 229.1 and 171.1; and embelin, with a precursor ion m/z value of 293.1758. Compounds identified via libraries were compared with spectral data from standards analyzed with similar instruments with the same type of ionization and fragmented with similar collision energies.

3.2. Effect of P. karwinskii Extract on Obesity, Insulin Resistance, and Oxidative Stress Parameters

Table 2 shows the weight record of the rats at weeks 0 and 20 of induction (before the experimental diet) and at the end of the experiment, along with indices of insulin resistance at week 20 as well as the total adipose tissue values of the rats at the end of the experimental diet. This information proves that the rats in the OG group were obese and presented insulin resistance before starting the experimental diet.

Figure 1 shows the effect of *P. karwinskii* extract on glucose, insulin, and triglyceride levels, as well as on HOMA-IR and TyG insulin resistance indices. All these parameters presented higher values in the OG group compared to the CG group. In the case of Figure 1a,c,e, only the OG group was different from the CG group. In the case of Figure 1b,d, the OG, PK, and MET groups were different from the CG group; however, PK differed from OG in both cases, while MET differed only in the case of Figure 1d. Administration of the extract and metformin reduced glucose and triglyceride levels and insulin resistance indices compared to the OG group. As for insulin, only the extract had an effect compared to OG. Table 3 shows the weekly liquid and solid diet and caloric intakes of the groups during the experimental diet period. Higher fluid intake was observed in the OG, PK, and MET groups compared to the CG group. Regarding the solid diet, consumption was higher in the CG group compared to the OG, PK, and MET groups. Although there were no significant differences between the total caloric intakes of the OG, PK, and MET groups at the end of week 4, there were significant differences in glucose and triglyceride levels and insulin resistance indices, as well as body weight and adipose tissue, evidencing the beneficial effect of the extract despite consumption of the same sucrose-rich diet.

Table 2. Body weight record for Wistar rats at weeks 0 and 20 of obesity induction and at the end of the experiment, along with total adipose tissue.

	CG	OG	PK	MET
Weight at week 0 (g)	74.3 ± 17.9	78.2 ± 15.5	-	-
Weight at week 20 (g)	368.5 ± 38.1	456.2 ± 42.6 *	-	-
TyG index at week 20	7.73 ± 0.48	8.85 ± 0.53 *	-	-
HOMA-IR at week 20	1.73 ± 0.25	5.01 ± 0.47 *	-	-
Weight at the end of the experiment (g)	423.1 ± 30.2	527.0 ± 45.3 *	487.0 ± 42.8 #	489.0 ± 55.2 #
Total adipose tissue at the end of the experiment (g)	14.3 ± 3.1	37.92 ± 4.4 *	29.38 ± 3.5 * **	33.35 ± 4.9 *

Values expressed as means ± SDs (n = 6). * Indicates a significant difference ($p < 0.05$) with respect to CG; ** indicates a significant difference ($p < 0.05$) with respect to OG; # indicates that the value does not differ significantly with respect to either CG or OG. CG: control group, OG: obese rats, PK: obese rats that received extract, and MET: obese rats that received metformin.

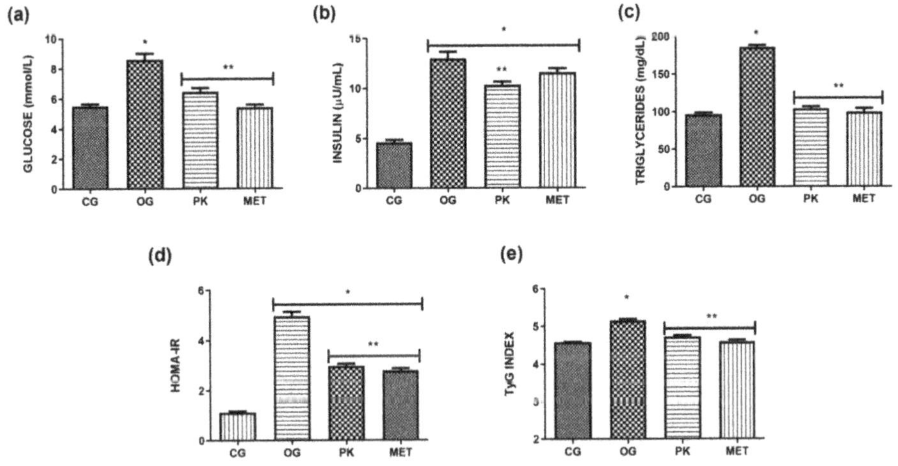

Figure 1. Glucose (**a**), insulin (**b**), and (**c**) triglyceride levels and HOMA-IR (**d**) and TyG (**e**) insulin resistance indices. Values expressed as means ± SDs. * Indicates a significant difference ($p < 0.05$) with respect to CG; ** indicates a significant difference ($p < 0.05$) with respect to OG. CG: control group, OG: obese rats, PK: obese rats that received extract, and MET: obese rats that received metformin.

Table 3. Weekly record of weight, liquid, and food consumption, as well as caloric intake, for each of the experimental diet groups.

Week 1	CG	OG	PK	MET
Weight (g)	389.00 ± 28.70	479.00 ± 35.97 *	472.00 ± 32.49 *	494.00 ± 67.77 *
Liquid consumption (mL/day)	30.60 ± 4.59	17.42 ± 2.61 *	13.85 ± 2.07 *	9.85 ± 1.47 * **
Liquid consumption (mL/day/100 g)	7.86 ± 0.58	3.63 ± 0.27 *	2.93 ± 0.20 * **	1.99 ± 0.27 * **
Equivalent in kcal in drinkable water	0.00	13.58 ± 1.02 *	10.96 ± 0.75 * **	7.44 ± 1.02 * **
Feed consumption (g/day)	24.28 ± 3.64	17.57 ± 2.63 *	17.71 ± 2.65 *	18.42 ± 2.76 *
Feed consumption (g/day/100 g)	6.24 ± 0.46	3.66 ± 0.27 *	3.75 ± 0.25 *	3.73 ± 0.51 *
Equivalent in kcal in feed	19.35 ± 1.42	11.37 ± 0.85 *	11.63 ± 0.80 *	11.56 ± 1.58 *
Total Kcal/day/100 g body weight	19.35 ± 1.42	24.95 ± 1.87 *	22.59 ± 1.55 *	19.01 ± 2.60 **

Table 3. *Cont.*

Week 2	**CG**	**OG**	**PK**	**MET**
Weight (g)	398.00 ± 25.21	482.00 ± 33.55 *	459.00 ± 35.41 *	482.00 ± 55.16 *
Liquid consumption (mL/day)	32.57 ± 4.88	28.14 ± 4.22	33.60 ± 5.04	30.77 ± 4.61
Liquid consumption (mL/day/100 g)	8.18 ± 1.10	5.83 ± 0.95 *	7.32 ± 0.38 **	6.38 ± 0.06 *
Equivalent in kcal in drinkable water	0.00	65.39 ± 4.55 *	76.80 ± 5.92 * **	75.08 ± 10.15 * **
Feed consumption (g/day)	27.28 ± 4.09	16.00 ± 2.4 *	11.85 ± 1.77 * **	12.71 ± 1.90 * **
Feed consumption (g/day/100 g)	6.85 ± 0.44	3.31 ± 0.23 *	2.58 ± 0.19 * **	2.63 ± 0.35 * **
Equivalent in kcal in feed	21.25 ± 1.36	10.29 ± 0.71 *	8.00 ± 0.61 * **	8.17 ± 1.10 * **
Total Kcal/day/100 g body weight	21.25 ± 1.36	75.68 ± 5.26 *	84.80 ± 6.54 *	83.26 ± 11.25 *
Week 3	**CG**	**OG**	**PK**	**MET**
Weight (g)	402.00 ± 30.75	492.00 ± 37.76 *	468.00 ± 29.42 *	486.00 ± 54.94 *
Liquid consumption (mL/day)	30.42 ± 4.56	44.85 ± 6.72 *	46.28 ± 6.94 *	31.00 ± 4.650 **
Liquid consumption (mL/day/100 g)	7.56 ± 0.57	9.11 ± 0.69 *	9.89 ± 0.62 *	6.37 ± 0.90 **
Equivalent in kcal in drinkable water	0.00	102.10 ± 7.83 *	110.70 ± 6.96 *	71.44 ± 10.13 * **
Feed consumption (g/day)	23.14 ± 3.47	12.14 ± 1.82 *	9.57 ± 1.43 *	9.57 ± 1.43 *
Feed consumption (g/day/100 g)	17.84 ± 1.36	7.65 ± 0.58 *	6.34 ± 0.39 * **	6.10 ± 0.86 * **
Equivalent in kcal in feed	5.75 ± 0.44	2.46 ± 0.18 *	2.04 ± 0.12 *	1.96 ± 0.27 *
Total Kcal/day/100 g body weight	17.84 ± 1.36	109.7 ± 8.42 *	117.10 ± 7.36 *	77.54 ± 11.00 * **
Week 4	**CG**	**OG**	**PK**	**MET**
Weight (g)	423.10 ± 30.20	527.00 ± 45.30 *	487.00 ± 42.80 #	489.00 ± 55.20 #
Liquid consumption (mL/day)	33.11 ± 4.96	48.91 ± 7.33 *	44.57 ± 6.68 *	52.97 ± 8.54 *
Liquid consumption (mL/day/100 g)	7.82 ± 0.63	9.28 ± 0.75 *	9.15 ± 0.61 *	10.83 ± 1.61 *
Equivalent in kcal in drinkable water	0.00	103.93 ± 8.40 *	102.55 ± 6.85 *	121.32 ± 18.09 *
Feed consumption (g/day)	24.42 ± 3.66	11.71 ± 1.75 *	10.71 ± 1.60 *	9.57 ± 1.43 *
Feed consumption (g/day/100 g)	5.77 ± 0.46	2.22 ± 0.17 *	2.19 ± 0.14 *	1.96 ± 0.27 *
Equivalent in kcal in feed	17.89 ± 1.44	6.88 ± 0.55 *	6.81 ± 0.45 *	6.09 ± 0.84 *
Total Kcal/day/100 g body weight	17.89 ± 1.44	110.81 ± 8.95 *	109.36 ± 10.34 *	127.41 ± 12.94 *

Values expressed as means ± SDs. (n = 6). * Indicates a significant difference ($p < 0.05$) with respect to CG; ** indicates a significant difference ($p < 0.05$) with respect to OG; # indicates that the value does not differ significantly with respect to either CG or OG. CG: control group, OG: obese rats, PK: obese rats that received extract, and MET: obese rats that received metformin.

Other studies have reported similar results, such as that of Paunovic et al. [18], where rats were fed a high-fat, high-fructose diet supplemented with blackcurrant juice and there was no significant difference in the caloric intake between the different groups; however, the supplemented diet did reduce adipose tissue and triglyceride levels and improved glucose tolerance, though, unlike our extract, it did not affect glucose or insulin levels. The study by Mostafa et al. [19] evaluated *Vitis vinifera* seed extract in rats fed a diet rich in fat and carbohydrates, and the extract reduced glucose and triglyceride levels and weight gain compared to the untreated group.

P. karwinskii leaves extract was able to decrease glucose levels and insulin resistance. The compounds identified in the extract that could have been responsible for the effect of the extract on parameters related to glucose metabolism were chlorogenic acid [20] and embelin [21], which regulate glucose metabolism and decrease IR [20], and rutin, which decreases carbohydrate absorption, inhibits gluconeogenesis, increases tissue glucose uptake, increases insulin secretion by β-pancreatic cells, and protects the islets of Langerhans [22]. Given the relationship of IR with various metabolic disorders and the ability of *P. karwinskii* extract to attenuate them, it has the potential to help treat diseases in which IR is implicated.

Figure 2 shows the activity of the antioxidant enzymes SOD and CAT, which were increased in the OG group compared to the GC group. However, in the PK and MET groups, the activity of these enzymes was downregulated to values that did not differ significantly from those recorded for the GC group.

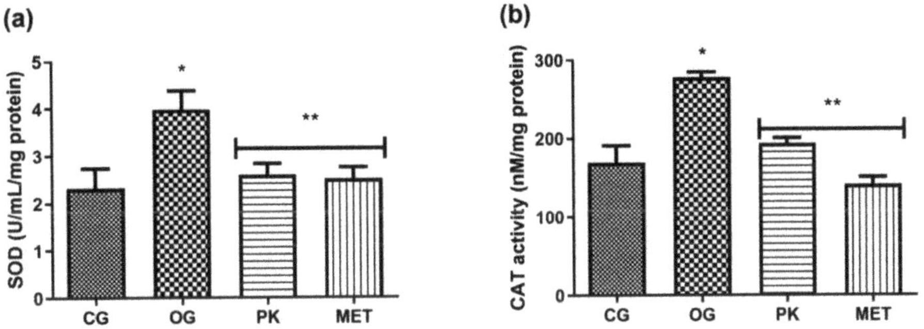

Figure 2. Superoxide dismutase enzyme activity (SOD) (**a**) and catalase enzyme activity (CAT) (**b**). Values expressed as means ± SDs. * Indicates a significant difference ($p < 0.05$) with respect to GC; ** indicates a significant difference ($p < 0.05$) with respect to OG. CG: control group, OG: obese rats, PK: obese rats that received extract, and MET: obese rats that received metformin.

Hyperglycemia can cause altered mitochondrial morphology and increase cellular oxidation of glucose and production of nicotinamide adenine dinucleotide (NADH), which contributes to oxidation in cellular processes, thereby increasing ROS generation [1]. ROS are produced in cells during normal aerobic metabolism; the generated ROS are removed by antioxidant enzymes, including SOD and CAT, which protect cells against oxidative stress by cellular detoxification of O_{2-} and H_2O_2, as excessive production of ROS and/or deficient antioxidant capacity can generate oxidative stress [2]. Hyperglycemia also activates different metabolic signaling pathways, triggering the activation of the antioxidant system to reduce ROS production and oxidative stress [1].

In the present study, the activity of the evaluated antioxidant enzymes increased in the obese group under oxidative stress compared to the control group. Although this behavior differs from the common behavior of antioxidant enzyme activity under oxidative stress, similar behavior has been reported in other investigations, such as Chen et al.'s [23] review on oxidative stress in non-alcoholic fatty liver disease and Barbosa et al.'s review [24] on oxidative stress in hypertensive diseases of pregnancy, where, in some cases, an increase in the activity of the antioxidant enzymes CAT, SOD, and glutathione peroxidase was observed, suggesting a compensatory mechanism for the increase in ROS due to oxidative stress.

Similarly, the research of Zeng et al. [25] on the effect of cadmium stress on the antioxidant system of crabs showed that SOD activity increased in both the hepatopancreas and the intestine in the cadmium-stressed group compared to the control group, while in the case of CAT, its activity decreased in the hepatopancreas but increased in the intestine in the Cd-stressed group. So, they consider antioxidant enzymes as non-specific adaptive mechanisms that protect against oxidative damage and serve as the first line of defense against oxidative stress.

According to Aouacheri et al. [26], whose results showed that SOD activity in diabetic patients was increased compared to the control group, increased antioxidant enzyme activity stimulates the cellular ability to remove and limit damage caused by ROS, acting as an adaptive response and a compensatory mechanism to detoxify harmful metabolites related to oxidative stress [23,26]. Leghi et al. [27] also observed that SOD enzyme activity was increased in a group with non-alcoholic steatohepatitis compared to those without, suggesting that the increased SOD activity counteracted the overproduction of O_{2-} radicals.

The increase in CAT activity could occur in response to high SOD activity by the synergistic action of both enzymes [2], as it increases H_2O_2 production by SOD, which is subsequently catalyzed by CAT. The extract downregulated the activity of SOD and CAT enzymes, which increased in OG to values like those presented by CG, possibly through a decrease in ROS, the levels of which were increased by hyperglycemia and IR. In addition, a previous study showed the ability of *P. karwinskii* leaf extract to inhibit ROS in cells [6].

This suggests that the compounds acted to reduce ROS, maintaining redox balance, without the physiological need to increase antioxidant enzyme activity. In other words, they can neutralize ROS and produce more stable radicals [28].

There have been other investigations where the effects of evaluated compounds were found to be similar to those reported in this research, such as that of Barbosa et al. [28], where they evaluated the effect of acai pulp in pregnant rats fed a high-fat diet and the activities of the antioxidant enzymes SOD and CAT were increased in a group under oxidative stress compared to the control group and, similar to what happened with our extract, the activity of antioxidant enzymes was modulated and biomarkers of oxidative stress which had been altered by the high-fat diet were reduced, thus improving the oxidative balance.

Something similar also occurred in okra plants: when they were stressed with heavy metals such as cadmium, ROS production increased [29]. It was observed that the activity of the antioxidant enzymes SOD and CAT increased in Cd-stressed plants compared to non-stressed plants and that receiving treatments with chelating agents, such as malic acid and EDTA, increased the resistance of plants to oxidative stress by decreasing H_2O_2 and the activity of SOD and CAT enzymes. Possibly something like what happens in plants occurred in our model, since, according to Hussain et al. [30], they develop defensive mechanisms to minimize or protect themselves from the toxic effects of ROS, maintaining a balance between their synthesis and decomposition to preserve cellular redox hemostasis. Antioxidant enzyme activity is a key indicator of the antioxidant defense mechanism.

The compounds identified in the extract with antioxidant properties are rutin [31], kaempferol-3-O-ruthinoside [32], embelin [21,33], and guanosine [34]. According to reports, these compounds decrease oxidative stress.

4. Conclusions

The activity of the antioxidant enzymes SOD and CAT was increased in the obese group as an adaptive response or compensatory mechanism to minimize the overproduction of ROS due to hyperglycemia and IR, which generated oxidative stress. Administration of the extract downregulated antioxidant enzyme activity to values like those of the control group, possibly by decreasing ROS formation by regulating glucose and IR levels, without increasing antioxidant enzyme activity to reduce oxidative stress. The orchid *P. karwinskii* has nutraceutical potential to treat conditions whose pathophysiology involves oxidative stress and/or insulin resistance. Thus, it can be considered a potential ingredient in the formulation of dietary supplements and functional foods to help treat diseases related to insulin resistance and oxidative stress. Therefore, future research concerning *P. karwinskii* should be directed towards the study of the properties of the extract as an ingredient, as well as the design of functional foods and nutraceuticals that share its properties.

Author Contributions: Conceptualization, G.S.B.-Z., L.L.-R. and A.A.-A.; methodology, G.S.B.-Z. and A.A.-A.; writing—original draft preparation, G.S.B.-Z.; writing—review and editing, L.L.-R., A.A.-A., R.S. and G.V.; visualization, G.S.B.-Z., L.L.-R., A.A.-A., R.S. and G.V.; supervision, L.L.-R., A.A.-A., R.S. and G.V.; funding acquisition, L.L.-R., A.A.-A. and R.S. All authors have read and agreed to the published version of the manuscript.

Funding: This research was funded by CONACYT (project 270428, postdoctoral grant 392827) and the Instituto Politécnico Nacional (projects SIP-20161270, SIP-2016RE/50, SIP-20220649 and SIP-20240944).

Institutional Review Board Statement: The animal study protocol was approved by the Cuerpo académico Genómica y Salud UVE CA-317 of Facultad de Bioanalisis from the Universidad Veracruzana with registration GS-317-1-2010 on August 2022.

Informed Consent Statement: Not applicable.

Data Availability Statement: The original contributions presented in the study are included in the article. Further inquiries can be directed to the corresponding authors.

Acknowledgments: We thank the students who helped in the implementation of the animal model and the organizers of the Semana Santa festivities in Zaachila, Oaxaca, for providing the plant material.

Conflicts of Interest: The authors declare no conflicts of interest.

Abbreviations

SOD: superoxide dismutase, CAT: catalase, HOMA-IR: Homeostatic Model Assessment of Insulin Resistance, TyG: triglyceride–glucose index, UHPLC-ESI-qTOF-MS/MS: ultra-high-performance liquid chromatography system combined with a mass spectrometer with electrospray ionization and quadrupole time of flight, RI: insulin resistance, ROS: reactive oxygen species, NADPH oxidase: nicotinamide adenine dinucleotide phosphate oxidase, NO: nitric oxide, PBS: phosphate-buffered saline, OG: obese group, CG: control group, PK: group administered *P. karwinskii* extract at a dose of 300 mg/kg, MET: group administered metformin at a dose of 200 mg/kg, ANOVA: analysis of variance, NADH: nicotinamide adenine dinucleotide.

References

1. Andreadi, A.; Bellia, A.; Di Daniele, N.; Meloni, M.; Lauro, R.; Della-Morte, D.; Lauro, D. The Molecular Link between Oxidative Stress, Insulin Resistance, and Type 2 Diabetes: A Target for New Therapies against Cardiovascular Diseases. *Curr. Opin. Pharmacol.* **2022**, *62*, 85–96. [CrossRef] [PubMed]
2. Hsouna, A.B.; Dhibi, S.; Dhifi, W.; Saad, R.B.; Brini, F.; Hfaidh, N.; Mnif, W. Essential Oil from Halophyte: Lobularia Maritima: Protective Effects against CCl4-Induced Hepatic Oxidative Damage in Rats and Inhibition of the Production of Proinflammatory Gene Expression by Lipopolysaccharide-Stimulated RAW 264.7 Macrophages. *RSC Adv.* **2019**, *9*, 36758–36770. [CrossRef] [PubMed]
3. Cruz-Garcia, G.; Solano-Gomez, R.; Lagunez-Rivera, L. Documentation of the Medicinal Knowledge of Prosthechea Karwinskii in a Mixtec Community in Mexico. *Rev. Bras. Farmacogn.* **2014**, *24*, 153–158. [CrossRef]
4. Rojas-Olivos, A.; Solano-Gómez, R.; Alexander-Aguilera, A.; Jiménez-Estrada, M.; Zilli-Hernández, S.; Lagunez-Rivera, L. Effect of Prosthechea Karwinskii (Orchidaceae) on Obesity and Dyslipidemia in Wistar Rats. *Alex. J. Med.* **2017**, *53*, 311–315. [CrossRef]
5. Barragán-Zarate, G.S.; Alexander-Aguilera, A.; Lagunez-Rivera, L.; Solano, R.; Soto-Rodríguez, I. Bioactive Compounds from Prosthechea Karwinskii Decrease Obesity, Insulin Resistance, pro-Inflammatory Status, and Cardiovascular Risk in Wistar Rats with Metabolic Syndrome. *J. Ethnopharmacol.* **2021**, *279*, 114376. [CrossRef]
6. Barragán-Zarate, G.S.; Lagunez-Rivera, L.; Solano, R.; Pineda-Peña, E.A.; Landa-Juárez, A.Y.; Chávez-Piña, A.E.; Carranza-Álvarez, C.; Hernández-Benavides, D.M. Prosthechea Karwinskii, an Orchid Used as Traditional Medicine, Exerts Anti-Inflammatory Activity and Inhibits ROS. *J. Ethnopharmacol.* **2020**, *253*, 112632. [CrossRef] [PubMed]
7. Salinas-Nolasco, C.; Barragán-Zarate, G.S.; Lagunez-Rivera, L.; Solano, R.; Favari, L.; Jiménez-Andrade, J.M.; Chávez-Piña, A.E. Role of LTB4 and Nitric Oxide in the Gastroprotective Effect of Prosthechea Karwinskii Leaves Extract in the Indomethacin-Induced Gastric Injury in the Rat. *Nat. Prod. Res.* **2022**, *37*, 819–822. [CrossRef] [PubMed]
8. Pridgeon, A.M.; Cribb, P.J.; Chase, M.W.; Rasmussen, F.N. *Genera Orchidacearum Volume 4: Epidendroideae (Part One)*; Oxford University Press: New York, NY, USA, 2005.
9. Soto, M.A.; Hágsater, E.; Jiménez, R.; Salazar, G.A.; Solano, R.; Flores, R.; Ruiz, I. *Orchids of Mexico. Digital Catalogue*; Herbario AMO: México City, Mexico, 2007.
10. Villaseñor, J.L. Checklist of the native vascular plants of Mexico. *Rev. Mex. De Biodivers.* **2016**, *87*, 559–902. [CrossRef]
11. Solano, R.; Salazar-Chávez, G.; Jiménez-Machorro, R.; Hagsater, E.; Cruz-García, G. *Actualización Del Catálogo de Autoridades Taxonómicas de Orchidaceae de México*; Data Base SNIB-CONABIO, Project KT005; Instituto Politécnico Nacional, Centro Interdisciplinario de Investigación para el Desarrollo Integral Regional Unidad Oaxaca: Mexico City, Mexico, 2020.
12. *NORMA Oficial Mexicana NOM-062-ZOO-1999*; Especificaciones Técnicas Para la Producción, Cuidado y Uso de Los Animales de Laboratorio 2001; Diario Oficial de la Federación. SAGARPA (Secretaría de Agricultura, Ganadería, Desarrollo Rural, Pesca y Alimentación): Diario Oficial de la Federación, Mexico, 2011.
13. Matthews, D.R.; Hosker, J.P.; Rudenski, A.S.; Naylor, B.A.; Treacher, D.F.; Turner, R.C. Homeostasis Model Assessment: Insulin Resistance and β-Cell Function from Fasting Plasma Glucose and Insulin Concentrations in Man. *Diabetologia* **1985**, *28*, 412–419. [CrossRef]
14. Simental-Mendía, L.E.; Guerrero-Romero, F. The Correct Formula for the Triglycerides and Glucose Index. *Eur. J. Pediatr.* **2020**, *179*, 1171. [CrossRef]
15. De Souza, M.P.; Bataglion, G.A.; da Silva, F.M.A.; de Almeida, R.A.; Paz, W.H.P.; Nobre, T.A.; Marinho, J.V.N.; Salvador, M.J.; Fidelis, C.H.V.; Acho, L.D.R.; et al. Phenolic and Aroma Compositions of Pitomba Fruit (Talisia Esculenta Radlk.) Assessed by LC-MS/MS and HS-SPME/GC-MS. *Food Res. Int.* **2016**, *83*, 87–94. [CrossRef]
16. Ma, T.; Lin, J.; Gan, A.; Sun, Y.; Sun, Y.; Wang, M.; Wan, M.; Yan, T.; Jia, Y. Qualitative and Quantitative Analysis of the Components in Flowers of Hemerocallis Citrina Baroni by UHPLC–Q-TOF-MS/MS and UHPLC–QQQ-MS/MS and Evaluation of Their Antioxidant Activities. *J. Food Compos. Anal.* **2023**, *120*, 105329. [CrossRef]

17. Dahibhate, N.L.; Dwivedi, P.; Kumar, K. GC–MS and UHPLC-HRMS Based Metabolite Profiling of Bruguiera Gymnorhiza Reveals Key Bioactive Compounds. *South Afr. J. Bot.* **2022**, *149*, 1044–1048. [CrossRef]
18. Paunovic, M.; Milosevic, M.; Mitrovic-ajtic, O.; Velickovic, N. Heliyon Polyphenol-Rich Black Currant and Cornelian Cherry Juices Ameliorate Metabolic Syndrome Induced by a High-Fat High-Fructose Diet in Wistar Rats. *Heliyon* **2024**, *10*, e27709. [CrossRef] [PubMed]
19. Mostafa, S.; Abbas, M.; El-shiekh, R.A.; Bakr, A.F.; Rashad, M.M.; Emam, S.R.; El-banna, H.A. *Vitis vinifera* L. Seed Standardized Extract; a Promising Therapeutic against Metabolic Syndrome Induced by High-Fat/High-Carbohydrate Diet and Streptozotocin in Rats. *S. Afr. J. Bot.* **2024**, *167*, 476–486. [CrossRef]
20. Ma, Y.; Gao, M.; Liu, D. Chlorogenic Acid Improves High Fat Diet-Induced Hepatic Steatosis and Insulin Resistance in Mice. *Pharm. Res.* **2015**, *32*, 1200–1209. [CrossRef] [PubMed]
21. Durg, S.; Veerapur, V.P.; Neelima, S.; Dhadde, S.B. Antidiabetic Activity of Embelia Ribes, Embelin and Its Derivatives: A Systematic Review and Meta-Analysis. *Biomed. Pharmacother.* **2017**, *86*, 195–204. [CrossRef]
22. Ghorbani, A. Biomedicine & Pharmacotherapy Mechanisms of Antidiabetic efects of Flavonoid Rutin. *Biomed. Pharmacother. J.* **2017**, *96*, 305–312. [CrossRef]
23. Chen, Z.; Tian, R.; She, Z.; Cai, J.; Li, H. Role of Oxidative Stress in the Pathogenesis of Nonalcoholic Fatty Liver Disease. *Free Radic. Biol. Med.* **2020**, *152*, 116–141. [CrossRef]
24. Barbosa, P.O.; Tanus-santos, J.E.; Cavalli, R.D.C.; Bengtsson, T.; Barbosa, P.O.; Tanus-santos, J.E.; Cavalli, R.D.C.; Bengtsson, T. The Nitrate-Nitrite-Nitric Oxide Pathway: Potential Role in Mitigating Oxidative Stress in Hypertensive Disorders of Pregnancy. *Nutrients* **2024**, *16*, 1475. [CrossRef]
25. Zeng, L.; Wang, Y.H.; Ai, C.X.; Zhang, B.; Zhang, H.; Liu, Z.M.; Yu, M.H.; Hu, B. Differential Effects of Oxytetracycline on Detoxification and Antioxidant Defense in the Hepatopancreas and Intestine of Chinese Mitten Crab under Cadmium Stress. *Sci. Total Environ.* **2024**, *930*, 172633. [CrossRef] [PubMed]
26. Aouacheri, O.; Saka, S.; Krim, M.; Messaadia, A.; Maidi, I. The Investigation of the Oxidative Stress-Related Parameters in Type2 Diabetes Mellitus. *Can. J. Diabetes* **2015**, *39*, 44–49. [CrossRef]
27. Leghi, G.E.; Domenici, F.A.; Vannucchi, H. Influence of Oxidative Stress and Obesity in Patients with Nonalcoholic Steatohepatitis. *Arq. Gastroenterol.* **2015**, *52*, 228–233. [CrossRef] [PubMed]
28. Barbosa, P.O.; Souza, M.O.; Silva, M.P.S.; Santos, G.T.; Silva, M.E.; Bermano, G.; Freitas, R.N. Açaí (Euterpe Oleracea Martius) Supplementation Improves Oxidative Stress Biomarkers in Liver Tissue of Dams Fed a High-Fat Diet and Increases Antioxidant Enzymes' Gene Expression in Offspring. *Biomed. Pharmacother.* **2021**, *139*, 111627. [CrossRef] [PubMed]
29. Mousavi, A.; Pourakbar, L.; Siavash Moghaddam, S. Effects of Malic Acid and EDTA on Oxidative Stress and Antioxidant Enzymes of Okra (*Abelmoschus Esculentus* L.) Exposed to Cadmium Stress. *Ecotoxicol. Environ. Saf.* **2022**, *248*, 114320. [CrossRef]
30. Hussain, I.; Siddique, A.; Ashraf, M.A.; Rasheed, R.; Ibrahim, M.; Iqbal, M.; Akbar, S.; Imran, M. Does Exogenous Application of Ascorbic Acid Modulate Growth, Photosynthetic Pigments and Oxidative Defense in Okra (*Abelmoschus esculentus* (L.) Moench) under Lead Stress? *Acta Physiol. Plant.* **2017**, *39*, 144. [CrossRef]
31. Londero, É.P.; Bressan, C.A.; Pês, T.S.; Saccol, E.M.H.; Baldisserotto, B.; Finamor, I.A.; Pavanato, M.A. Rutin-Added Diet Protects Silver Catfish Liver against Oxytetracycline-Induced Oxidative Stress and Apoptosis. *Comp. Biochem. Physiol. Part C Toxicol. Pharmacol.* **2021**, *239*, 108848. [CrossRef]
32. Wang, Y.; Tang, C.; Zhang, H. Hepatoprotective Effects of Kaempferol 3-O-Rutinoside and Kaempferol 3-O-Glucoside from *Carthamus Tinctorius* L. on CCl4-Induced Oxidative Liver Injury in Mice. *J. Food Drug Anal.* **2015**, *23*, 310–317. [CrossRef]
33. Gupta, R.; Sharma, A.K.; Sharma, M.C.; Gupta, R.S. Antioxidant Activity and Protection of Pancreatic b -Cells by Embelin in Streptozotocin-Induced Diabetes. *J. Diabetes* **2012**, *4*, 248–256. [CrossRef]
34. Nonose, Y.; Pieper, L.Z.; da Silva, J.S.; Longoni, A.; Apel, R.V.; Meira-Martins, L.A.; Grings, M.; Leipnitz, G.; Souza, D.O.; de Assis, A.M. Guanosine Enhances Glutamate Uptake and Oxidation, Preventing Oxidative Stress in Mouse Hippocampal Slices Submitted to High Glutamate Levels. *Brain Res.* **2020**, *1748*, 147080. [CrossRef]

foods

MDPI

Review

Recent Insights into Eco-Friendly Extraction Techniques for Obtaining Bioactive Compounds from Fruit Seed Oils

Sandra Rodríguez-Blázquez *, Esther Gómez-Mejía, Noelia Rosales-Conrado and María Eugenia León-González *

Department of Analytical Chemistry, Faculty of Chemistry Sciences, Complutense University of Madrid, 28040 Madrid, Spain; egomez03@ucm.es (E.G.-M.); nrosales@ucm.es (N.R.-C.)
* Correspondence: sandro08@ucm.es (S.R.-B.); leongon@ucm.es (M.E.L.-G.)

Abstract: The valorization of agri-food waste has emerged as a global priority. In this context, fruit seed waste is being investigated for oil extraction due to its richness in bioactive compounds with remarkable health benefits. This review (2020–2025) focuses on the current state of eco-friendly extraction techniques for obtaining high-yield oils enriched with compounds such as tocopherols, polyphenols, fatty acids, phytosterols, and carotenoids. A comparison of the present method with conventional extraction techniques reveals several notable distinctions. Conventional methods are generally characterized by prolonged extraction times, elevated temperatures, and high amounts of solvents and/or energy. The findings of this review suggest that the extraction methodologies employed exerts a substantial influence on the yield and bioactive composition of the oil, which in turn affects its health-promoting properties. Furthermore, the results have demonstrated that alternative methodologies (microwave-assisted extraction, ultrasound-assisted extraction, pressurized liquid extraction, electric pulse extraction, enzyme-assisted extraction, subcritical extraction, and combinations thereof) have analogous oil yields in comparison with conventional methods. In addition, these oils present a superior bioactive profile with feasible potential in industrial and health applications. The novelty of this work lies in its emphasis on the valorization of fruit seed waste, as well as its sustainable approach. This sustainable approach utilizes experimental design strategies, the implementation of developments that employ comprehensive ecological metrics, and the latest trends in the application of artificial intelligence.

Keywords: agri-food residues; fruit seed oils; bioactive compounds; green extraction techniques; design of experiments; green metrics; artificial intelligence; industrial applications

1. Introduction

Vegetable oils are natural extracts derived from seeds, fruit pulps, pits, sprouts, or tubers of various plants, which serve as nutritionally dense sources of energy [1]. Beyond their caloric value, these oils are rich in bioactive compounds that offer a wide range of health-promoting properties, including antioxidant, antibacterial, antifungal, anticancer, anti-inflammatory, and neuroprotective activities [2,3]. Furthermore, their versatility has led to the widespread use of vegetables oils across multiple industries, from nutrition and healthcare to cosmetics and everyday products such as varnishes, adhesives, synthetic resins, lubricants, soaps, paints, and bactericides [1,4].

Driven by global population growth, rising economic standards, and increasing awareness of the health benefits of vegetable oils, global demand continues to surge. In 2019–2020 alone, global consumption reached 200.27 million metric tons [1]. Meeting this growing demand, both for industrial applications and human consumption, requires the production of high-quality oils in substantial quantities. At the same time, consumers are increasingly seeking foods that not only satisfy sensory and nutritional expectations but also produced through environmentally responsible processes [4]. Preserving a greater proportion of bioactive compounds in the final product thus presents a significant challenge for the industry [1,4].

Among the various stages involved in vegetable oil processing, conventional techniques such as cold mechanical pressing are still widely used [1]. However, these methods often yield low oil recovery, as a significant portion remains trapped in the solid residue. To improve yields, organic solvents are commonly employed to extract the remaining oil. Yet, there is growing interest in reducing solvent use due to concerns over high removal costs, toxicity, safety risks, and environmental pollution [1]. In response, innovative extraction methods have been explored, particularly those utilizing waste materials from commonly consumed fruits such as avocados, cherries, plums, peaches, *citrus* fruits, and tropical varieties. These by-products represent an abundant and cost-effective source of bioactive-rich vegetable oils. Converting food waste into valuable bioproducts like these oils is a cornerstone of advancing global sustainability and fostering a circular economy. This waste management strategy also aligns with the United Nations' Sustainable Development Goals (SDGs) outlined in the 2030 Agenda, promoting sustainable practices and reducing environmental impact [5,6].

This has spurred the development of innovative and environmentally friendly extraction technologies, including ultrasound-assisted extraction (UAE) [7], microwave-assisted extraction (MAE) [8], pressurized liquid extraction (PLE) [7], pulsed electric fields (PEFs) [9], subcritical extraction [10], and enzyme-assisted extraction (EAE) [11], as well as hybrid and sequential approaches, such as the combination of supercritical fluid extraction and ultrasound-assisted extraction (SFE-UAE) [12] or the complex pulsed electric field and supercritical fluid extraction (PEF-SFE) [9] or ultrasound-microwave assisted extraction (UMAE) [13]. These techniques aim to maximize the recovery of bioactive compounds while minimizing environmental impact, by shortening extraction times, reducing solvent use, and optimizing solvent nature, to name just a few.

Currently, many of these methods are being optimized or developed at the laboratory scale. As such, it is essential not only to assess their extraction efficiency but also to evaluate their overall sustainability. In this regard, chemometric tools such as experimental design, along with modeling and predictive tools based on artificial intelligence, can play a crucial role in determining the practical feasibility of large scale implementation and identifying optimal operating conditions.

Accordingly, in light of the aforementioned considerations, the primary objective of this review is to provide a critical and comprehensive analysis of green and emerging technologies for the extraction of bioactive compounds from fruit seeds, an agri-food waste that holds significant untapped potential. The scope encompasses a detailed evaluation of current eco-friendly extraction strategies, with particular attention given to hybrid techniques, the integration of simulation tools for assessing the sustainability of developed processes, and the application of artificial intelligence as both a predictive and transformative tool to enhance process scalability and maximize the extraction efficiency. From this targeted perspective, this review aims to underscore both the technological advances and sustainability implications of these approaches, offering insights into their practical applicability and limitations.

The selection of studies was primarily conducted using the Scopus and ScienceDirect databases. The search strategy employed the following sequence of keywords: (seed oil), (waste or residue), (eco-friendly extraction or method), and (bioactive). Only articles and review papers published in English between 2020 and 2025 were considered. Additionally, the subject area was restricted to 'Chemistry', and duplicate entries were removed to avoid redundancy (Figure 1a). Studies offering substantial contributions to the field were selected for in-depth analysis, with a particular focus on those involving the extraction of oils from food waste sources such as plum seeds, citrus seeds, and avocado seeds, among others. As shown in Figure 1b, the annual distribution of publications retrieved from Scopus demonstrates a clear upward trajectory in research activity within this field since 2012. Although minor fluctuations are observed, the overall trend remains positive, culminating in 31 publications by 2024.

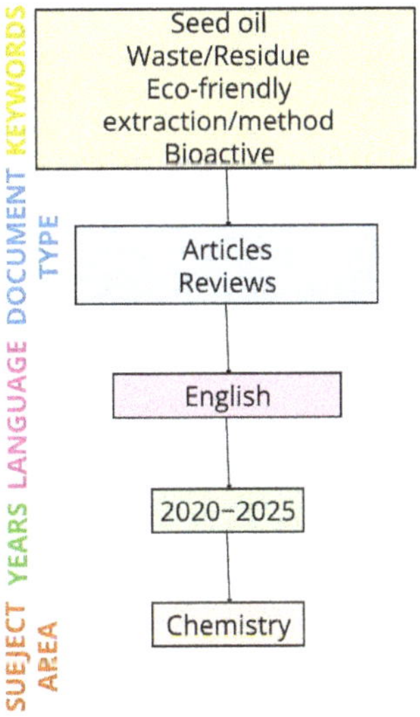

(a)

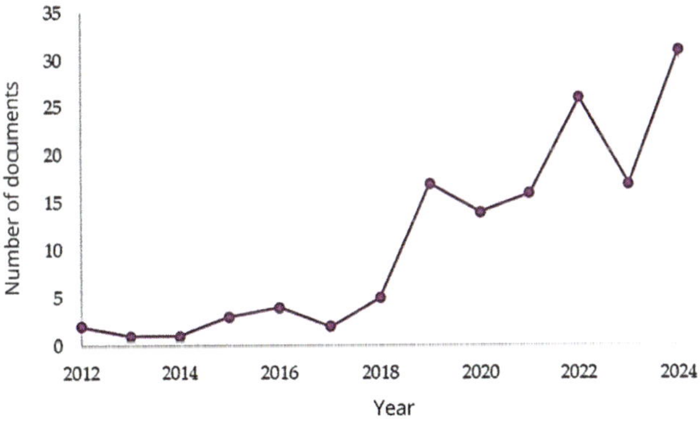

(b)

Figure 1. (**a**) Flow diagram illustrating the bibliographic selection process employed in this review. (**b**) Annual distribution of documents retrieved from Scopus using the keyword sequence outlined in the Methodology section.

As a complementary element, Figure 2 displays a network of keywords identified based on co-occurrence within the analyzed literature, considering a minimum threshold of three occurrences for the selected terms ("seed oil", "waste or residue", "eco-friendly extraction or method", and "bioactive") between 2012 and 2025. The connecting lines illustrate the relationships among the keywords, while the size of each circle indicates the relative frequency or weight of the term. Greater distances between circles reflect a lower degree of correlation or thematic association. As can be seen, it reveals significant thematic intersections among terms such as "plant seed" (red cluster), "lipid composition" (green cluster), "flavour compounds" (blue cluster), and "food chemistry" (yellow cluster). Moreover, the term "extraction" appears in distinct configurations—specifically as "extraction techniques" and as "extraction"—within two separate clusters, represented in yellow and green, respectively. This distinction underscores the influence of the employed technique on both the chemical composition of the extracted compounds and the sustainability of the process.

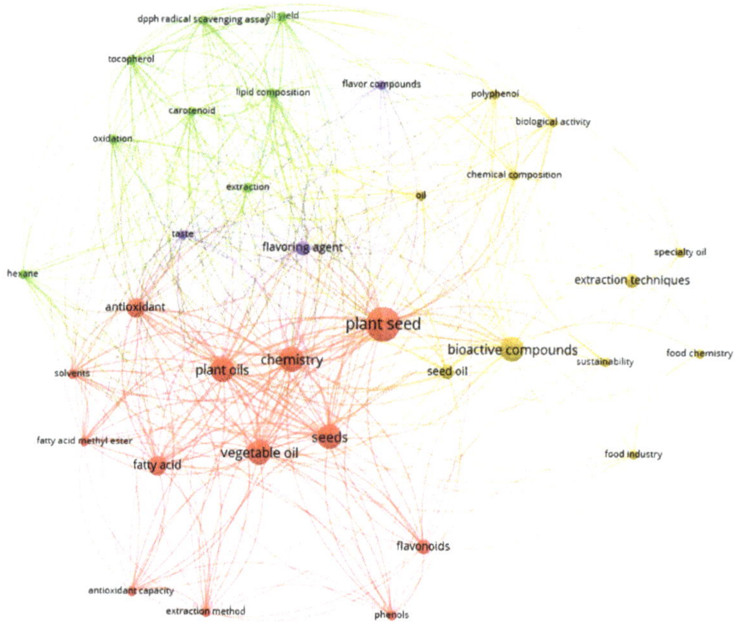

Figure 2. Keyword co-occurrence map generated using VOSviewer software (version 1.6.20), based on 150 documents retrieved from Scopus between 2012 and 2025. Clusters and their corresponding colors were automatically assigned according to a minimum threshold of three co-occurrences among the selected keywords ("seed oil", "waste or residue", "eco-friendly extraction or method", and "bioactive").

On the other hand, Figure 2 highlights the conceptual interconnectivity and thematic clustering of these domains, reflecting an interdisciplinary approach to the study of bioactive compound and plant-derived oil extraction. Notably, these topics frequently co-occur with keywords such as "vegetable oil", "plant seed", "seed oil", and "plant oils" within a more specific context denoting various bioactive compounds including fatty acids, phenols, flavonoids, tocopherols, and carotenoids (see red and green clusters). Furthermore, these terms are interconnected with themes related to extraction techniques, "chemistry", "food industry/chemistry", and "sustainability", underscoring the pivotal role of analytical chemistry in the development and optimization of sustainable and innovative extraction methodologies that can significantly impact and contribute to the functioning of the agri-food industry.

2. Valorization of Fruit Seeds as Agri-Food Residues

The fruit and vegetable industry generates a substantial amount of waste, including peels, seeds, pomace, and pulp, which can account for between 25% and 30% of the total product volume. This waste is often classified as industrial waste, and its improper disposal can have a negative environmental impact. Furthermore, it should be noted that many of these products contain valuable bioactive substances. In the case of seeds, lipids, phytosterols, tocopherols, phenolic compounds, and carotenoids, among others, can be found [14]. In many cases these fruit seeds are discarded, and that is an economic loss and an environmental issue. The utilization of edible seeds for oil extraction has a long historical precedent, primarily due to their high lipid content. However, many seeds, including those from grapes and pomegranates, have not been employed in a commercial capacity for oil extraction until sufficient information regarding their potential health benefits has become available [15]. The extraction of bioactive compounds from agri-food waste has applications in various industrial sectors, including food, pharmaceuticals, and cosmetics. However, the sustainable extraction of these compounds presents various challenges. These include selective extraction to avoid the co-extraction of undesirable substances, the possibility of degradation of the compounds subject to extraction, or the consumption of solvents in some extraction procedures and the need for a balance between the costs and benefits of the overall extraction process [5]. To obtain extracts that can be used at an industrial level, conventional extractions include Soxhlet extraction, cold pressing, or supercritical CO_2 extraction. These complete extracts have been used as such in the pharmaceutical industry in herbal preparations, as food supplements, or as ingredients in cosmetics. Not only do conventional techniques require a long extraction time and high energy consumption, but also a large volume of solvent in the case of Soxhlet extraction. To obtain higher bioactive recoveries, to use smaller amounts of solvents, or even to use so-called green solvents, unconventional extraction techniques have been employed both on a laboratory scale and on an industrial scale [16,17].

Each method of oil extraction has advantages and disadvantages from an environmental, practical, and economic point of view. Information on the environmental impact between extraction methods is limited, although there is scientific evidence on the adverse effects of solvents such as *n*-hexane. Another aspect that must be considered when implementing an extraction technique is economic viability, especially when it is desired to develop on an industrial scale. A techno-economic analysis on benefits and costs will include the valorization of by-products and the added value that can be derived from a higher quality of the oil by co-extraction of other compounds. It is imperative to consider the preservation of easily degradable compounds in certain experimental conditions that are indispensable to the extraction process. Such conditions may encompass elevated temperatures or pressure, among others [18]. For all these reasons, further research is required to study its practical and economic feasibility at an industrial level.

The efficacy of the extraction technique used in the isolation of compounds of interest from biological matrices is paramount. Consequently, a range of technologies have been developed for the extraction of bioactive compounds from diverse types of seeds that have not traditionally been utilized for oil extraction. A notable illustration of this phenomenon is the extraction of oil from the seeds of passion fruit by means of cold pressing [19,20], by industrial-scaled hydraulic pressing [21], by low-temperature continuous phase transition [22], by ultrasound-assisted extraction [11], by continuous expeller press [23], by supercritical CO_2 extraction assisted by ultrasound [12], by subcritical fluid extraction [24], or by surfactant assisted aqueous extraction [25]. Other seeds widely studied for oil extraction and valorization are pomegranate [26,27] and citrus seeds [28–30], using various methods such as MAE, UAE, extraction in supercritical fluid or subcritical fluid and/or combined with EAE. As an example, Table 1 includes different pomegranate seed oil extraction procedures. It is evident that the extraction yield and the time required for extraction can vary significantly. This variation is attributed to the type of solvent or solvents employed, as well as the geographical origin of the sample.

Table 1. Effect of the extraction method on the oil yield of pomegranate seeds.

Extraction Method	Solvent	Extraction Time	Oil Recovery (wt%)	References
Soxhlet	*n*-hexane	8 h	15.66	[25]
Soxhlet	Petroleum ether	8 h	14.92	[25]
Shaking extraction	*n*-hexane	2 h	12.10	[25]
Shaking extraction	Petroleum ether	2 h	11.32	[25]
Mechanical pressing	-	8 min	9.00	[31]
MAE	*n*-hexane	9 min	12.47	[25]
MAE	Petroleum ether	9 min	11.72	[25]
MAE	*n*-hexane or petroleum ether	9 min	11.72–12.47	[32]
MAE	*n*-hexane	5 min	35.19	[33]
UAE	*n*-hexane	18 min	13.63	[25]
UAE	Petroleum ether	18 min	12.82	[25]
UAE	water	2–8 h	4.29–25.11	[26]
SubE	*n*-butane	40 min	14.5	[25]
Sc-CO_2	CO_2	2 h	16	[34]

Other oils have been extracted using supercritical fluids under different conditions from seeds such as *Sapucaia* (*Lecythis pisonis*) nuts [35], *Pracaxi* (*Pentaclethra macroloba*) seed [36], blackberry seeds, [32,37], *Butia odorata* seed [38], apple seeds [39], cherry seed [40] and raspberry seed [41].

In all cases, before choosing the extraction method, the complex nature of the seeds must be considered, since the selectivity of the extraction will be given both by the selected extraction method and its extraction conditions and by the nature of the matrix on which the extraction is applied. A considerable number of the methods employed are founded, at least in part, on the principles of green chemistry, which include the elimination of toxic solvents and minimal energy consumption. In numerous instances, the outcomes are contrasted with those attained by conventional extraction techniques. Nevertheless, the evaluation of the activity of the extracts and their chemical composition are the key to developing industrial applications that allow for the valorization of these wastes (Figure 3).

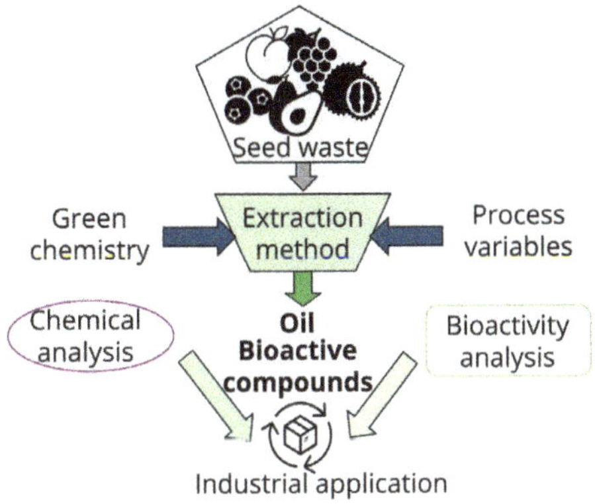

Figure 3. Waste seed recovery scheme.

3. Bioactive Compounds in Fruit Seed Oils: Nutritional Value, Application Potential, and Determination Methods

Many fruit seed oils are nutritionally rich, containing high levels of bioactive compounds including antioxidants, vitamins, essential fatty acids, and phytonutrients, thereby rendering them suitable for diverse applications in food, nutraceuticals, cosmetics, and functional products (Figure 4) [14].

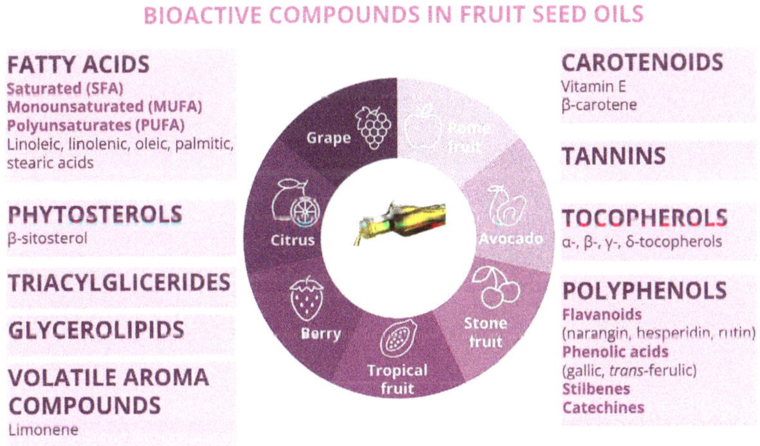

Figure 4. Representative bioactive compounds present in fruit seed oils.

Several factors have been reported to influence the quality, yield, and potential applications of fruit seed oil. These include preharvest conditions—such as the cultivar and the fruit's growing region—as well as the processing steps involved in oil production, including seed drying, pretreatment, and extraction methods. Together, these factors have a significant impact on the oil's chemical composition and functional properties [42].

Grape seeds, a by-product of winemaking and juice production, are among the most valuable components derived from grapes. They contain approximately 7% to 20% of oil, depending on the variety and processing methods. Grape seed oil is notably rich in saturated (SFAs), monounsaturated (MUFAs), and polyunsaturated fatty acids (PUFAs), along with lipophilic compounds such as vitamin E and phytosterols. In addition, it contains a wide range of hydrophilic bioactives, including polyphenols (such as flavonoids, phenolic acids, stilbenes, catechins, and epicatechins), carotenoids, and tannins. These diverse constituents contribute to the oil's potent antioxidant, anti-inflammatory, anticancer, and antibacterial properties, as well as its potential in preventing hypertension and supporting cardiovascular health. As a result, grape seed oil holds great promise for use in dietary supplements, anti-aging skincare products, therapeutic formulations, and other health-promoting applications [14]. Grapefruit seed oil is particularly notable for its high content of linoleic acid, β-sitosterol, and α-tocopherol, as well as flavonoids such as naringin, hesperidin, neohesperidin, and rutin. It also contains phenolic acids, with *trans*-ferulic acid and gallic acid being the most prominent. Additionally, a total of 26 volatile aromatic compounds have been identified in grapefruit seed oil, including d-limonene, furfural, and 3-methylbutanal [43].

Citrus fruit seeds are also valuable sources of edible oil, particularly rich in unsaturated fatty acids, including essential omega-6 and omega-3 fatty acids. In addition to these, *citrus* seed oils contain significant amounts of bioactive compounds such as tocopherols (vitamin E), limonoids, carotenoids, and phenolic compounds—primarily flavonoids. The high tocopherol content, combined with these phenolics, contributes to the oil's strong antioxidant capacity, helping stabilize unsaturated fatty acids and providing effective free radical scavenging activity. As a result, *citrus* seed oils offer various health benefits, including cardiovascular disease prevention and immune modulation in inflammatory conditions. Notably, lemon seed oil contains a high concentration of limonene, a compound that has shown promise in early-stage cancer treatment [14], while sweet orange seed oil is a promising nutritional source with prominent antioxidant potential. It contains high levels of unsaturated fatty acids, particularly MUFA and PUFA types such as oleic, linoleic, and α-linolenic acids, all of which are associated with potential health benefits in humans [30]. Mandarin (*Citrus reticulata*) and lemon (*Citrus limon*) seed oils are characterized by the presence of key fatty acids, including linoleic, oleic, palmitic, α-linolenic, and stearic acids. Furthermore, the citrus oils also contain triacylglycerols, notably monolinolein and dilinolein derivatives [29].

Regarding tropical and exotic fruit seeds, they emerge as promising sources of functional and edible vegetable oils due to their high lipid content, offering valuable opportunities for selective use as industrial raw materials. The dominant fatty acids typically found in these oils are palmitic, oleic, and linoleic acids. Lipidomic studies have shown that passion fruit, cherimoya, and durian seeds are particularly rich in glycerolipids, sphingolipids, and fatty acyls, respectively, while litchi seeds contain high levels of glycerophospholipids and saccharolipids. Moreover, oils extracted from passion fruit and cherimoya seeds are especially rich in UFA, making them well suited for daily dietary lipid intake in alignment with internationally recognized dietary patterns such as the Mediterranean diet [44].Vitamin E and β-carotene have been also determined in passion fruit seed oil [22], as well as oleamides (particularly 9-octadecenamide) [32] and volatile aroma compounds [20]. Oil extracted from yellow passion fruit (*Passiflora edulis var. flavicarpa*) seeds is rich in unsaturated fatty acids and tocopherols. It also exhibits strong antioxidant properties and notable antibacterial activity against *Escherichia coli, Salmonella enteritidis, Staphylococcus aureus*, and *Bacillus cereus* [24].

Furthermore, dragon fruit seeds are highly nutritious, particularly in terms of essential fatty acids, and the use of seed oils has become increasingly popular. The oil primarily contains PUFAs, with linoleic acid being the most abundant. It also includes various tocopherol groups, such as α-tocopherol, γ-tocopherol, and δ-tocopherol. Dragon fruit seed oil is known for its anti-aging properties, helping to smooth and firm the skin. Rich in antioxidants, it is a valuable addition to creams, lotions, and moisturizers. In addition, due to its high levels of polyphenols and unsaturated fats, this oil holds significant value and can be microencapsulated for use in a variety of products [8]. Conversely, seed oil from cactus tropical fruit has also exhibited important amounts of linoleic acid, α-tocopherol, and canolol phenolic acid [45].

On the other hand, berry seeds are an important by-product, with their oils offering a rich source of essential nutrients. The primary fatty acids found in most berry seed oils are linoleic acid, linolenic acid, and oleic acid, resulting in low n−6/n−3 ratios. These oils typically contain around 90% UFAs, with a high concentration of PUFAs. Following oil extraction, significant amounts of antioxidants, phenolic compounds (such as phenolic acids and flavonoids), vitamins, phytosterols, and carotenoids remain in the oils. Berry seed oils are also rich in tocopherols, particularly α- and δ-tocopherol, with β-sitosterol being the primary sterol. Additionally, some berry seed oils are excellent sources of vitamins C and E, enhancing their value in the food, pharmaceutical, and cosmetic industries. These oils also exhibit antioxidant, antimicrobial, and anti-carcinogenic properties [14]. Particularly, blackberry seed oils have exhibited high contents of unsaturated fatty acid (mainly linoleic acid), β-carotene [37], polyphenols (mainly vanillin), and oleamides (9-octadecenamide) [32].

Stone fruits, belonging to the Rosaceae family, are distinguished by a hard pit or endocarp shell that surrounds the seeds or kernels. The most common stone fruits include almonds (*Prunus dulcis*), peaches (*Prunus persica*), nectarines (*Prunus persica* var. *nucipersica*), plums (*Prunus domestica*), cherries (Sweet: *Prunus avium* L; Sour: *Prunus cerasus* L.), and apricots (*Prunus armeniaca*). Seed shells contain high concentrations of toxic amygdalin, particularly in apricots. This toxicity limits the use of stone fruit seed oils in the food industry, driving interest in their potential applications in biodiesel production and cosmetics. The oils are rich in fatty acids and γ-tocopherol, along with sterols, primarily β-sitosterol, although some stone fruit seed oils are also distinguished for their high levels of potassium, magnesium, and B-complex vitamins [14]. In general, stone fruit oils and, particularly, plum seed oils, have exhibited good antioxidant activity and oxidative stability, thus showing potential application in the cosmetic and nutraceutical industries [2,46]. Additionally, cherry seed oil, which is rich in fatty acids, particularly linoleic, oleic, palmitic, and stearic acids, as well as γ-tocopherol, has demonstrated strong antioxidant potential, making it also suitable for use in the food and pharmaceutical industries [40].

In contrast, research on pome fruit seed oils has been limited, primarily due to the small size of the seeds. However, apple and pear seed oils have been found to be rich in fatty acids, particularly oleic and linoleic acids, as well as tocopherols (α-, β-, γ-, and δ-tocopherols) and carotenoids. These oils exhibit notable antibacterial and antioxidant properties, along with potential cytotoxic effects against certain cancer cells [14]. Specifically, apple seed oils are rich in antioxidant phenolic compounds such as phlorizin, phloretin, naringenin, vanillin, ferulic acid, and 4-hydroxybenzoic acid [39]. Pear seed oil is also viable and useful for sustainable biodiesel production [47], while pomegranate seed oil has been abundant in PUFAs (especially punicic acid), triacylglycerol molecules, squalene, β-sitosterol, and tocopherols (α, γ, and δ), thus exhibiting high nutritional properties [26].

Furthermore, the avocado processing industry generates a substantial amount of by-products, including avocado seeds. Avocados are well known for their rich nutritional profile, particularly due to their high content of MUFAs, especially oleic acid. The oil content of avocados is also of great significance, as they are rich in lipophilic components such as fatty acids (palmitic, oleic, and linoleic), tocopherols, and sterols. The hydrophilic components of avocados, including polyphenols and carotenoids, provide numerous health benefits and have proven effective as natural preservatives and antibacterial agents.

Moreover, avocado oil has shown antidiabetic and hypotensive effects, mainly attributed to key compounds like stigmasterol and campestanol [48].

Otherwise, when determining the bioactive compounds in fruit seed oils, both the spectrophotometric and chromatographic methods are commonly used. Thus, the total polyphenol content (TPC) is often estimated using the simple and rapid Folin–Ciocalteu assay, by measuring the absorption of a blue complex at 765 nm, after an incubation period, and by employing gallic acid as standard [22,29,30,39]. However, this method does not differentiate between individual polyphenolic compounds and can be affected by other reducing substances present in the sample. In contrast, profiling the phenolic compounds involves identifying and quantifying individual polyphenols, typically achieved through high-performance liquid chromatography (HPLC) or ultra-HPLC coupled to diode-array (DAD) [39,43] or mass spectrometry (MS) [45] detectors. Separations are mainly conducted in the reversed phase under gradient elution mode, using non-polar C18 stationary phases and mixtures of organic solvents (acetonitrile or methanol) with additives such as ammonium acetate [45], formic acid [45] and sulphuric acid [43], or mixtures of methanol and water [39], as the mobile phase. While more specific and sensitive than spectrophotometric assays, HPLC methods require more advanced techniques and expertise.

α-, β, γ-, and δ-Tocopherol levels are usually determined using liquid chromatography in both the reversed phase [24,43,45] and normal phase modes [26,40], often with DAD [24,43] and fluorescence [8,26,40,45] detection. Common mobile phases are constituted with tetrahydrofurane–hexane [40], dioxane–hexane [8], and isopropanol–hexane [26] mixtures in the normal phase and methanol–water [43,45] and methanol–isopropanol [24] in the reversed phase mode.

Spectrophotometric methods are also used to assess the β-carotene content in fruit seed oils by measuring the absorbance of diluted samples in *n*-hexane [37,49] or dichloromethane [22] at 450 nm.

For fatty acid determination, gas chromatography (GC) is employed after a derivatization step to obtain fatty acid methyl esters (FAMEs), followed by temperature programming and detection through MS [22,28,32,50] or flame ionization detection (FID) [8,26,29,37,40,43–45]. Several reagents have been employed to obtain derivatized fatty acids, such as sodium hydroxide–methanol [44,50], boron–trifluoride methanol [40], BF_3 methanolic [30,44], sodium methoxide [26], and KOH methanolic [28] solutions. A wide range of columns have been also used for the determination of FAMEs, including SH-Rtx-Wax (30 m × 0.32 mm ID × 0.25 µm) [37], SP-2560 (100 m × 0.25 mm ID, 0.20 µm) [40], RTX-wax (30 m × 0.25 mm ID × 0.25 µm) [30], CP-Sil 88 (60 m × 0.25 mm ID × 0.20 µm) [26] (100 m × 0.25 mm ID × 0.20 µm) [24], Agilent HP-50 (30 m × 0.32 mm ID × 0.25 µm) [8,45], cyanopropyl methylpolysiloxane HP-23 [29], HP-5MS UltraInert (30 m × 0.25 mm ID × 0.25 µm) [22], DB 5 ms (30 m × 0.25 mm × 0.25 µm) [32], HP-88 (100 m × 0.25 mm ID × 0.20 µm) [43], and SUPELCOWAX (30 m × 0.25 mm ID × 0.25 µm) [28] fused silica capillary columns.

4. Conventional Extraction Methods: Limitations and Challenges

The extraction of fruit seed oils is typically accomplished through conventional methodologies, encompassing mechanical pressing, solvent extraction, and supercritical CO_2 extraction. It is evident that these methodologies have attained a high level of industry-wide adoption; nevertheless, it is imperative to acknowledge their inherent limitations [1,51]. Hence, this section discusses the principles of each extraction method, the factors influencing extraction yield, and the effect of the extraction method used on the bioactive composition of the oils, as well as the limitations and challenges of each extraction method.

4.1. Mechanical Pressing

Mechanical pressing is a solvent-free method that is often used for the extraction of oils from seed samples. The conventional mechanical extraction technique for oil production offers numerous benefits, including the superior quality of the extracted oil, the expeditious nature of the extraction process, the absence of organic solvents, which promotes process sustainability, and its straightforward scalability for industrial implementation. Nonetheless, the process is accompanied by several disadvantages such as low capacity per batch, low oil recovery percentage, high energy consumption, and high dependence on moisture content [1,51].

The process relies on the mechanical pressing of seeds, divided into two main stages: material preparation and subsequent oil extraction (Figure 5). The preparation of the seed material consists of cleaning, preheating, breaking, and/or crushing [18,51]. Subsequently, as shown in Figure 5, seeds are collecting and pressing into the barrel surrounding the screw and the resulting oil is collected in a basin under the screw, and the defatted seeds is discharged at the end of the screw. Finally, due to the high content of impurities in the oil (such as pieces of non-extracted seeds) the oil is centrifuged and separated from the pellet [1].

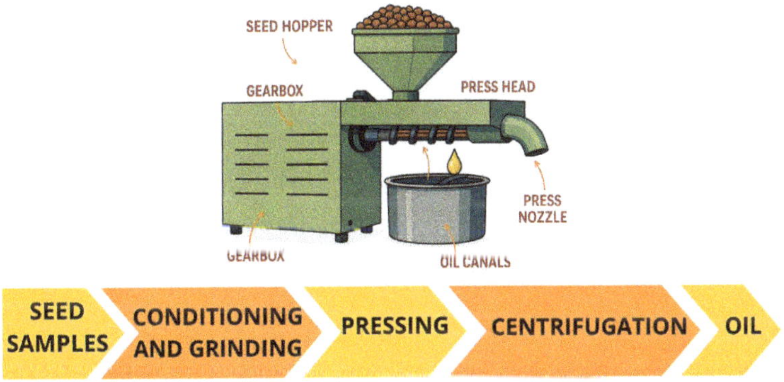

Figure 5. Flow chart of oil extraction through mechanical pressing.

According to mechanical pressing, different factors directly affect the oil recovery, such as the moisture content of the seed samples, the pretreatment of the seeds (e.g., roasting), the pressure, and the temperature. With regard to moisture content, normally an increase in moisture content has different effects on oil yield depending on the seed species studied. In general, an increase generally leads to a decrease in oil yield. Hasanov et al. [31] showed that by increasing flaxseed moisture content from 6 to 11–15.57% the oil recovery decreased moderately from 13.17 to 4.34%. They demonstrated this finding on the basis that higher moisture content in the raw material could exert a lubricating effect and significantly reduce the friction inside the press, thus resulting in lower oil yield. Another crucial factor when controlling the mechanical pressing is the pressure of the system. In general, the more pressure applied to the system, the higher the oil extraction yield [52]. Niu et al. [53] demonstrated that roasting tomato seeds previously at 130 to 170 °C significantly increased the oil content, increasing from 18.72 to 22.16%. This phenomenon was explained by the fact that the heat treatment process promoted oil release by decreasing enzymatic activity, coagulating proteins, and inducing structural changes in the seeds. The final factor to be considered in mechanical seed pressing is the selected extraction temperature. According to this factor, two distinct types of mechanical pressing are typically employed in the extraction of seed oils. These methods are referred to as cold pressing and hot pressing, respectively. The process of hot pressing involves the extraction of the desired substance at temperatures that typically range from 90 to 100 °C [54]. Cold-pressed oils are well known to be healthier because they have higher concentrations of phenolic compounds, tocopherols, phytosterols, fatty acids, and carotenoids, whereas bioactive compounds in hot-pressed oils are more prone to degradation due to the elevated temperatures used in this technique [55]. However, hot pressing correlates with higher oil extraction yields, with up to 80% of the oil present in the oilseed being recovered [54]. Specifically, Zhen et al. [56], found that the oil yield in flaxseed oil was significantly different using cold pressing (22.2%) and hot pressing, with the highest oil yield being obtained in hot pressing (32.7%).

Cold pressing has been widely employed for the extraction of fruit seed oils, such as pomegranate seeds, cherry seeds, passion fruit seeds, and lemon seeds, among others. The related information taken from the literature data is shown in Table 2.

Table 2. Literature data on the application of hot and cold pressing in the extraction of fruit seed oils.

Food Waste Material	Seed Mass (g)	Mechanical Pressing Method	Temperature Pressing (°C)	Oil Recovery (wt%)	Content of Principal Bioactive Compounds	References
Pomegranate seeds (*Punica granatum* L.)	200	Cold pressing YODA YD-ZY-02A (Yoda, Warsaw, Poland)	28–32	9	Linolenic acid: 80% Oleic acid: 5% Linoleic acid: 5%	[57]
Cherry seeds	6000	Cold pressing (SZR "Mikron", Temerin, Serbia)	70	4	Linoleic acid: 46.82% Oleic acid: 41.92% α-tocopherol: 6.39 $mg \cdot 100\ g^{-1}$ β-tocopherol: 1.07 $mg \cdot 100\ g^{-1}$ γ-tocopherol: 25.22 $mg \cdot 100\ g^{-1}$ δ-tocopherol: 5.41 $mg \cdot 100\ g^{-1}$	[40]
Passion fruit seeds	200	Cold pressing (grain type MKS-J03)	62	6–12	Linoleic acid: 38–70% Oleic acid: 9–17%	[58]
Muskmelon (*Cucumis bisexualis*) seeds	100	Hot pressing (DH-50 screw oil press)	90	23	Oleic acid: 49.60% Linoleic acid: 26.60%	[59]
Lemon (*Citrus limon* L.) seeds	3000	Cold pressing (Koçmaksan, ESM 3710, İzmir, Turkey)	40	17	Palmitic acid: 20.88% Oleic acid: 30.86% Linoleic acid: 30.77% Beta-sitosterol: 76.55% α-tocopherol: 155 $mg \cdot kg^{-1}$	[60]
Berry seeds	n.d.	Cold pressing (IBG Momforts Oekotec, DD85G, Mönchengladbach, Germany)	50	n.d.	TPC: 8.9–19.3 mg GAE·100 g^{-1} Brassicasterol: <0.1–465.7 $mg \cdot kg^{-1}$ Campesterol: <0.1–200 $mg \cdot kg^{-1}$ Stigmasterol: <0.1–34 $mg \cdot kg^{-1}$	[61]

n.d.: not declared.

4.2. Solvent Extraction

Solvent extraction is another traditional seed oil extraction. When comparing with the mechanical pressing, it is less sustainable, although this extraction technique allows for higher oil yields (oil recovery rate from 90 to 98%). However, this technique necessitates a substantial solvent demand and significant time and energy expenditure. The fundamental principle underlying this extraction technique involves the utilization of a solvent that comes into direct contact with the seed to extract the oil. In certain instances, the application of heat and agitation may be employed to enhance the extraction process. Once the solid–liquid extraction is completed, the mixture is centrifuged to remove the defatted seed from the oil phase, and then the organic solvent is evaporated from the oil (Figure 6) [1,62]. Based on this principle, there are several methodologies that can be employed (e.g., maceration, Soxhlet extraction, shaking extraction, and distillation) [1].

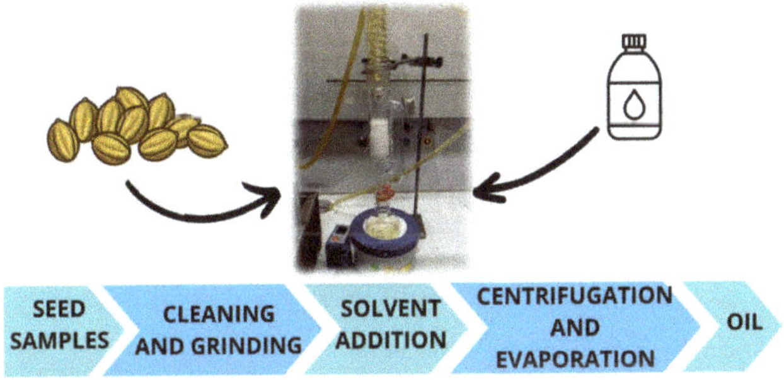

Figure 6. Flow chart of solvent oil extraction.

Oil recovery by solvent extraction mostly depends on the particle size of the oilseed, as well as the extraction temperature, the type and nature of the solvent (polar and non-polar), the solid–solvent ratio, and the extraction time [55].

Soxhlet extraction is the main classical methodology for obtaining vegetable oils from the oleaginous matter of seeds. This technique allows for the extraction of oils with high efficiency and reproducibility, especially in laboratory and research contexts, but its use is limited on an industrial scale, being replaced by continuous extraction methods that are more efficient and compatible with large-scale production. The solvent selection for this methodology is mainly based on the highest leaching properties of the required solute substrate. Among the main solvents used in seed oil extraction are *n*-hexane, ethanol, methanol, chloroform, diethyl ether, petroleum ether, and acetone. Gonfa Keneni et al. [63] studied different solvents (*n*-hexane, ethanol, diethyl ether, and heptane) for the Soxhlet extraction of oils from Jatropha seeds and observed that the highest extraction yield was obtained using *n*-hexane (41.24%). Liu et al. [26] evaluated the effect of *n*-hexane or petroleum ether on the recovery of pomegranate (*Punica granatum* L.) seed oil by Soxhlet extraction and observed that *n*-hexane was the most effective. Giuffrè et al. [64] used various solvents (acetone, ethyl acetate, chloroform, and petroleum ether) for tomato seed oil recovery by Soxhlet extraction and demonstrated that the highest oil yield was achieved using petroleum ether (21.2%).

Hexane is the most widely used solvent for oil extraction globally, due to its advantageous properties such as easy recovery of the extract, low latent heat of vaporization (330 kJ·kg^{-1}), a narrow boiling point range (63–69 °C), excellent solubilizing capacity, high extraction efficiency, and its non-polar character. Notwithstanding, this solvent has many disadvantages such as its flammable properties with potential health hazards [18].

Owing to its high efficiency in lipid solubilization, *n*-hexane has been established by the Association of Official Analytical Collaboration (AOAC) as the standard solvent in Method 960.39 for the determination of oils and fats from seeds by conventional extraction [65]. In addition, Table 3 summarizes some data from the literature using this methodology to obtain fruit seed oils.

Table 3. Summary of literature data using Soxhlet extraction with *n*-hexane as a solvent for obtaining fruit seed oils.

Food Waste Material	Mass Seed to Solvent Ratio (*w*/*v*)	Time of Extraction (h)	Oil Recovery (wt%)	Content of Principal Bioactive Compounds	References
Pomegranate seeds (*Punica granatum* L.)	1:6	8	15.66	Punicic acid: 762.2 $mg \cdot g^{-1}$ Oleic acid: 55.4 $mg \cdot g^{-1}$ Linoleic acid: 57.5 $mg \cdot g^{-1}$ Squalene: 1.20 $mg \cdot g^{-1}$ β-Sitosterol: 3.23 $mg \cdot g^{-1}$ α-tocopherol: 17.66 $mg \cdot 100\ g^{-1}$ γ-tocopherol: 325.3 $mg \cdot 100\ g^{-1}$ δ-tocopherol: 8.33 $mg \cdot 100\ g^{-1}$	[26]
Passion seeds	1:20	6	26.12	Palmitic acid: 10.22% Oleic acid: 17.11% Linoleic acid: 68.97% α-tocopherol: 1.68 $mg \cdot 100\ g^{-1}$ γ-tocopherol: 1.62 $mg \cdot 100\ g^{-1}$ δ-tocopherol: 4.93 $mg \cdot 100\ g^{-1}$	[24]
Citrus seeds	1:20	20	24.94	Palmitic acid: 21.72% Oleic acid: 26.00% Linoleic acid: 34.32% β-Sitosterol: 32.25% Limonene: 5.1% Δ-Avenasterol: 5.69% Campesterol: 7.55% 1,2-Benzenedicarboxylic acid, mono(2-ethylhexyl) ester: 18.66%	[66]
Avocado seeds (*Persea Americana* L.)	1:10	8	41.3	TPC: 25.78 mg $GAE \cdot g^{-1}$ TFC: 72.69 mg $QE \cdot 100\ mg^{-1}$ Myristic acid: 33.33% Linoleic acid: 20.12% Oleic acid: 11.60% Palmitic acid: 7.07%	[67]
Prunus seeds (peach, apricot, plum, and cherry)	1:10	6	Peach: 30% Apricot: 38% Plum: 37.4% Cherry: 36.0%	Palmitic acid: 5.71–8.1% Oleic acid: 48.6–72.7% Linoleic acid: 16.4–39.3%	[46]
Apple seeds	1:15	10	22.5	Quercetin: 0.01 $\mu g \cdot g^{-1}$ *p*-coumaric: 0.098 $\mu g \cdot g^{-1}$ *Trans*-ferulic acid: 0.104 $\mu g \cdot g^{-1}$ Phloridzin dihydrate: 0.01 $\mu g \cdot g^{-1}$ Naringenin: 0.31 $\mu g \cdot g^{-1}$ Palmitic acid: 6.24% Stearic acid: 2.97% Oleic acid: 24.15% Linoleic acid: 49.03% Linolenic acid: 1.87% Arachidic acid: 1.22%	[68]
Cherry seeds	1:4	6	3.19	Linoleic acid: 41.44% Oleic acid: 47.17% α-tocopherol: 7.13 $mg \cdot 100\ g^{-1}$ β-tocopherol: 1.112 $mg \cdot 100\ g^{-1}$ γ-tocopherol: 31.47 $mg \cdot 100\ g^{-1}$ δ-tocopherol: 6.64 $mg \cdot 100\ g^{-1}$	[40]

4.3. Supercritical CO_2 Extraction

Supercritical CO_2 extraction ($ScCO_2$) is widely used today for extracting oils from seeds and is considered a sustainable method. This method uses variations in temperature and pressure to bring a gas to its supercritical state—a phase in which the properties of gases and liquids merge, making the two phases indistinguishable [1].

According to Versteeg et al. [69], ScCO_2 is a separation technique based on mass transfer, where convection within the supercritical solvent phase (CO_2) typically serves as the dominant mechanism for transporting solutes. This methodology has several strengths in comparison with solvent extraction. The absence of residual solvents, the low temperature applied during the separation of the seed materials, the enhancement of diffusivity, which allows for greater selectivity, and the reduction in extraction times are all factors that must be considered [70]. In addition, Morya et al. [55] exposed that supercritical CO_2 extraction enables obtaining an oil quality similar to that of the mechanical pressing process explained above.

The typical system comprises a CO_2 tank, a pump, an oven containing a sample vessel, an extractor, and a collector. The liquid CO_2 is pumped to the extractor containing the seed, where the oil is dissolved. The resulting mixture then passes into a low-pressure chamber where the oil is separated from the CO_2. The CO_2 is converted into a gas, which can then be recycled or released [69,70]. The flow chart of the system is represented in Figure 7.

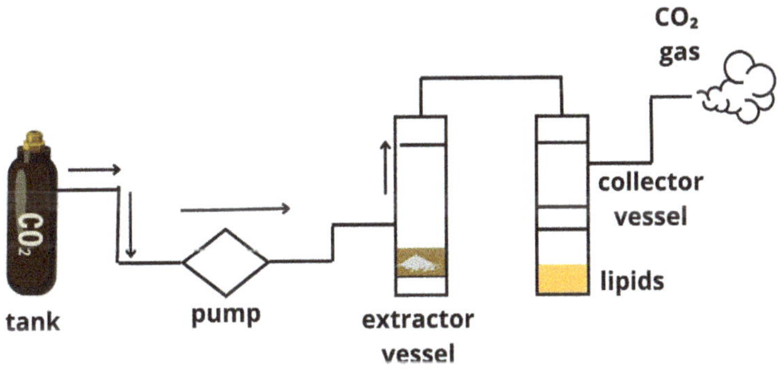

Figure 7. Flow chart of supercritical CO_2 extraction.

The efficiency of the oil extraction process using this system depends on the temperature, the pressure, and the solubility of the oil in the extraction fluid, as well as the contact time between the oil material and the extraction fluid. Particularly, Pavlić et al. [41] used a Box–Behnken design to optimize the raspberry oil extraction method using kinetic models of equal extraction and showed that the highest extraction efficiency with the maximum initial rate of matter transfer was achieved at higher levels of pressure (35 MPa) and CO_2 flow rate (0.4 kg $CO_2 \cdot h^{-1}$). Concurrently, it was ascertained that temperature (40 °C) and particle size (200–400 μm) levels required maintenance at low levels.

In addition, a quite common practice to improve extraction efficiency is the use of co-solvents such as ethanol or *n*-hexane. The observed effect can be attributed to the enhanced solubility of lipids. This increase in solubility may result from a higher density in the ScCO_2 and co-solvent mixture, or from specific intermolecular interactions between the co-solvent and certain solutes. Using co-solvent can be beneficial in separation processes, especially when it improves selectivity by interacting specifically with target compounds. Even so, the overall solubility of all components in the mixture also tends to increase due to the rise in fluid density [70]. Specially, Souza Correa et al. [71], demonstrated that the use of ethanol as a co-solvent during supercritical CO_2 oil extraction with berry seeds provided a higher oil yield, as well as greater efficiency in the extraction of bioactive components (e.g., fatty acids and phenolic compounds) as well as enhanced antioxidant activity. In other study, Macawile et al. [72] proved that the efficiency of oil extraction from *Gliricidia sepium* seeds using supercritical CO_2 extraction with *n*-hexane as a co-solvent significantly improved due to the increased solubility of lipid components.

Furthermore, recent studies to optimize the supercritical extraction process tend to resort to artificial intelligence simulators. These include artificial neural networks (ANNs). Specifically, Dimić et al. [34] carried out a study to develop and optimize the supercritical CO_2 extraction of cherry seed oil using sequential extraction kinetics modeling and an ANN. The resulting Box–Behnken design demonstrated that the highest oil recovery was obtained at 350 bar, 50 °C, and 0.4 kg CO_2. In addition, an ANN was applied to calculate the initial mass transfer rate, as it allowed for the simulation of the initial phase and to control the extraction process.

Nowadays, this methodology has emerged as a traditional method for extracting oils from fruit seeds. Therefore, Table 4 shows a summary of the literature data that apply this methodology in the extraction of fruit seed oils as well as the composition of the bioactive compounds present in them.

Table 4. Literature review of the application of supercritical CO_2 extraction in seed oil recovery.

Food Waste Material	Pressure (bar)	Temperature (°C)	CO_2 Flow Rate ($kg \cdot h^{-1}$)	Particle Size (μm)	Oil Recovery (wt%)	Content of Principal Bioactive Compounds	References
Cherry seeds	200–350 bar	40–70	0.2–0.4	800	3.62–13.02	Oleic acid: 40.89–41.65% Linoleic acid: 46.32–47.37% α-tocopherol: 1.75–10.54 $mg \cdot 100\ g^{-1}$ β-tocopherol: 0.78–1.11 $mg \cdot 100\ g^{-1}$ γ-tocopherol: 6.23–40.60 $mg \cdot 100\ g^{-1}$ δ-tocopherol: 2.08–8.42 $mg \cdot 100\ g^{-1}$	[40]
Citrus seeds	200–250	45–60	1.62	n.d.	15.2–19.6	Palmitic acid: 17.68–20.01% Oleic acid: 29.51–30.67% Linoleic acid: 36–39.44% Limonene: 5.73–6.39% β-sitosterol: 26.63–28.37% 1,2-Benzenedicarboxylic acid, mono(2-ethylhexyl) ester: 26.55–30.34%	[66]
Berry seeds	1000	100	240	n.d.	12–18	α-tocopherol: 21.7 176.1 $mg \cdot kg^{-1}$ β + γ-tocopherol: 50.6–1027.7 $mg \cdot kg^{-1}$ δ-tocopherol: 2.1–1104.8 $mg \cdot kg^{-1}$ α-linolenic acid: 3.2–38.0% Linoleic acid: 47.9–72.9% Palmitic acid: 2.1–4.3% Oleic acid: 10.5–18.0%	[73]

Table 4. *Cont.*

Food Waste Material	Pressure (bar)	Temperature (°C)	CO_2 Flow Rate (kg·h^{-1})	Particle Size (μm)	Oil Recovery (wt%)	Content of Principal Bioactive Compounds	References
Apple seeds	75–350	25–90	0.12–7.6	n.d.	20.5	Quercetin: 0.03 μg·g^{-1} *p*-coumaric: 0.17 μg·g^{-1} Trans-ferulic acid: 0.12 μg·g^{-1} Phloridzin dihydrate: 2.97 μg·g^{-1} Naringenin: 0.63 μg·g^{-1} Palimitic acid: 13.39% Stearic acid: 7.69% Oleic acid: 34.84% Linoleic acid: 63.76% Linolenic acid: 1.61% Arachidic acid: 2.04%	[68]
Pomegranate seeds (*Punica granatum* L.)	200	40	0.108	n.d.	16	Punicic acid: 80.7% Oleic acid: 1.58% Linoleic acid: 3.43% 2,4-Dihydroxyphenylacetic acid: 0.14 mg·100 g^{-1} Ferulic acid: 0.29 mg·100 g^{-1} *Trans*-Cinnamic acid: 0.14 mg·100 g^{-1} *p*-hydroxibenzoic acid acid: 0.11 mg·100 g^{-1} *p*-Coumaric acid: 0.07 mg·100 g^{-1} Naringenin: 0.31 mg·100 g^{-1} Vanillic acid: 0.07 mg·100 g^{-1} β-Carotene: 0.02 mg·100 g^{-1} β-Sitosterol≈ 1000 mg·100 g^{-1} Stigmasterol≈ 1000–1500 mg·100 g^{-1} Campesterol≈ 50 mg·100 g^{-1} α-tocopherol≈ 5 mg·100 g^{-1} γ-tocopherol≈ 2000 mg·100 g^{-1} δ-tocopherol≈ 5 mg·100 g^{-1}	[74]
Passion seeds	284	56	1.02	n.d.	2.67	Isovitexin: 90.78 mg·g^{-1} Ferulic acid: 31.08 mg·g^{-1} *p*-coumaric: 24.74 mg·g^{-1} Gallic acid: 13.99 mg·g^{-1} Isorhmnetin: 7.40 mg·g^{-1} Quercetin: 3.05 mg·g^{-1}	[75]

n.d.: not declared.

However, despite the advantages of this technique such as the selectivity and sustainability of the process, this technique requires a high application cost of $ScCO_2$ which restricts its use mainly to oils produced in small quantities [69]. Consequently, the development of new sustainable alternatives for oil extraction is required.

5. Recent Advances in Eco-Friendly Extraction Methods and Factors Affecting Extraction Efficiency and Oil Quality

As previously reported, traditional extraction methods often involve the use of harmful solvents, long processing times, and high temperatures, which can degrade heat-sensitive compounds. To address these challenges, advanced extraction techniques such as UAE, MAE, EAE, PLE, and subcritical extraction, as well as the novel PEF technique have been employed to efficiently recover bioactive compounds from fruit seeds (Figure 8).

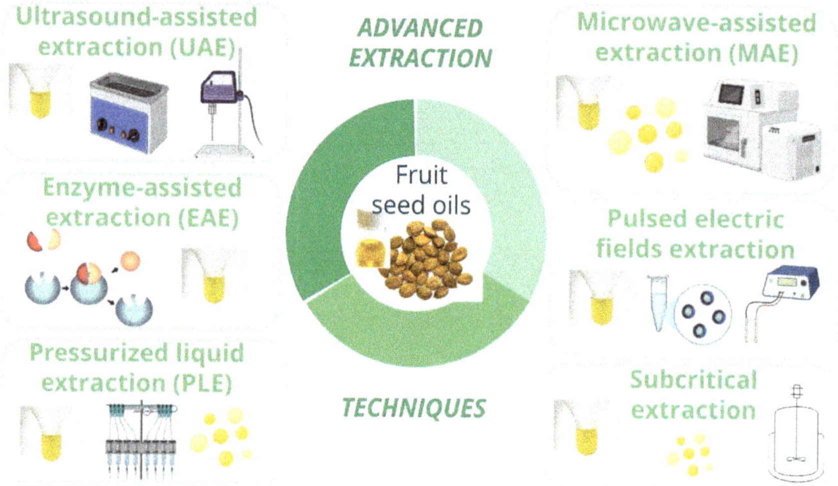

Figure 8. Advanced sustainable extraction methods for fruit seed oil recovery.

In many cases, it is necessary to use hybrid or combined extraction techniques, which involve the coupling of two or even three methods, to enhance extraction efficiency.

5.1. Ultrasound-Assisted Extraction (UAE)

The utilization of ultrasound-assisted extraction in the extraction of bioactive compounds is consistent with several principles inherent to green chemistry. These principles include the enhancement of operator safety, the reduction in energy consumption, the effective management of waste, and the minimization of the use of toxic or environmentally harmful solvents [76].

UAE is distinguished by the implementation of high-frequency waves, which are utilized to induce the formation of cavitation bubbles within the solvent. The subsequent collapse of these bubbles results in the generation of high temperatures and localized pressures, which in turn modify the sample matrix and enhance the extraction efficiency. Once the extraction is completed, the oil mixture requires a centrifugation or filtration step in order to eliminate the defatted seeds and then the solvent is evaporated using a rotary evaporator or under vacuum at a low temperature.

The main advantages of UAE are focused on reducing solvent consumption, the time needed to extract, and improving extraction efficiency [77]. The extraction conditions using UAE depend on the bioactive compounds to be extracted and the material from which they must be extracted. A number of authors have evaluated experimental designs and response surface analyses with the objective of obtaining optimal extraction conditions. They have also sought to derive equations that would allow for the relating of the experimental parameters used with oil extraction performance or other factors, such as oxidative stability [11,12]. The breakdown of cell structures and cell walls in plant samples produces an improvement in the extraction of bioactive compounds so that the efficiency of the extraction depends on the power of the ultrasound, the frequency, the duty cycle, the extraction time, and the solvent selected, so the optimal extraction conditions must be evaluated for each type of bioactive compound and each type of sample. Deep eutectic solvent extraction and ultrasound-assisted extraction have been performed in order to improve the oil and amygdalin extraction performance in bitter apricot kernels. The combined effect of UAE and the mixture of choline chloride–lactic acid facilitate the breakage of the apricot kernels, which would explain the increase in extraction yield [78]. In other study, Marinaccio L. et al. [79] used a natural volatile deep eutectic solvent (NADES), specifically menthol–thymol 1:1, for ultrasound-assisted grape seed oil extraction and compared it with UAE using *n*-hexane as a solvent. Their findings demonstrated that the phenolic content increased significantly (18.65 mg GAE·g^{-1}), as well as the α-linolenic acid content (5.18%). Meanwhile, the oleic and linoleic acid content decreased in the oil extracted using NADES.

In the optimization process, it is important to consider that increasing the power of the ultrasound can enhance the extraction of some compounds, but it may also have a negative impact on others. While an increase in acoustic power and extraction time can facilitate mass transfer, it can also potentially lead to an increase in the formation of free radicals [45] These free radicals, in turn, have the potential to degrade certain bioactive compounds, such as polyphenols, or to induce the extraction of unwanted compounds, like amygdalin [39]. It is important to note that the UAE also depends on the frequency used. Flexible plant materials are best suited to a frequency range of 20 to 40 kHz. However, for more rigid materials, a frequency range of up to 500 kHz may be necessary [77]. The extraction of fatty acids in oil is strongly influenced by their affinity for the solvent, but it is also affected by the time of exposure to ultrasound and by the content of SFAs and UFAs in the sample [39].

The ultrasound waves can be generated in a water bath or by using a probe immersed in the sample. While the ultrasonic bath can be used for multiple samples, the probe sonicator can only be applied to one sample at a time. It should also be considered that the intensity provided by the bath ultrasound [30] is lower than that provided for the probe [12,27,80].

Compared to oil extraction by conventional methods such as stem distillation or mechanical pressing, ultrasonically assisted extraction allows for lower energy consumption and a reduction in solvent or substitution with green solvents and an extraction yield of the same order of magnitude. Table 5 shows the conditions used to extract oil with bioactive compounds from various seed residues. In a more advanced situation, ultrasound can be combined with other technologies like microwave [13], enzymatic [27], and supercritical fluid extractions [12] to be used directly in the process or could be used as a pretreatment process. These hybrid techniques are useful in increasing the extraction efficiency for oil [26,27,81].

Table 5. Applications of UAE for extraction of fruit seed oils.

Food Waste	Solvent	Processing Conditions	Oil Recovery (wt%)	Bioactive Compounds	References
Cactus fruit seeds	*n*-hexane	Ultrasound intensity: 150–600 W Time: 60 min; Temperature: 40 °C Solid–solvent ratio 1:10 (*w*/*v*)	20.28 ± 0.68	α-tocopherol: 20.01–36.94 mg·100 g^{-1} γ-tocopherol: 12.73–19.15 mg·100 g^{-1} δ-tocopherol: 0.88–2.95 mg·100 g^{-1} Sinapic acid: 0.87–1.31 μg·g^{-1} Syringic acid: 0.14–0.15 μg·g^{-1} Salicylic acid: 0.21–0.22 μg·g^{-1} Palmitic acid: 10.01–12.71% Stearic acid: 5.14–6.33% Oleic acid: 10.04–11.46% Linoleic acid: 57.46–61.93% Linolenic acid: 5.41–7.20%	[45]
Passion seeds	ethanol	Power: 165 W; Frequency: 25 kHz Time: 10–50 min; Temperature: 30–60 °C Solid–solvent ratio 2–4 mL·g^{-1}	12.32–21.76	Palmitic acid: 12.31% Palmitoleic acid: 0.19% Oleic acid: 17.03 Stearic acid: 2.44 Linoleic acid: 61.07 Linolenic acid: 0.12 Stigmasterol: 75.67 mg·100 g^{-1} Campesterol: 21.55 mg·100 g^{-1} β-Sitosterol: 69.11 mg·100 g^{-1}	[11]
Apple seeds	*n*-hexane	Ultrasound power: 18–70 W Amplitude: 20–80%, Temperature: 30 °C Solid–solvent ratio 1:15 (*w*/*v*)	17.20 ± 2.3	Phloretin and phloridzin	[39]
Passion seeds	supercritical CO_2	Ultrasound power: 160 W Pressure: 16 MPa, Temperature: 40 (*w*/*v*)	29	γ -tocopherol: 6.7–10.6 mg·100 g^{-1} γ-tocotrienol: 24–36 mg·100 g^{-1} δ-tocotrienol: 33–45 mg·100 g^{-1} Palmitic acid: 10.9–11.8% Oleic acid: 15.89–19% Linoleic acid: 63–67.91%	[12]
Mango seeds	*n*-hexane	Power: 70 W; Frequency: 42 kHz Time: 20–60 min; Temperature: 35–55 (*w*/*v*) Solid–solvent ratio 1:5 (*w*/*v*)	96.6 ± 1.3	Palmitic acid: 4.95% Stearic acid: 39.59% Oleic acid: 32.38% Linoleic acid: 3.31% Eicosadienoic acid: 2.32% Docosahexaenoic acid: 15.74%	[13]

Table 5. *Cont.*

Food Waste	Solvent	Processing Conditions	Oil Recovery (wt%)	Bioactive Compounds	References
Pomegranate seeds	water	Power: 130 W; Frequency: 20 kHz Temperature: 35–55 °C Time: 2–8 h	4.29–25.11	n.d.	[27]
Citrus seeds	*n*-hexane	Temperature: 25 °C Time: 90 min	Lime seeds: 22.07–22.08 Sweet orange seeds: 22.84–23.11	Lime seed TPC: ≈ 0.6–0.8 mg GAE·g^{-1} Palmitic acid: 19.83–20.79% Oleic acid: 17.27–18.07% Linoleic acid: 34.07–34.25% α-Linolenic acid: 10.96–11.45% Orange sweet seed TPC: ≈0.6–1 mg GAE·g^{-1} Palmitic acid: 23.56–23.73% Oleic acid: 20.95–21.58% Linoleic acid: 35.74–36.32% α-Linolenic acid: 3.67–3.98%	[30]
Bitter apricot kernels	natural deep eutectic solvent (choline chloride–lactic acid)	Liquid-to-solid ratio: 18 mL·g^{-1}; Ultrasonic power: 420 W; Temperature: 42 °C; Extraction time: 32 min	45.64 ± 0.24	Palmitic acid:16.63 ±0.52 mg·g^{-1} Palmitoleic acid: 3.15 ± 0.16 mg·g^{-1} Ginkgolic acid: 0.62 ± 0.02 mg·g^{-1} Stearic acid: 3.86 ± 0.20 mg·g^{-1} Oleic acid: 309.18 ± 9.47 mg·g^{-1} Linoleic acid: 109.40 ± 8.35 mg·g^{-1} Arachidic acid: 0.36 ± 0.01 mg·g^{-1} *cis*-11-eicosenoic acid: 0.42 ± 0.02 mg·g^{-1} Octadecatrienoic acid: 0.60 ± 0.01 mg·g^{-1}	[78]
Grape seeds	natural deep eutectic solvent (menthol–thymol)	Sample-to-solvent ratio: 1:4 Ultrasound sonicator (20 kHz, 400 W, and 70% amplitude) Temperature: 70 °C Time extraction: 20 min		Palmitic acid: 14.47% Stearic acid: 5.48% Oleic acid: 15.49% Linoleic acid: 56.17% α-linolenic acid: 5.18% TPC: 18.65 mg GAE·g^{-1} TFC: 0.73 mg RE·g^{-1}	[79]

n.d.: not declared.

5.2. *Microwave-Assisted Extraction (MAE)*

Microwave-assisted extraction has emerged as a compelling and increasingly adopted non-conventional technology for oil extraction. It operates through the application of non-ionizing electromagnetic radiation (ranging from 300 MHz to 300 GHz) directed at polarizable materials, including both the solvent and the oil-bearing matrix. This radiation induces internal heating via ionic conduction and dipolar rotation, leading to the rupture of plant cell walls as a result of vapor pressure buildup. Such structural disruption facilitates the rapid release and dissolution of oil into the solvent, thereby enhancing extraction efficiency [14,82]. The rapid heating characteristic of MAE enables the production of high-quality oils in significantly reduced timeframes and with minimal solvent usage, an advantage that positions it favorably against conventional solvent-based techniques such as Soxhlet extraction [1,3,14].

Unlike traditional methods, MAE does not necessitate prior thermal treatment of the raw material, offering both economic and energy-saving benefits for industrial applications. Although microwave-assisted extraction has demonstrated considerable potential in the valorization of agro-industrial seed residues—such as grape seeds [83,84], pomegranate seeds [26], and dragon fruit seeds—its application remains relatively scarce. This underutilization persists despite promising results in terms of oil yield and bioactive compound recovery, as summarized in Table 6, which outlines the extraction conditions and key outcomes for various seed types.

Be that as it may, microwave energy can serve as a pretreatment step in other extraction processes [1,14]. For instance, the use of microwave drying and heating as a prior step for oil recovery from avocado seeds was reported by Satriana et al. [85], and Bose et al. [33] reported a twofold increase in oil yield from mechanically pressed tropical seeds (*Simarouba glauca*) following microwave pretreatment. Nevertheless, as previously noted, the use of microwaves solely as a pretreatment is generally less efficient, both economically and energetically, than their direct application as extraction technique.

Table 6. MAE conditions for the extraction of agri-food seed oils.

Food Waste	Solvent	Processing conditions	Oil Recovery (wt%)	Bioactive Compounds	References
Pomegranate seeds	*n*-hexane or petroleum ether	Power: 500 W Frequency: 40 kHz Time: 9 min; Sample-to-solvent ratio: 1:6 (*w*/*v*)	11.72–12.47	Palmitic acid: 29.3% Oleic acid: 49.5% Stearic acid: 19.9% Linoleic acid: 53.6% Punicic acid: 776.5% α-Eleoestearic acid: 5.3% β-Eleoestearic acid: 33.0% Catalpic acid: 10.2% Arachidic acid: 6.2% Arachidonic acid: 7.6% β-Sitosterol: 3.47 mg·g^{-1} Squalene: 1.59 mg·g^{-1} α-tocopherol: 17.24 mg·100 g^{-1} γ-tocopherol: 322.7 mg·100 g^{-1} δ-tocopherol: 8.23 mg·100 g^{-1}	[38]
	n-hexane	Power: 220 W, Time: 5 min; Sample-to-solvent ratio: 1:10 (*w*/*v*)	35.19	Palmitic acid: 2.04% Oleic acid: 4.10% Stearic acid: 1.71% Linoleic acid: 3.84% α-Linolenic acid: 86.53% Margaric acid: 0.04% Arachidic acid: 0.37% Gadoleic acid: 0.69.0% TPC: 7.42 mg·g^{-1}	[86]

Table 6. *Cont.*

Food Waste	Solvent	Processing conditions	Oil Recovery (wt%)	Bioactive Compounds	References
Dragon fruit seeds	-	Power: 450 W; Time: 10 min	34.3	Palmitic acid: 14.69% Stearic acid: 7.30% Oleic acid: 18.66% Linolenic acid: 52.14% Arachidonic acid:0.04% TPC: 96.71 mg GAE·g^{-1} α-tocopherol: 52.19 mg·kg^{-1} γ-tocopherol: 921.58 mg·kg^{-1} δ-tocopherol: 76.6 mg·kg^{-1}	[8]

In summary, MAE offers numerous advantages, including shorter processing times, lower operational costs, compact equipment requirements, reduced solvent consumption, and higher yields. However, despite its many benefits, MAE may not be universally suitable for all plant matrices. Excessive microwave energy can compromise plant tissue integrity, and the elevated temperatures reached during the process may degrade thermolabile bioactive compounds, thereby diminishing oil quality and bioactivity. Consequently, MAE is not recommended for the extraction of oils rich in PUFAs, which are particularly prone to oxidative degradation [5,14].

5.3. *Enzyme-Assisted Extraction (EAE)*

Enzyme-assisted enzyme extraction relies on the enzymatic pretreatment of seeds to release substances that bind to cell walls in order to increase the extraction of compounds of interest. The extraction of oil from seeds is a challenging process due to the complexity of the cell wall composition, which consists of cellulose, hemicellulose, lignin, and pectin. The application of specific enzymes, such as carbohydrates, that are capable of degrading the cell membrane, has been shown to facilitate the release of oil. Proteases have been demonstrated to play a pivotal role in the hydrolysis of cell membrane proteins, thereby enabling the efficient extraction of oil. From an ecological perspective, it is imperative to acknowledge that enzymatic treatment is conducted at relatively low temperatures, thereby eliminating the generation of solvent residues. A potential disadvantage of enzyme use is the cost, which can be significant. However, this can be mitigated by increasing the extraction yield or by reusing immobilized enzymes. In the extraction process, it is crucial to consider the working pH of the enzyme and the isoelectric point of the proteins (pI) that constitute the seed, since, on the one hand, if the pH is not adequate, the enzyme can be deactivated, and, on the other hand, if the pH is close to pI, precipitation of the seed proteins can occur [14,18].

In most cases, enzyme-assisted extraction consists of the stages described in Figure 9 [87]. Some of these steps are common with other extraction methods, but in this case pH and temperature control are essential parameters in process optimization.

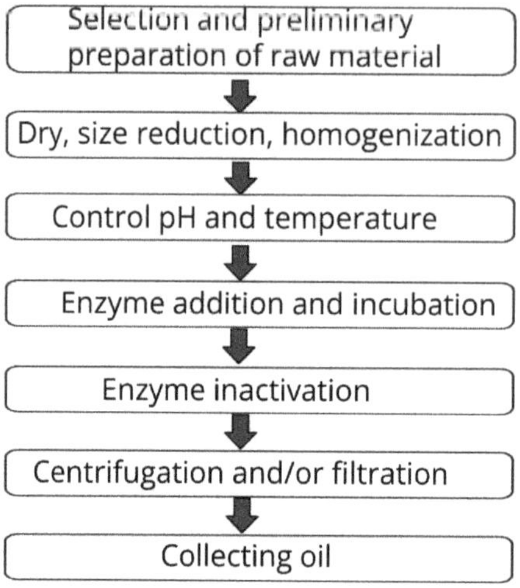

Figure 9. Enzyme-assisted extraction main steps.

The use of proteases for the joint extraction of oils and proteins in seeds/grains of fruits such as mango, lemon, or pumpkin has been evaluated to improve the performance of protein and fat extraction. The oil extracts showed a high nutritional value as they had a high content of unsaturated fatty acids. In the procedure, an enzyme–substrate ratio of 1:100 (w/w) was used in a phosphate buffer at a pH of 7.5 and hydrolyzed at 60 °C for 16 h. To deactivate the enzyme, it was heated for 10 min at 90 °C and finally centrifuged to separate the different phases [28]. With a focus on increasing the extraction yield and decreasing the time of extraction, enzyme-assisted extraction has been combined with ultrasound-assisted extraction using enzyme treatment as a pre-step or using ultrasound stirring during the incubation process [27,77].

5.4. Pressurized Liquid Extraction (PLE)

PLE is a technique that uses a solvent in its liquid state under elevated temperatures (typically between 27 and 227 °C) and pressures (from 5 to 20 MPa) but below the solvent's critical point. These subcritical conditions enhance solvent penetration into the solid plant tissue and disrupt the binding interactions between the target compounds and the matrix, thereby accelerating the extraction process.

Considered a sustainable method, PLE minimizes solvent use, reduces energy consumption, and is easily automated, all of which contribute to its low environmental impact. Ethanol and hydroalcoholic mixtures are commonly used as solvents because they are considered green solvents and are easy to remove. Their adjustable polarity makes them versatile solvent systems for various extraction purposes.

In addition, PLE can be conducted in either intermittent or dynamic modes. In intermittent extraction, the system is pressurized for a single cycle, which is often repeated several times to improve extraction efficiency. This approach offers significant benefits, such as shorter extraction times and reduced solvent consumption, leading to energy and solvent savings. In contrast, dynamic extraction involves the continuous flow of solvent through the raw material bed, enabling the constant recovery of extracts and the collection of different fractions during the process. However, continuous solvent flow can lead to unnecessary dilution of the extract if the extraction time is extended. Increasing the solvent flow rate can reduce extraction time, which is particularly advantageous for thermo-sensitive compounds, as it minimizes their exposure to heat.

On the other hand, key process variables—such as temperature, pressure, solvent flow rate, extraction time, solvent choice, solvent-to-feed mass ratio (S/F), moisture content, and particle size—are crucial to the efficiency of the technique. Thermal effects impact the process by reducing solvent viscosity, weakening matrix–compound interactions, and altering solvent properties. Moreover, continuous solvent flow plays a vital role in maintaining constant interaction between the solvent and the processed material, which facilitates better mass transfer and extraction kinetics. Increasing solvent flow rates can significantly enhance the extraction yield. In terms of the extraction solvent, higher water content in hydroalcoholic mixtures improves the recovery of hydrophilic compounds but may reduce the extractability of non-polar lipids. It is important to note that higher moisture content leads to greater interaction with water, improving extraction efficiency when using water or hydroalcoholic mixtures as solvents. The S/F ratio also plays a critical role in balancing extraction efficiency with economic feasibility. Regarding particle size, reducing it increases the surface area available for extraction, enhances solvent penetration, and improves mass transfer. However, smaller particle sizes may lead to compaction, which can reduce solvent flow and decrease extraction efficiency.

The findings demonstrate that PLE is highly effective in extracting a wide range of bioactive compounds, including polyphenols, unsaturated fatty acids, tocopherols, and carotenoids. The resulting extracts typically exhibit significant anticancer, antidiabetic, antioxidant, and antimicrobial activities [48,88].

Compared to conventional methods, PLE offers several advantages, including shorter extraction times, reduced solvent consumption, high extraction efficiency, moderate environmental impact, and the possibility of solvent recycling. However, there are some drawbacks. These include potential analyte degradation under high pressure and temperature, the lack of miniaturized commercial devices that hinder its integration with other techniques, and the strong dependence of extraction efficiency on factors such as matrix type and the optimization of parameters like time and temperature. Additionally, the high cost of equipment and the removal of solvents remains a significant challenge for industrial-scale implementation [89,90].

However, PLE has been applied to recover valuable polyphenolic compounds from avocado seeds at temperatures up to 200 °C and pressures ranging from 11 to 22.5 MPa, using ethanol or hydroalcoholic mixtures. The major polyphenolic compounds recovered include procyanidins, and flavonoids (such as epicatechin, catechin, quercetin, and rutin) and their derivatives, as well as phenolic acids like chlorogenic, vanillic, and gallic acids [48].

Furthermore, Figueroa et al. [7] have employed an ethanol–water 50:50 (v/v) mixture at 200 °C and 11 MPa to efficiently extract organic acids, phenolic acids, flavonoids, catechins, and condensed tannins from avocado seeds. This method enabled obtaining extracts rich in 4-hydroxybenzoic acid with high antioxidant properties [7].

PLE has also been utilized to extract oils from the seeds of various berry fruits, including cranberries, black currants, red currants, strawberries, and chokeberries. The seed samples were mixed with diatomaceous earth, and the extraction was carried out in four 5 min cycles, using n-hexane as the solvent, at 150 °C and a pressure of 10.34 MPa. The primary fatty acids identified in the extracted oils were linoleic, α-linolenic, and oleic acids, which also exhibited high oxidative stability. Furthermore, the oils displayed low atherogenicity and thrombogenicity indexes, and, combined with high hypocholesterolemic index values, they are considered oils of high nutritional quality [91].

In order to summarize the PLE methods for the extraction of fruit seed oils, a summary of the literature data is shown in Table 7.

Table 7. Overview of the literature data on the application of PLE for obtaining fruit seed oils.

Extraction Method	Food Waste Material	Processing Conditions	Oil Recovery (wt%)	Content of Principal Bioactive Compounds	References
PLE	Avocado seeds	200 °C 11 MPa ethanol 20 min 5 g	39	Total polyphenols: 88 mg GAE·100 g^{-1}	[48]
PLE	Avocado seeds	200 °C 11 MPa Ethanol–water 50:50 (*v*/*v*)	n.d	Organic acids, phenolic acids, flavonoids, catechins, and condensed tannins	[81]
Hot PLE	Grape pomace seeds	160 °C 10 atm 60% ethanol	n.d	Gallic acid: 21.66 μg·g^{-1} Ellagic acid: 4.55 μg·g^{-1} Protocatechuic acid: 5.31 μg·g^{-1} Caffeic acid: 4.92 μg·g^{-1} Chlorogenic acid: 6.66 μg·g^{-1} *p*-oumaric acid: 2.24 μg·g^{-1} Catechin: 85.91 μg·g^{-1} Epicatechin: 28.53 μg·g^{-1} Epigallocatechin: 13.21 μg·g^{-1} Gallocatechin: 3.27 μg·g^{-1}	[87]
PLE	Berry fruit seeds	150 °C 10.34 MPa hexane 4 cycles × 5 min	16.00 (cranberry) 13.53 (strawberry) 5.08 (chokeberry) 7.60 (black currant) 1.70 (red currant)	Palmitic acid: 5.54–9.14% Stearic acid: 1.23–2.62% Palmitoleic acid: 0.11–0.36% Elaidic acid: 13.73–23.52% Linoleic acid: 36.35–67.23% γ-Linoleic acid: 0.12–8.72% Linolenic acid: 1.99–31.80%	[88]

n.d.: not declared.

5.5. Pulsed Electric Fields (PEF) Extraction

Pulsed electric field technology has emerged as a cutting-edge, non-thermal, environmentally sustainable, and energy-efficient method for the extraction of bioactive compounds from biological matrices, making it particularly suitable for the recovery of oils from residual seeds. This technique is grounded in the principle of electroporation, wherein a high-intensity electric field is applied between two electrodes immersed in a solution containing the target material. The application of short-duration electric pulses, ranging from microseconds to milliseconds, induces electropermeabilization of cellular membranes, thereby facilitating the release of valuable intracellular compounds without the need for elevated temperatures. This characteristic is especially advantageous for preserving thermolabile substances. Unlike conventional thermal methods, PEF treatment maintains the organoleptic and nutritional qualities of food products, including the color, aroma, flavor, texture, and nutritional value, rendering it an appealing option for oil recovery processes intended for the food industry [4,5].

PEF systems typically comprise a high-voltage pulse generator, a monitoring and control unit, and a treatment chamber, the geometry and electrode configuration of which vary according to the processing mode. The effectiveness of PEF treatment is governed by several critical parameters, including the electric field strength (determined by the ratio of applied voltage to the distance between electrodes), the shape, number, and duration of pulses, the pulse frequency, and the specific energy, defined as the energy delivered per unit mass of the treated material. Collectively, these parameters not only enhance the efficiency and speed of the extraction process but also minimize waste generation and enable rapid processing. These attributes underscore the potential of PEF as a pivotal tool in advancing sustainable, high-quality processing technologies within the modern food industry [4].

Moreover, the non-destructive nature of PEF technology, coupled with its low energy demands, significantly enhances its suitability for industrial-scale applications [5]. However, its broader adoption continues to be limited by the substantial initial capital investment required for implementation. In this context, PEF has predominantly been employed as a pretreatment strategy for seeds, aiming to disrupt the cellular structure and thereby facilitate more efficient oil extraction in subsequent stages [92]. These subsequent extraction methods may include conventional techniques such as Soxhlet extraction and mechanical pressing, as well as alternative approaches like UAE [4]. For instance, [93] applied PEF to recover oils from berry seeds, including blackcurrant, redcurrant, raspberry, and chokeberry, using 2 g of sample material, electrode voltages of 8 or 10 kV, a pulse width of 7 μs, a frequency of 20 Hz, and a specific energy input of 50 $kV \cdot kg^{-1}$ over a 30 min treatment period. Among the tested samples, PEF proved most effective for enhancing oil yield from raspberry seeds, while no significant improvement was observed in the case of blackcurrant seeds. Notably, PEF treatment was found to improve the oxidative stability of the extracted oils without adversely affecting their thermal properties or fatty acid profiles, underscoring its potential as a gentle yet effective technique for high-quality oil recovery [93].

5.6. *Subcritical Extraction*

Subcritical fluid extraction is predicated on the utilization of solvents at temperatures between the boiling point and the critical temperature, and at a pressure below the critical point, with the capacity to maintain the solvent in a liquid state. Extractions with compressed solvents under subcritical conditions are considered green because the solvent can be reused and recovered. In the various subcritical fluids that have been utilized in the literature, *n*-butane and propane appear to be the most commonly employed in oil extraction processes. This preference is likely attributed to their colorless nature and cost-effectiveness. In addition to their excellent solubility of lipophilic compounds, they have the potential to be used in shorter extraction times and at lower critical pressures and temperatures, which may result in a higher-quality product [10].

Subcritical extraction is a simple process which has many astounding advantages in comparison with the traditional extraction techniques, including low pressure and temperature, a reduction in the extraction time, improved selectivity, and environmental compatibility, as well as eliminating toxic solvents. The parameters that govern the extraction process, including the extraction temperature, time, solvent ratio, particle size, and co-solvents, have been demonstrated to modulate the efficiency of subcritical extraction at low temperatures [94].

This emerging extraction technique has been used as a method for extracting fruit seed oil. In this particular case, Liu et al. [26] employed subcritical extraction with *n*-butanol to extract oil from pomegranate seeds. The experimental conditions encompassed a solid–solvent ratio of 1:6, wt.%, a temperature of 45 °C, an extraction time of 40 min, and a pressure of 0.6 MPa. A comparative analysis was conducted between this methodology and Soxhlet extraction as well as the sustainable MAE and UAE techniques. The findings indicated that this methodology resulted in the greatest increase in oil percentage (14.50%) compared to UAE and MAE. In addition, the oil extraction percentage obtained in this study was comparable to that achieved through Soxhlet extraction (15.66%) with *n*-hexane as a solvent. Furthermore, the oils exhibited comparable oleic (52.4 $mg \cdot g^{-1}$) and linoleic acid (53.7 $mg \cdot g^{-1}$) content to that obtained through Soxhlet extraction with *n*-hexane (55.4 and 57.5 $mg \cdot g^{-1}$, respectively). Furthermore, these samples exhibited higher concentrations of punicic acid (772.3 vs. 762.1 $mg \cdot g^{-1}$). Conversely, this method yielded higher levels of squalene, β-sitosterol, and α-, β-, γ-, and δ-tocopherols compared to Soxhlet extraction.

In addition, the use of water as a solvent for the subcritical extraction of essential oils from seeds has also been reported in the literature. In this case, this solvent allows for the extraction of higher amounts of polyphenols, terpenes, and other polar components [95].

In short, this emerging technique is currently being investigated for application with seed oils as an alternative to conventional methods.

5.7. Hybrid Extraction Technologies

The use of two or three techniques for the extraction of seed oils has been applied with two fundamental objectives: either to increase the efficiency of the extraction and/or to improve the extraction of bioactive products, which may be affected by the pressure conditions and/or temperatures used in various extraction techniques and which may lead to losses of antioxidant capacity or stability of the oil, among others [96].

Usually, to improve the extraction efficiency of bioactive compounds from seeds containing low levels of oil, a combination of conventional and alternative processes is required. In the case of passion fruit seed oil extraction, combining supercritical fluid extraction (at 40 °C and 16 MPa) with ultrasonic agitation (160 W) increased the overall yield to 20.9%, compared to 12.3% for SFE (at 40 °C and 16 MPa) without ultrasound-assisted extraction. This fact can be explained on the basis of the possible cavitation produced by the ultrasonic waves in the extraction bed, which favored the rupture of the sample matrix and the more favorable release of the oily phase. This combination of techniques must be previously optimized to find the conditions that produce a greater release of the extractable material from the seeds without increasing by too much the energy cost involved in the application of ultrasound. Although the oil yield increased, a decrease in γ-tocopherol (7.3 vs. 10.6 $mg \cdot 100\ g^{-1}$), γ-tocotrienol (27 vs. 36 $mg \cdot 100\ g^{-1}$), and δ-tocotrienol (34 vs. 45 $mg \cdot 100\ g^{-1}$) content was observed under these conditions. Meanwhile, an increase in oleic acid (19 vs. 15.9%) content and a decrease in linoleic acid (65 vs. 67.9%) was observed when ultrasound was applied in combination with SFE. This could be explained by the breakdown of the linoleic double bond when using a previous cavitation process by ultrasound [12].

In other study, Ćurko et al. [9] optimized a supercritical CO_2 oil extraction method in combination with an electric pulse pretreatment from grape seeds. A Box–Behnken design was applied by studying the pressure, temperature, and flow rate factors of the SFE. The PEF pretreatments were performed at 5 $kV \cdot cm^{-1}$ and 120 hz for 5 and 1 min of extraction. The findings revealed that the PEF pretreatment in supercritical oil extraction significantly increased the oil yield to a value of 81.8 $kg \cdot g^{-1}$. Other authors in the literature concluded that the utilization of PEF as a pretreatment before the extraction process leads to the electrical decomposition of the cells and therefore to their increased permeability, thus favoring the extraction of the oil from the seed samples [92]. Furthermore, this research demonstrated that this hybrid technique significantly increased the content of sterols, tocopherols, polyunsaturated fatty acids, and phenolic acids compared to the untreated procedure.

As previously reported, the mechanical pressing technique is characterized by low extraction yields of the oils obtained. Therefore, the use of enzymes as a pretreatment phase could be a sustainable alternative, as they tend to soften and/or destroy the cellular structures, thus aiding in oil extraction [97]. In one research study, Candan et al. [98] developed a hybrid methodology with the combination of an enzymatic pretreatment prior to oil extraction by mechanical pressing in the recovery of grape, apricot, and flaxseed oils. Their results showed that enzymatic pretreatment and cold-press extraction showed positive results in terms of total carotenoids, tocopherols, oil yield, phenolics, antioxidant activity, and oxidative stability.

In other work, Bakhshabadi et al. [92] employed two pretreatment methodologies for black cumin seeds before cold pressing was employed, PEF (electric field strength of 3.25 $kV{\cdot}cm^{-1}$ and number of pulses of 30) and MAE (540 W for 180 s), to study the oil extraction efficiency. The achieved findings demonstrated that both of the pretreatments increased the oil extraction efficiency and its oxidative stability.

In addition to the combination of conventional and alternative methods. Recently, oil extraction processes have been developed by combining alternative sustainable methods (e.g., UMAE; aqueous enzymatic pretreatment microwave-assisted extraction (HCUEPM)) with each other.

In this line, ultrasound stirring in combination with microwave-assisted extraction (UMAE) has also been used to improve the extraction of kernel power oil. Specifically, Mwaurah et al. [13] noted that this combination resulted in an increase in extraction yield of 8.13%. This increase can be explained by the extraction method requiring two stages, the first involving the penetration of the solvent into the cell matrix and the second involving the diffusion of the extractable compounds through the porous cell wall to the solvent used for extraction. The first stage seems to be favored by ultrasound agitation. Furthermore, this hybrid technique allowed for the maintenance of the bioactive composition of the oils compared to the MAE methodology in the absence of UAE [13].

Alternatively, with a focus on enhancing oil extraction from *Litsea cubeba* seeds, Yang et al. [50] used homogenization circulating ultrasound in combination with an aqueous enzymatic pretreatment microwave-assisted extraction (HCUEPM). The enzymes, hemicellulose, pectinase, and neutral protease, yielded oils with a similar fatty acid composition, but the use of enzymes resulted in a higher yield in oil extraction using shorter extraction times.

In other research study, Hu et al. [99] optimized a simultaneous ultrasonic–microwave-assisted aqueous enzymatic extraction (UMAEE) for the extraction of cherry seed oils. Hence, a Plackett–Burman design and Box–Behnken design were employed, and the factors considered in this study were the ultrasonic power, the microwave power, the extraction time, and the enzyme concentration. Under the optimal extraction conditions (560 W, 323 W, 38 min, and 2.7%, respectively), the findings achieved an oil recovery rate of 83.85%. In addition, the fatty acid composition of the oil was not significantly modified in this new methodology in comparison to solvent extraction. However, it was characterized by better physicochemical properties and a higher content of bioactive compounds. It should be noted that the UMAEE technique is an efficient and environmentally friendly method.

Concluding all the results discussed above, Table 8 shows a summary of the bibliographic data applying hybrid techniques in the extraction of fruit seed oils as well as the composition of the bioactive compounds present in them.

Table 8. Overview of the hybrid techniques applied for obtaining fruit seed oils.

Extraction Method	Food Waste Material	Processing Conditions	Oil Recovery	Content of Principal Bioactive Compounds	References
SFE-UAE	Passion seeds	Temperature: 40–50 °C Pressure: 16–29 MPa CO_2 flow rate: 0.63 kg·h^{-1} Solvent-seed mass ratio: 210 kg CO_2·g^{-1} Ultrasound power: 160–800 W Time: 20 min	13.5–20.9%	γ-tocopherol: 6.7–9.4 mg·100 g^{-1} γ-tocotrienol: 24–34 mg·100 g^{-1} δ-tocotrienol: 33–44 mg·100 g^{-1} Palmitic acid: 10.9–11.6% Stearic acid: 3.01–5% Oleic acid: 15.89–19% Linoleic acid: 63–68.2%	[12]
PEF-SFE	Grape seeds	Solvent-seed mass ratio: 1/2 Temperature: 35–45 °C Pressure: 35–50 MPa CO_2 flow rate: 0.9–2.7 kg·h^{-1}	70–80 g·kg^{-1}	Campesterol: 370–395.6 mg·kg^{-1} Stigmasterol: 508.8–527.2 mg·kg^{-1} β-Sitosterol: 3784.4–3973.9 mg·kg^{-1} α-tocopherol: 63.1–73.6 mg·kg^{-1} α-tocotrienol: 60.9–77.8 mg·kg^{-1} γ-tocopherol: 16.1–20.5 mg·kg^{-1} γ-tocotrienol: 88.2–108.1 mg·kg^{-1} Oleic acid: 210.8–211.1 g·kg^{-1} Linoleic acid: 645.2–646.1 g·kg^{-1} Procyanidin dimer B1: 912.1–1338.0 μg·kg^{-1} (+)-Catequin: 293.8–450.3 μg·kg^{-1} Procyanidin dimer B1: 912.1–1338.0 μg·kg^{-1} p-coumaric acid: 130.6–282.0 μg·kg^{-1} Gallic acid: 130.0–582.5 μg·kg^{-1} Dihydroxibenzoic acid: 191.5–267.4 μg·kg^{-1}	[9]

Table 8. *Cont.*

Extraction Method	Food Waste Material	Processing Conditions	Oil Recovery	Content of Principal Bioactive Compounds	References
Enzyme-assisted pressing	Apricot seeds, flaxseeds, and grape seeds	Types of enzymes: pectolytic, cellulotic, and hemicellulotic; time and temperature during incubation: 60 °C and 3 h; pH of enzyme solution: 6	11.05–30.85%	Total carotenoids: 0.037–0.731 mg·kg^{-1} Total phenolic compounds: 31.08–127.92 mg·kg^{-1} α-tocopherol: 111.60–156.59 mg·L^{-1} β-tocopherol: 0.147–154.29 mg·L^{-1} γ-tocopherol: 413.00–619.04 mg·L^{-1} δ-tocopherol: 0.147–150.97 mg·L^{-1} Oleic acid: 18.10–69.28% Linoleic acid: 14.32–68.33% α-linoleic acid: 0.08–57.49%	[98]
UMAE	Mango seeds	Bath of ultrasonicator (42 kHz, 70 WUltrasound); Ultrasound extraction time: 20–60 min Ultrasound extraction temperature: 35–55 °C Type of solvent: *n*-hexane Mass–solvent ratio: 1:3–1:8 Power MAE: 250–350 W MAE time: 5–15 min	11.60%	Stearic acid: 39.59% Oleic acid: 32.33% Docosahexaenoic acid: 15.74%	[13]
HCUEPM	*Litsea cubeba* seeds	Types of enzymes: hemicellulose, pectinase, and neutral protease Incubation temperature: 30–50 °C	204.23 mL·kg^{-1}	*D*-Limonene: ND-0.48% Lauric acid: 48.72–57.47% Decanoic acid: 10.89–11.67% Oleic acid: 8.93–14.37% Linoleic acid: 4.28–7.69%	[50]
UMAEE	Cherry seeds	Ultrasonic power: 360 W, Microwave power: 323 W for 38 min at 40 °C Enzyme: 2.7% (cellulase, hemicellulase, and pectinase)	83.85%	Palmitic acid: 5.68% Oleic acid: 53.82% Linoleic acid: 38.29%	[99]

6. Green Assessment of Extraction Techniques

In recent years, given the increasing importance of sustainability in chemical processes, the design and application of Green Analytical Metrics (GAMs) has gained significant relevance. These metrics are essential tools for evaluating the environmental impact of analytical methods in alignment with the principles of Green Analytical Chemistry (GAC), as illustrated in Figure 10 [100].

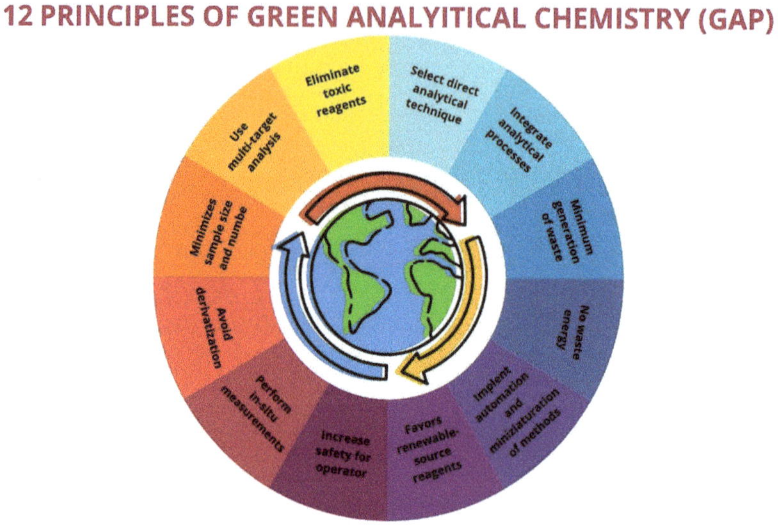

Figure 10. The 12 principles of Green Analytical Chemistry (GAC).

Along these lines, to estimate the environmental impacts of chemical processes, several metrics are currently used that are distinguished by a different degree of sophistication, known as GAMs. These metrics consider relevant aspects such as waste generation, energy consumption, reagent toxicity, process efficiency, and operator safety [101]. These sustainability metrics in analytical chemistry have undergone significant advances since the introduction of the preliminary National Index of Environmental Methods (NEMI) in 2002. In recent years, several updated tools have been proposed in the literature, reflecting a broader and more refined approach to assessing ecology. These include the Eco-scale, the Green Motion tool, the E-factor, the Green Certificate, the Green Analytical Procedure Index (GAPI), its complementary version ComplexGAPI, the Analytical Greenness Metric Approach (AGREE), the Blue Applicability Grade Index (BAGI), and the Analytical Greenness Metric for Sample Preparation (AGREEprep) [102].

Among all these metric methodologies, the most widely used due to its inclusiveness in covering all GAC principles, as well as offering a comprehensive, flexible, and simple evaluation approach that provides easy-to-interpret and informative results, is AGREEprep. This methodology is widely used as a guide for the development of new methodologies that are more responsible and aligned with current sustainability objectives [103]. AGREEprep was created by Wojciech Wojnowski et al. [103] in 2022. This metric evaluates the sustainability of sample preparation methods in analytical chemistry using a 10-principle approach based on the 12 principles of green chemistry. Each principle is rated from 0 to 1, with 1 indicating the highest level of sustainability. Finally, an overall sustainability index (from 0 to 1) is generated that summarizes the overall performance of the method and is represented in a visual chart. In the charts generated by AGREEprep, the green areas represent high sustainability across the evaluated criteria, while the red zones highlight critical aspects that need improvement from an environmental perspective [101].

Specifically, Ferrara et al. [104] used the AGREEprep metric to compare different extraction and derivatization workflows for FAMEs in different classes of food samples. They demonstrated that this methodology enables the rapid and simple development of new approaches aligned with sustainability. They highlighted the factors of safety, ease of use, effectiveness, and compatibility of various sample matrices during sustainable oil extraction.

In other study, Rabiej-Kozioł et al. [105], two conventional methods of oil extraction from *Nigella sativa* L. seeds, namely cold pressing and solvent extraction, were evaluated in terms of their ecological factor using the AGREEprep method. The green scores calculated for oil extraction by cold pressing and extraction with *n*-hexane as a solvent were 0.79 and 0.25, respectively. This demonstrated the effectiveness of cold pressing as a sustainable method of oil extraction.

In a different research study, Costa et al. [106] performed the AGREEprep methodology to evaluate the methods applied in the preparation of vegetable oil samples: microwave-assisted extraction based on emulsion breaking (MAEB); emulsion breaking-induced extraction; microemulsion; and ultrasound-assisted extraction. Their results noted that both MAEB and microemulsion contributed significantly to the sustainability of the oil extraction process.

In this review, to assess the environmental performance of the advanced sustainable techniques reviewed, as compared to conventional extraction methods, a theoretical evaluation using AGREEprep was carried out based on the methodological information available in the literature. The results of this evaluation are presented in Table 9, which summarizes the performance of various oil extraction methodologies according to the ten AGREEprep criteria selected for analysis.

Table 9. AGREEprep theoretical simulation of fruit oil extraction techniques.

Extraction Techniques	Food Waste Material	Analytes	Processing Conditions	Analytical Techniques	Criterion	Metrics	References
UAE	Passion seeds	Sterols and fatty acids	Power 165 W, frequency 25 kHz, time 10–50 min, temperature 30–60 °C, and sample–solvent ratio 2–4 mL·g^{-1}	GC-MS	Criterion 1: ex situ Criterion 2: non-hazardous materials Criterion 3: only sustainable materials Criterion 4: no waste Criterion 5: 1 g of sample Criterion 6: 4 sample throughput Criterion 7: 2 steps or fewer and semi-automated Criterion 8: 82.5 W·h^{-1} Criterion 9: liquid chromatography; gas chromatography with quadrupole detection Criterion 10: non-hazardous or no exposure	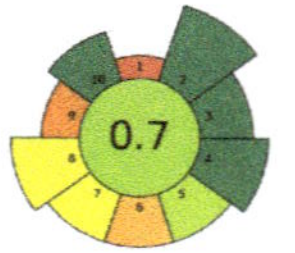	[11]
MAE	Pomegranate seeds	Fatty acids, sterols, and tocopherols	Sample–solvent ratio: 1:6 (w/v) 500 W, and frequency of 40 kHz for 9 min	GC-FID HPLC-FLD	Criterion 1: ex situ Criterion 2: non-hazardous materials Criterion 3: only sustainable materials Criterion 4: no waste Criterion 5: 1 g of sample Criterion 6: 10 sample throughput Criterion 7: 3 steps or fewer and semi-automated Criterion 8: 41.65 W·h^{-1} Criterion 9: liquid chromatography; gas chromatography with quadrupole detection Criterion 10: 1 hazard	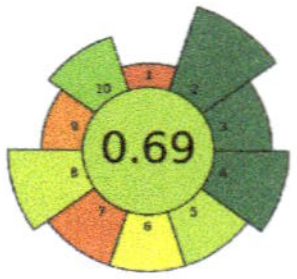	[26]
EAE	Mango seeds	Fatty acid	15 g of seeds Sample–water ratio: 1:10 (w/v) 5 min 2% of enzymes	GC-FID	Criterion 1: ex situ Criterion 2: non-hazardous materials Criterion 3: only sustainable materials Criterion 4: no waste Criterion 5: 15 g of sample Criterion 6: 12 sample throughput Criterion 7: 2 steps or fewer and manual systems Criterion 8: 69.4 W·h^{-1} Criterion 9: gas chromatography with non-MS detection Criterion 10: non-hazardous or no exposure	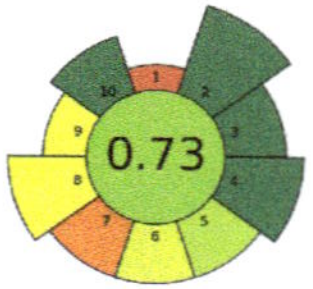	[107]

Table 9. *Cont.*

Extraction Techniques	Food Waste Material	Analytes	Processing Conditions	Analytical Techniques	Criterion	Metrics	References
PLE	Radish seeds	Phytosterols and tocopherols	5.2 g of seeds 120, 135, or 150 °C 5, 10, or 15 MPa 10 or 30 min	GC-MS	Criterion 1: ex situ Criterion 2: non-hazardous materials Criterion 3: only sustainable materials Criterion 4: no waste Criterion 5: 15 g of sample Criterion 6: 12 samples throughout Criterion 7: 2 steps or fewer and manual systems Criterion 8: 250 W·h^{-1} Criterion 9: liquid chromatography; gas chromatography with quadrupole detection Criterion 10: non-hazardous or no exposure	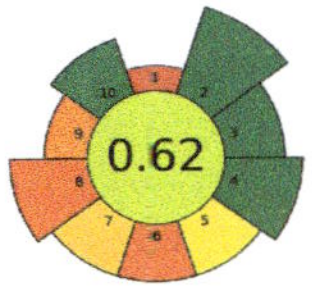	[108]
PEF	Berry seeds	Fatty acids	2 g of seeds 8 or 10 kV electrode voltage, 7 µs pulse width, 20 Hz frequency, and an adequate number of pulses to achieve energy intake of 50 kV kg^{-1} 30 min	GC-FID	Criterion 1: ex situ Criterion 2: non-hazardous materials Criterion 3: only sustainable materials Criterion 4: no waste Criterion 5: 2 g of sample Criterion 6: 2 sample throughput Criterion 7: 2 steps or fewer and semi-automated Criterion 8: 13.89 W·h^{-1} Criterion 9: gas chromatography with non-MS detection Criterion 10: non-hazardous or no exposure	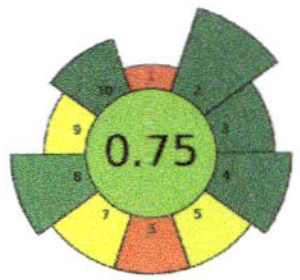	[93]
Cold pressing	Cherry seeds	Fatty acids and tocopherols	6000 g of seed 70 °C	GC-MS and HPLC-FLD	Criterion 1: ex situ Criterion 2: non-hazardous materials Criterion 3: only sustainable materials Criterion 4: no waste Criterion 5: 2000 g of sample Criterion 6: 6 sample throughput Criterion 7: 2 steps or fewer and semi-automated Criterion 8: 200 W·h^{-1} Criterion 9: gas chromatography with non-MS detection Criterion 10: non-hazardous or no exposure	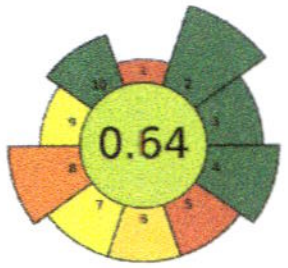	[40]

Table 9. *Cont.*

Extraction Techniques	Food Waste Material	Analytes	Processing Conditions	Analytical Techniques	Criterion	Metrics	References
Soxhlet extraction	Passion seeds	Fatty acids and tocopherols	10 g of seed Sample–solvent ratio: 1:20 (*w*/*v*) 4 h 60 °C	GC-MS and HPLC-DAD	Criterion 1: ex situ Criterion 2: 200 mL of hazardous materials Criterion 3: >75% sustainable materials Criterion 4: 200 mL of waste Criterion 5: 10 g of sample Criterion 6: 0.25 sample throughput Criterion 7: 3 steps or fewer and semi-automated Criterion 8: 2000 W·h^{-1} Criterion 9: liquid chromatography; gas chromatography with quadrupole detection Criterion 10: 1 hazard	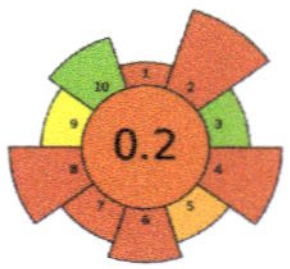	[24]
Supercritical CO_2 extraction	Berry seeds	Fatty acids and tocopherols	14 kg of seed sample 1000 bar 100 °C 240 kg·h^{-1}	RP-HPLC-FLD	Criterion 1: ex situ Criterion 2: non-hazardous materials Criterion 3: only sustainable materials Criterion 4: no waste Criterion 5: 14,000 g of sample Criterion 6: 1 sample throughput Criterion 7: 3 steps and semi-automated Criterion 8: 3000 W·h^{-1} Criterion 9: liquid chromatography; gas chromatography with quadrupole detection Criterion 10: no hazards	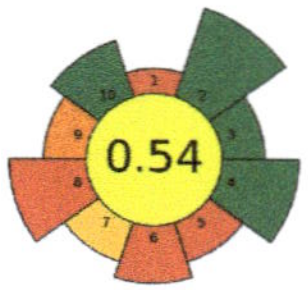	[73]

To emphasize the sustainability aspect, a weighting system was applied to each criterion, assigning the highest weight (score of 5) to criterion 2, which addresses the use of hazardous materials, followed by a score of 4 for both criterion 4, related to the non-generation of waste during the process, and criterion 8, concerning energy consumption.

A score of 3 was assigned to criterion 3 on sustainability, renewability, and reusability of materials, as well as to criterion 5 on sample size economy, criterion 6 on sample throughput, criterion 7 on method integration and automation, and criterion 10 on operator safety.

The lowest scores were given to criterion 1, which relates to the location of the sample analysis, and criterion 9, which evaluates the configuration of the sample for post-preparation analysis, as these were considered to have less influence on the overall environmental impact of the methodology.

As shown in the results presented in Table 9, all the advanced sustainable techniques examined in this review demonstrated a notably higher degree of sustainability, with AGREEprep scores ranging from 0.62 to 0.75, compared to conventional extraction methodologies, which exhibited lower scores between 0.20 and 0.64.

Although AGREEprep can significantly aid in evaluating the potential for industrial scalability of various methodologies, it should be viewed as a preliminary tool rather than a definitive solution. Its application offers valuable insight into how certain processes might perform at a larger scale and under industrial conditions. However, given that many of these methodologies are relatively recent and still under development, it is essential to continue conducting in-depth research [103].

7. Integrating Artificial Intelligence into Waste Valorization Processes

Food waste is a serious global issue, with significant economic, social, and environmental repercussions. The extensive waste generated across the entire food supply chain—from production to consumption—highlights the urgent demand for innovative and efficient management strategies. In this line, artificial intelligence (AI) has emerged as a powerful ally, thanks to its ability to process and interpret large volumes of data, recognize patterns, and forecast outcomes. Technologies such as machine learning and advanced data analytics can be leveraged to examine detailed information on waste production, composition, and valorization pathways. Specifically, there is a growing need to apply AI-driven tools in the treatment of seed-derived food waste to optimize oil extraction processes, enhancing resource recovery and promoting circular economy practices [109].

Machine learning represents a methodological shift towards systems that learn from data. In traditional programming, data and rules are provided to the computer, and then the expected response is calculated using the rules provided. However, in machine learning, both the data and the expected responses are provided to the model, and the rules are the result. These learned rules are then applied to predict the response for new data, thus enabling faster machine learning. There are two different ways in which the machine can learn, unsupervised and supervised learnings. On the one hand, supervised learning is a widely used machine learning approach in which the algorithm is trained with labelled data, e.g., data that includes both the input features and the correct output. The goal is to learn the correlations between the inputs and the corresponding outputs. This method is commonly used in tasks such as classification and regression. On the other hand, unsupervised learning is used when the samples do not contain labels. In this type of learning, the machine is allowed to discover hidden patterns in the data on its own, without the help of labels. The most common categories are clustering and dimension reduction [110]. Specifically, Iweka et al. [111] employed machine learning to optimize and predict the extraction of sustainable green oil from mature palm kernel seeds using the Soxhlet technique and observed that the value predicted by the model for the maximum yield was 38.0354% (wt) at 40 min of extraction time, 200 mL of solvent volume, and 60 g of sample weight.

Another type of artificial intelligence widely applied in the literature is artificial neural networks (ANNs), inspired by the structure and functioning of the human brain. In particular, an ANN also provides a robust approach to predicting oil extraction performance using various extraction techniques. By mimicking the way the human brain processes information, ANNs can analyze complex, non-linear relationships between input and output variables. This makes them particularly valuable for modeling and comparing different extraction methods, thereby assessing the sustainability of the process. ANNs can learn from experimental data to accurately predict performance and optimize process conditions with a focus on improving the efficiency of oil recovery [112]. Agu et al. [113] employed the response surface methodology (RSM) and an ANN to optimize the oil yield of *Terminalia catappa* L. seeds through Soxhlet extraction with n-hexane as the solvent. Under the optimum conditions for temperature, particle size, and extraction time, the RSM predicted an oil yield of 62.92%, which was validated as 60.34%, whereas the ANN-predicted yield was 60.39%, which was validated as 60.40%. The obtained optimization results noted that the ANN was a more effective approach than the RSM.

The current state of these artificial intelligence prediction methodologies is very new, so further study is required. The implementation of AI in the valorization of seed waste to obtain bioactive oils can be a powerful tool for predicting the oil recovery that would be obtained with different extraction methods, as well as predicting the bioactive compounds that would be obtained. However, its widespread adoption is currently limited. For large-scale operations, this can be a barrier due to the high cost of implementation. Furthermore, the complexity of integrating AI systems into existing agri-food infrastructure can also pose a significant challenge. In addition, ensuring data privacy and security is crucial for the ethical and responsible implementation of AI in the agri-food sector. Hence, further research and development are essential to overcome these limitations and enable broader adoption of AI in seed waste valorization and other sustainable practices. Collaborative efforts between industry and policymakers will be key to establishing standardized protocols, improving data accessibility, and reducing implementation costs. By addressing these challenges, AI has the potential to significantly enhance the efficiency and environmental performance of bioactive oil production, ultimately contributing to a more circular and sustainable agri-food system [109].

8. Industrial Applications and Future Perspectives

Although numerous authors have proposed potential applications for oils extracted via sustainable methods—emphasizing their bioactive properties, such as antioxidant and antibacterial activities, as well as their capacity to avoid the extraction of toxic or anti-nutritional compounds [20,21,35]—there is a limited number of studies that have developed these applications at an industrial or semi-industrial scale. Additionally, economic evaluations regarding the feasibility of their industrial-scale implementation are still limited [17].

Among studies evaluating the economic impact, the production of biodiesel from oil-rich waste, such as pear seeds via transesterification processes, stands out [47]. Other authors, using the circular economy criteria, have also obtained biodiesel by converting *Citrus aurantium* seed oil into methyl esters using recyclable zirconium oxide nanoparticles [114].

It is also worth noting the applications in the development of fungal control materials that extend the shelf life of foods such as strawberries, proposed by Silva et al. [19], using passion fruit seed oil together with other materials. Another novel application that may have an industrial application is to use oil from avocado seeds in a cold thermal storage system for food preservation [115].

In addition, the use of seed oils in the formulation of sunscreens is also noteworthy. Along these lines, Chu et al. [116] reported that pumpkin seed oil, due to its high content of fatty acids and squalene, could be used in the formulation of sunscreens. Moreover, Tsiapali et al. [117] pointed out that grape seed oil, due to its high content of omega-6 fatty acids, polyphenols, flavonoids, and tannins, is suitable for use in the formulation of sunscreens thanks to its added ability to delay skin ageing.

Another possible application for seed oils is their potential use as a functional food component. Afzal et al. [118] noted that sweet cherry oil obtained from food industry waste is a rich source of bioactive compounds, carotenoids, unsaturated fatty acids, and antioxidants. It is widely used as a preservative and functional food ingredient.

As for the future perspective for the valorization of oils produced from seed residues, work is underway to design new metrics that will facilitate informed decision-making for various industries, enabling them to prioritize more sustainable extraction practices. In this context, Path2Green was created by de Souza Mesquita et al. [119] to provide a holistic assessment of the sustainability of an extraction method, considering factors such as resource depletion, energy consumption, waste generation, and biodiversity preservation. However, it is necessary to further develop and refine these metrics to align with future needs and evolving regulatory frameworks.

Therefore, there is no doubt that the study of the properties of seed residue oils in the food industry can favor their potential application in industry, but there are still studies to be carried out on a semi-industrial scale to favor the integration of these oils in different industries. All of these alternative extraction technologies are currently known to be mature, offering significant intellectual property opportunities for specific applications. However, they are underutilized due to a lack of information on return on investment. In this line, initiatives are currently underway to conduct a technical–economic analysis (TEA) in order to predict the economic impact of each extraction technique [47]. Additionally, among the modeling options, Consequential Life Cycle Assessments (CLCAs) and Attributional Life Cycle Assessments (ALCAs) are of particular interest. These assessments predict the economic and environmental impact of each extraction process, thereby assisting in the design of appropriate strategic decisions.

Other relevant difficulties and barriers include a lack of technical experts and start-ups in the industry, poor competition among equipment suppliers, and a shortage of appropriate pilot facilities [17]. Furthermore, the entry of these technologies into the market is delayed due to the exhaustive safety, harmlessness, and quality testing required by the EFSA (European Food Safety Authority) or the FDA (Food and Drug Administration). Another major barrier to its industrial implementation is consumer acceptance. Although there is currently a growing trend toward natural products, the cost of accessing sustainable products is holding this trend back [120].

Notwithstanding, existing regulations increasingly emphasize the recognition and promotion of waste recycling and the enhancement of food system sustainability, aiming to facilitate its practical application across industries. Specifically, European Union policies support the development of new carbon-neutral labeling schemes, thereby strengthening the credibility and impact of sustainability initiatives (European commission 2024, [121]). Consequently, should these trends persist, it is anticipated that detailed information regarding each process for extracting oil from seed residues will become available in the future, thereby facilitating the implementation of these methodologies on an industrial scale.

9. Conclusions

This review underscores a prevailing tendency among researchers to devise more sustainable extraction methodologies and systematically evaluate the extracts obtained, with a view to correlating the bioactivity of the oils with the nature of the compounds extracted. A considerable proportion of the experimental developments incorporate optimization processes or the utilization of neural networks, which facilitate the establishment of experimental conditions conducive to achieving elevated oil extraction yields. Furthermore, these processes encompass the consideration of additional properties, including stability against oxidation and the presence of compounds exhibiting cardioprotective and antioxidant activity. These characteristics render these residues more appealing to industry. In the field of environmental science, there have been concerted efforts to develop processes that are more environmentally friendly. These efforts have involved the use of various programs that facilitate the scaling up of extraction procedures. The programs take into account multiple parameters related to environmental impact, thereby ensuring that the development of more environmentally friendly processes is informed by a comprehensive understanding of ecological considerations. Despite the existence of certain economic methodologies that facilitate the evaluation of the prospective industrial application, the majority of research endeavors concentrate on the optimization of extraction yield, the analysis of bioactive substances, and the comparison of these with conventional extraction techniques, eschewing economic comparisons between the disparate extraction techniques. In the preceding twelve-month period, the incorporation of artificial intelligence into customary extraction methodologies has been observed. Notwithstanding, it is anticipated that in the imminent future, the economic accessibility of generative artificial intelligence will be enhanced, rendering it a viable instrument for the assessment and refinement of more sustainable extraction techniques.

Author Contributions: S.R.-B.: Conceptualization, writing—original draft preparation, writing—review and editing. E.G.-M.: Conceptualization, writing—original draft preparation, writing—review and editing. N.R.-C.: Conceptualization, writing—original draft preparation, writing—review and editing. M.E.L.-G.: Conceptualization, writing—original draft preparation, writing—review and editing. All authors have read and agreed to the published version of the manuscript.

Funding: This work is part of the grant PID2023-149424OB-I00, funded by MCIU/AEI/10.13039/501100011033 and ESF+. Sandra Rodríguez-Blázquez acknowledges the Complutense University of Madrid and the Santander Bank for the pre-doctoral grant funding (CT25-24).

Institutional Review Board Statement: Not applicable.

Informed Consent Statement: Not applicable.

Data Availability Statement: No new data were created or analyzed in this study. Data sharing is not applicable to this article.

Conflicts of Interest: The authors declare no conflicts of interest.

Abbreviations

The following abbreviations are used in this manuscript:

AGREE	Analytical Greenness Metric Approach
AGREEprep	Analytical Greenness Metric for Sample Preparation
AI	Artificial intelligence
ANN	Artificial neural network
BAGI	Blue Applicability Grade Index
CLCAs	Life Cycle Assessments
ACLCAs	Attributional Life Cycle Assessments
DAD	Diode-array detector

EAE	Enzyme-assisted extraction
EFSA	European Food Safety Authority
FDA	Food and Drug Administration
FAMEs	Fatty acid methyl esters
FID	Flame ionization detection
GAC	Green Analytical Chemistry
GC	Gas chromatography
GAMs	Green Analytical Metrics
GAPI	Green Analytical Procedure Index
HCUEPM	Homogenization-circulating ultrasound with enzymatic pretreatment microwave-assisted extraction
HPLC	High-performance liquid chromatography
MAE	Microwave-assisted extraction
MAEB	Microwave-assisted extraction based on emulsion breaking
MS	Mass spectrometry
MUFAs	Monounsaturated fatty acids
NEMI	National Index of Environmental Methods
PEFs	Pulsed electric fields
PEF-SFE	Pulsed electric field–supercritical fluid extraction
PLE	Pressurized liquid extraction
PUFAs	Polyunsaturated fatty acids
RSM	Response surface methodology
$ScCO_2$	Supercritical carbon dioxide extraction
SDGs	Sustainable Development Goals
SFAs	Saturated fatty acids
S/F	Solvent-to-sample feed mass ratio
SFE	Supercritical fluid extraction
SFE-UAE	Supercritical fluid extraction–ultrasound-assisted extraction
TEA	Technical–economic analysis
TPC	Total polyphenol content
TFC	Total flavonoid content
UAE	Ultrasound-assisted extraction
UFAs	Unsaturated fatty acids
UMAE	Ultrasound–microwave assisted extraction
UMAEE	Ultrasound–microwave-assisted aqueous enzymatic extraction

References

1. Rani, H.; Sharma, S.; Bala, M. Technologies for Extraction of Oil from Oilseeds and Other Plant Sources in Retrospect and Prospects: A Review. *J. Food Process Eng.* **2021**, *44*, e13851. [CrossRef]
2. Rodríguez-Blázquez, S.; Pedrera-Cajas, L.; Gómez-Mejía, E.; Vicente-Zurdo, D.; Rosales-Conrado, N.; León-González, M.E.; Rodríguez-Bencomo, J.J.; Miranda, R. The Potential of Plum Seed Residue: Unraveling the Effect of Processing on Phytochemical Composition and Bioactive Properties. *Int. J. Mol. Sci.* **2024**, *25*, 1236. [CrossRef] [PubMed]
3. Gómez-Mejía, E.; Vicente-Zurdo, D.; Rosales-Conrado, N.; León-González, M.E. Unlocking the in Vitro Neuroprotection of Sloe Residues Phenolic Extracts by Bioanalytical and Chemometric Strategies. *Food Chem.* **2025**, *463*, 141208. [CrossRef] [PubMed]
4. Ferraz, L.P.; Silva, E.K. Pulsed Electric Field-Assisted Extraction Techniques for Obtaining Vegetable Oils and Essential Oils: Recent Progress and Opportunities for the Food Industry. *Sep. Purif. Technol.* **2025**, *354*, 128833. [CrossRef]
5. Vicente-Zurdo, D.; Gómez-Mejía, E.; Morante-Zarcero, S.; Rosales-Conrado, N.; Sierra, I. Analytical Strategies for Green Extraction, Characterization, and Bioactive Evaluation of Polyphenols, Tocopherols, Carotenoids, and Fatty Acids in Agri-Food Bio-Residues. *Molecules* **2025**, *30*, 1326. [CrossRef]
6. Dhiman, S.; Thakur, B.; Kaur, S.; Ahuja, M.; Gantayat, S.; Sarkar, S.; Singh, R.; Tripathi, M. Closing the Loop: Technological Innovations in Food Waste Valorisation for Global Sustainability. *Discov. Sustain.* **2025**, *6*, 258. [CrossRef]
7. Gil-Martín, E.; Forbes-Hernández, T.; Romero, A.; Cianciosi, D.; Giampieri, F.; Battino, M. Influence of the Extraction Method on the Recovery of Bioactive Phenolic Compounds from Food Industry By-Products. *Food Chem.* **2022**, *378*, 131918. [CrossRef]

8. Boyapati, T.; Rana, S.S.; Ghosh, P. Microwave-Assisted Extraction of Dragon Fruit Seed Oil: Fatty Acid Profile and Functional Properties. *J. Saudi Soc. Agric. Sci.* **2023**, *22*, 149–157. [CrossRef]
9. Ćurko, N.; Lukić, K.; Tušek, A.J.; Balbino, S.; Vukušić Pavičić, T.; Tomašević, M.; Redovniković, I.R.; Ganić, K.K. Effect of Cold Pressing and Supercritical CO_2 Extraction Assisted with Pulsed Electric Fields Pretreatment on Grape Seed Oil Yield, Composition and Antioxidant Characteristics. *LWT* **2023**, *184*, 114974. [CrossRef]
10. Wang, W.Y.; Yan, Y.Y.; Liu, H.M.; Qi, K.; Zhu, X.L.; Wang, X.D.; Qin, G.Y. Subcritical Low Temperature Extraction Technology and Its Application in Extracting Seed Oils. *J. Food Process Eng.* **2021**, *44*, e13805. [CrossRef]
11. Massaa, T.B.; Iwassaa, I.J.; Stevanatoa, N.; Garciab, V.A.S.; Silva, C. Passion Fruit Seed Oil: Extraction and Subsequent Transesterification Reaction. *Grasas Y Aceites* **2021**, *72*, 2. [CrossRef]
12. Barrales, F.M.; Rezende, C.A.; Martínez, J. Supercritical CO_2 Extraction of Passion Fruit (*Passiflora Edulis* sp.) Seed Oil Assisted by Ultrasound. *J. Supercrit. Fluids* **2015**, *104*, 183–192. [CrossRef]
13. Mwaurah, P.W.; Kumar, S.; Kumar, N.; Attkan, A.K.; Panghal, A.; Singh, V.K. Ultrasound and Microwave-Assisted Solvent Extraction of Mango Kernel Oil and Evaluation of Physicochemical Properties and Fatty Acid Profile. *J. Food Process Preserv.* **2021**, *45*, e16090. [CrossRef]
14. Lin, Z.; Grasso, S. Exploring Seed-Based Upcycled Oils: Types, Extraction Processes, and Emerging Applications. *Crit. Rev. Food Sci. Nutr.* **2025**, 1–20. [CrossRef]
15. Choe, U.; Childs, H.; Zeng, M.; Zheng, W.; Zhu, H.; Zhu, L.; Xie, Z.; Gao, B.; Yu, L. Value-Added Utilization of Fruit Seed Oils for Improving Human Health: A Progress Review. *ACS Food Sci. Technol.* **2023**, *3*, 528–538. [CrossRef]
16. Kumar, M.; Barbhai, M.D.; Puranik, S.; Radha; Natta, S.; Senapathy, M.; Dhumal, S.; Singh, S.; Kumar, S.; Deshmukh, V.P.; et al. Combination of Green Extraction Techniques and Smart Solvents for Bioactives Recovery. *TrAC Trends Anal. Chem.* **2023**, *169*, 117286. [CrossRef]
17. Belwal, T.; Chemat, F.; Venskutonis, P.R.; Cravotto, G.; Jaiswal, D.K.; Bhatt, I.D.; Devkota, H.P.; Luo, Z. Recent Advances in Scaling-up of Non-Conventional Extraction Techniques: Learning from Successes and Failures. *TrAC Trends Anal. Chem.* **2020**, *127*, 115895. [CrossRef]
18. Lavenburg, V.M.; Rosentrater, K.A.; Jung, S. Extraction Methods of Oils and Phytochemicals from Seeds and Their Environmental and Economic Impacts. *Processes* **2021**, *9*, 1839. [CrossRef]
19. Araújo, A.S.; de Lima, G.S.; Nunes, I.d.S.; Aguiar, J.C.R.d.O.F.d.; Navarro, D.M.D.A.F.; Melo, N.F.C.B.; Magalhães, N.S.S.; França, R.; Carvalho, R.d.S.F.; Stamford, T.C.M. Chitosan Hydrochloride-Gum Arabic-Passion Fruit Seed Oil Nanoparticle Edible Coating to Control Fungal Infection and Maintain Quality Parameters of Strawberries. *Food Control* **2024**, *161*, 110360. [CrossRef]
20. Zheng, L.; Wang, S.; Yang, Y.; Zheng, X.; Xiao, D.; Ai, B.; Sheng, Z. Volatile Aroma Compounds of Passion Fruit Seed Oils: HS-GC-IMS Analysis and Interpretation. *Food Chem. X* **2024**, *21*, 101212. [CrossRef]
21. Le, X.-T.; Ly, T.-T.; Tong, T.-D.; Luu, X.-C.; Thuy Nguyen Pham, D.; Hien Tran, T. Oil Extraction of Purple Passion Fruit Seeds Grown in Vietnam. *Mater. Today Proc.* **2023**. [CrossRef]
22. Chen, P.; Zhang, X.; Yang, Q.; Li, L.; Cao, Y.; Zhong, R.; Miao, J. Extraction and Characterization of Passion Fruit Seed Oil and Investigation of Its Hypolipidemic Activity. *J. Agric. Food Res.* **2025**, *19*, 101672. [CrossRef]
23. Antoniassi, R.; Wilhelm, A.E.; Rosa Reis, S.L.; Regis, S.A.; Faria-Machado, A.F.; Bizzo, H.R.; Cenci, S.A. Expeller Pressing of Passion Fruit Seed Oil: Pressing Efficiency and Quality of Oil. *Braz. J. Food Technol.* **2022**, *25*, e2021168. [CrossRef]
24. Pereira, M.G.; Maciel, G.M.; Haminiuk, C.W.I.; Bach, F.; Hamerski, F.; de Paula Scheer, A.; Corazza, M.L. Effect of Extraction Process on Composition, Antioxidant and Antibacterial Activity of Oil from Yellow Passion Fruit (*Passiflora Edulis* Var. *Flavicarpa*) Seeds. *Waste Biomass Valorization* **2019**, *10*, 2611–2625. [CrossRef]
25. Surlehan, H.F.; Azman, N.; Zakaria, R.; Amin, M. Extraction of Oil from Passion Fruit Seeds Using Surfactant-Assisted Aqueous Extraction. *Food Res.* **2019**, *3*, 348–356. [CrossRef]
26. Liu, N.; Ren, G.; Faiza, M.; Li, D.; Cui, J.; Zhang, K.; Yao, X.; Zhao, M. Comparison of Conventional and Green Extraction Methods on Oil Yield, Physicochemical Properties, and Lipid Compositions of Pomegranate Seed Oil. *J. Food Compos. Anal.* **2022**, *114*, 104747. [CrossRef]
27. Goula, A.M.; Papatheodorou, A.; Karasavva, S.; Kaderides, K. Ultrasound-Assisted Aqueous Enzymatic Extraction of Oil from Pomegranate Seeds. *Waste Biomass Valorization* **2018**, *9*, 1–11. [CrossRef]
28. Lolli, V.; Viscusi, P.; Bonzanini, F.; Conte, A.; Fuso, A.; Larocca, S.; Leni, G.; Caligiani, A. Oil and Protein Extraction from Fruit Seed and Kernel By-Products Using a One Pot Enzymatic-Assisted Mild Extraction. *Food Chem. X* **2023**, *19*, 100819. [CrossRef]
29. Rosa, A.; Era, B.; Masala, C.; Nieddu, M.; Scano, P.; Fais, A.; Porcedda, S.; Piras, A. Supercritical CO_2 Extraction of Waste Citrus Seeds: Chemical Composition, Nutritional and Biological Properties of Edible Fixed Oils. *Eur. J. Lipid Sci. Technol.* **2019**, *121*, 1800502. [CrossRef]
30. Garrido, G.; Chou, W.-H.; Vega, C.; Goïty, L.; Valdés, M. Influence of Extraction Methods on Fatty Acid Composition, Total Phenolic Content and Antioxidant Capacity of Citrus Seed Oils from the Atacama Desert, Chile. *J. Pharm. Pharmacogn. Res.* **2019**, *7*, 389–407. [CrossRef]

31. Hasanov, J.H.; Mirzaxmedov, S.D.; Sultonova, E.M.; Salikhov, S.I. Effect of Moisture Content on the Quality and Quantity of Screw-Pressed Flax Seed Oil. *Food Process. Tech. Technol.* **2023**, *53*, 309–315. [CrossRef]
32. Arturo-Perdomo, D.; Mora, J.P.J.; Ibáñez, E.; Cifuentes, A.; Hurtado-Benavides, A.; Montero, L. Extraction and Characterization of the Polar Lipid Fraction of Blackberry and Passion Fruit Seeds Oils Using Supercritical Fluid Extraction. *Food Anal. Methods* **2021**, *14*, 2026–2037. [CrossRef]
33. Bose, P.; Pradhan, D.; Sonawane, A.; Pathak, S.S.; Anupriya, S.; Pradhan, R.C. Optimization of Hot Air and Microwave Pretreatment for Enhancing the Mechanical Expression Yield of *Simarouba glauca* Oil. *J. Agric. Food Res.* **2020**, *2*, 100024. [CrossRef]
34. Dimić, I.; Pezo, L.; Rakić, D.; Teslić, N.; Zeković, Z.; Pavlić, B. Supercritical Fluid Extraction Kinetics of Cherry Seed Oil: Kinetics Modeling and Ann Optimization. *Foods* **2021**, *10*, 1513. [CrossRef]
35. Teixeira, G.L.; Ghazani, S.M.; Corazza, M.L.; Marangoni, A.G.; Ribani, R.H. Assessment of Subcritical Propane, Supercritical CO_2 and Soxhlet Extraction of Oil from Sapucaia (*Lecythis pisonis*) Nuts. *J. Supercrit. Fluids* **2018**, *133*, 122–132. [CrossRef]
36. Teixeira, G.L.; Maciel, L.G.; Mazzutti, S.; Gonçalves, C.B.; Ferreira, S.R.S.; Block, J.M. Composition, Thermal Behavior and Antioxidant Activity of Pracaxi (*Pentaclethra macroloba*) Seed Oil Obtained by Supercritical CO_2. *Biocatal. Agric. Biotechnol.* **2020**, *24*, 101521. [CrossRef]
37. Correa, M.S.; Fetzer, D.L.; Hamerski, F.; Corazza, M.L.; Scheer, A.P.; Ribani, R.H. Pressurized Extraction of High-Quality Blackberry (*Rubus* spp. Xavante Cultivar) Seed Oils. *J. Supercrit. Fluids* **2021**, *169*, 105101. [CrossRef]
38. Gallon, R.; Draszewski, C.P.; Santos, J.A.A.; Wagner, R.; Brondani, M.; Zabot, G.L.; Tres, M.V.; Hoffmann, R.; Castilhos, F.; Abaide, E.R.; et al. Obtaining Oil, Fermentable Sugars, and Platform Chemicals from *Butia odorata* Seed Using Supercritical Fluid Extraction and Subcritical Water Hydrolysis. *J. Supercrit. Fluids* **2023**, *203*, 106062. [CrossRef]
39. Gasparini, A.; Ferrentino, G.; Angeli, L.; Morozova, K.; Zatelli, D.; Scampicchio, M. Ultrasound Assisted Extraction of Oils from Apple Seeds: A Comparative Study with Supercritical Fluid and Conventional Solvent Extraction. *Innov. Food Sci. Emerg. Technol.* **2023**, *86*, 103370. [CrossRef]
40. Dimić, I.; Pavlić, B.; Rakita, S.; Cvetanović Kljakić, A.; Zeković, Z.; Teslić, N. Isolation of Cherry Seed Oil Using Conventional Techniques and Supercritical Fluid Extraction. *Foods* **2023**, *12*, 11. [CrossRef]
41. Pavlić, B.; Pezo, L.; Marić, B.; Tukuljac, L.P.; Zeković, Z.; Solarov, M.B.; Teslić, N. Supercritical Fluid Extraction of Raspberry Seed Oil: Experiments and Modelling. *J. Supercrit. Fluids* **2020**, *157*, 104687. [CrossRef]
42. Kaseke, T.; Opara, U.L.; Fawole, O.A. Fatty Acid Composition, Bioactive Phytochemicals, Antioxidant Properties and Oxidative Stability of Edible Fruit Seed Oil: Effect of Preharvest and Processing Factors. *Heliyon* **2020**, *6*, e04962. [CrossRef] [PubMed]
43. Yilmaz, E.; Aydeniz Guneser, B.; Ok, S. Valorization of Grapefruit Seeds: Cold Press Oil Production. *Waste Biomass Valorization* **2019**, *10*, 2713–2724. [CrossRef]
44. Cui, J.; Li, T.; Zhou, Y.; Wang, L.; Li, T.; Zhang, W. Comparative Lipidomics Analysis of Seed Oils from Nine Tropical Fruits: Emphasizing the Fatty Acid and Lipid Molecule Profiles. *Food Res. Int.* **2024**, *190*, 115004. [CrossRef]
45. Li, X.; Qi, B.; Zhang, S.; Li, Y. Foodomics Revealed the Effects of Ultrasonic Extraction on the Composition and Nutrition of Cactus Fruit (*Opuntia ficus-indica*) Seed Oil. *Ultrason. Sonochem* **2023**, *97*, 106459. [CrossRef]
46. Rodríguez-Blázquez, S.; Gómez-Mejía, E.; Rosales-Conrado, N.; León-González, M.E.; García-Sánchez, B.; Miranda, R. Valorization of Prunus Seed Oils: Fatty Acids Composition and Oxidative Stability. *Molecules* **2023**, *28*, 7045. [CrossRef]
47. Umeagukwu, O.E.; Onukwuli, D.O.; Ude, C.N.; Chizoo, E.; Nnamdi Ekwueme, B.; Asadu, C.O.; Okey-Onyesolu, F.C.; Ikenna, M.U.; Innocent Chukwudi, E.; Ugwele, F.O. Techno-Economic Analysis of the Statistically Optimized Biodiesel Production Process Using African Pear Seed Oil and Activated Empty Palm Fruit Bunch Biocatalyst. *Waste Manag. Bull.* **2024**, *2*, 95–112. [CrossRef]
48. García-Ramón, D.F.; Valdiviezo-Marcelo, J., Rodrigues, C.E.d.C.; Viganó, J.; de Oliveira, A.L. Advances and Trends in Compressed Fluid Extraction to Recovering Bioactive Compounds from Avocado: Effect of Extraction Parameters, Biological Properties, and Applications. *J. Supercrit. Fluids* **2025**, *218*, 106505. [CrossRef]
49. Chifomboti, L.P.; Chimphango, A.F.A. Development and Optimization of a Two-Step Co-Extraction Process for the Recovery of Pumpkin Seed Oil and in-Situ Enrichment with β-Carotene Compounds from Pumpkin Peel. *Food Bioprod. Process.* **2025**, *149*, 211–223. [CrossRef]
50. Yang, X.; Yang, Y.; Zhang, K.; Zhao, R.; Tian, H.; Yang, L., Zhao, X. Homogenization-Circulating Ultrasound in Combination with Aqueous Enzymatic Pretreatment for Microwave-Assisted Extraction of Kernel Oil and Essential Oil from the Fruit of *Litsea cubeba*. *Ultrason. Sonochem* **2024**, *111*, 107093. [CrossRef]
51. Geow, C.H.; Tan, M.C.; Yeap, S.P.; Chin, N.L. A Review on Extraction Techniques and Its Future Applications in Industry. *Eur. J. Lipid Sci. Technol.* **2021**, *123*, 2000302. [CrossRef]
52. Savoire, R.; Lanoisellé, J.L.; Vorobiev, E. Mechanical Continuous Oil Expression from Oilseeds: A Review. *Food Bioproc. Technol.* **2013**, *6*, 1–16. [CrossRef]

53. Niu, Z.; Zhu, Z.; Zhou, J.; Xu, C.; Wei, C.; Liu, W.; Liu, Z.; Wang, T.; Xiao, H. Effect of Roasting on the Chemical Composition and Oxidative Stability of Tomato (*Solanum lycopersicum* L.) Seed Oil. *Foods* **2024**, *13*, 1682. [CrossRef]
54. Hou, N.C.; Gao, H.H.; Qiu, Z.J.; Deng, Y.H.; Zhang, Y.T.; Yang, Z.C.; Gu, L.B.; Liu, H.M.; Zhu, X.L.; Qin, Z.; et al. Quality and Active Constituents of Safflower Seed Oil: A Comparison of Cold Pressing, Hot Pressing, Soxhlet Extraction and Subcritical Fluid Extraction. *LWT* **2024**, *200*, 116184. [CrossRef]
55. Morya, S.; Menaa, F.; Jiménez-López, C.; Lourenço-Lopes, C.; Binmowyna, M.N.; Alqahtani, A. Nutraceutical and Pharmaceutical Behavior of Bioactive Compounds of Miracle Oilseeds: An Overview. *Foods* **2022**, *11*, 1824. [CrossRef]
56. Zeng, J.; Xiao, T.; Ni, X.; Wei, T.; Liu, X.; Deng, Z.Y.; Li, J. The Comparative Analysis of Different Oil Extraction Methods Based on the Quality of Flaxseed Oil. *J. Food Compos. Anal.* **2022**, *107*, 104373. [CrossRef]
57. Siol, M.; Piasecka, I.; Mańko-Jurkowska, D.; Górska, A.; Bryś, J. Optimization and Impact of Ultrasound-Assisted Extraction on Pomegranate Seed Oil Quality: A Comparative Study of Bioactive Potential and Oxidation Parameters. *Molecules* **2025**, *30*, 1837. [CrossRef]
58. Andasuryani; Ifmalinda; Derosya, V.; Syukri, D. The Effect of Moisture on the Yield of Sweet Passion Fruit (*Passiflora lingularis* Juss Cv. *Gumanti*) Seed Oil from Indonesia by Mechanical Extraction. *IOP Conf. Ser. Earth Environ. Sci.* **2020**, *515*, 012021. [CrossRef]
59. Zhang, H.; Yuan, Y.; Zhu, X.; Xu, R.; Shen, H.; Zhang, Q.; Ge, X. The Effect of Different Extraction Methods on Extraction Yield, Physicochemical Properties, and Volatile Compounds from Field Muskmelon Seed Oil. *Foods* **2022**, *11*, 721. [CrossRef]
60. Yilmaz, E.; Güneşer, B.A. Cold Pressed versus Solvent Extracted Lemon (*Citrus limon* L.) Seed Oils: Yield and Properties. *J. Food Sci. Technol.* **2017**, *54*, 1891–1900. [CrossRef]
61. De Filette, M.; Schatteman, K.; Geuens, J. Characterization of Six Cold-Pressed Berry Seed Oils and Their Seed Meals. *Appl. Sci.* **2024**, *14*, 439. [CrossRef]
62. Kumar, S.P.J.; Prasad, S.R.; Banerjee, R.; Agarwal, D.K.; Kulkarni, K.S.; Ramesh, K.V. Green Solvents and Technologies for Oil Extraction from Oilseeds. *Chem. Cent. J.* **2017**, *11*, 9. [CrossRef] [PubMed]
63. Gonfa Keneni, Y.; Bahiru, A.; Marchetti, J.M.; Marchetti, J.M. Effects of Different Extraction Solvents on Oil Extracted from Jatropha Seeds and the Potential of Seed Residues as a Heat Provider. *Bioenergy Res.* **2021**, *14*, 1207–1222. [CrossRef]
64. Giuffrè, A.M.; Zappia, C.; Capocasale, M. Tomato Seed Oil: A Comparison of Extraction Systems and Solvents on Its Biodiesel and Edible Properties. *Riv. Ital. Delle Sostanze Grasse* **2017**, *94*, 140–160.
65. Association of Official Analytical Chemists (AOAC). *Method 960.39*, 17th ed.; AOAC International: Gaithersburg, MD, USA, 2000.
66. Ndayishimiye, J.; Getachew, A.T.; Chun, B.S. Comparison of Characteristics of Oils Extracted from a Mixture of Citrus Seeds and Peels Using Hexane and Supercritical Carbon Dioxide. *Waste Biomass Valorization* **2017**, *8*, 1205–1217. [CrossRef]
67. Onyedikachi, U.B.; Nkwocha, C.C.; Ejiofor, E.; Nnanna, C.C. Investigation of Chemical Constituents, Antioxidant, Anti-Inflammatory and Nutritional Properties of Oil of *Persea americana* (Avocado) Seeds. *Food Chem. Adv.* **2024**, *5*, 100770. [CrossRef]
68. Ferrentino, G.; Giampiccolo, S.; Morozova, K.; Haman, N.; Spilimbergo, S.; Scampicchio, M. Supercritical Fluid Extraction of Oils from Apple Seeds: Process Optimization, Chemical Characterization and Comparison with a Conventional Solvent Extraction. *Innov. Food Sci. Emerg. Technol.* **2020**, *64*, 102428. [CrossRef]
69. Versteeg, F.A.; Picchioni, F.; Versteeg, G.F. On the Mass Transfer of Supercritical Fluids, Specifically Super Critical CO_2: An Overview. *Chem. Eng. J.* **2024**, *493*, 152521. [CrossRef]
70. Dhara, O.; Rani, K.N.P.; Chakrabarti, P.P. Supercritical Carbon Dioxide Extraction of Vegetable Oils: Retrospective and Prospects. *Eur. J. Lipid Sci. Technol.* **2022**, *124*, 2200006. [CrossRef]
71. de Souza Correa, M.; Boschen, N.L.; Rodrigues, P.R.P.; Corazza, M.L.; de Paula Scheer, A.; Ribani, R.H. Supercritical CO_2 with Co-Solvent Extraction of Blackberry (*Rubus* spp. Xavante Cultivar) Seeds. *J. Supercrit. Fluids* **2022**, *189*, 105702. [CrossRef]
72. Macawile, M.C.; Auresenia, J. Utilization of Supercritical Carbon Dioxide and Co-Solvent n-Hexane to Optimize Oil Extraction from Gliricidia Sepium Seeds for Biodiesel Production. *Appl. Sci. Eng. Prog.* **2022**, *15*, 1. [CrossRef]
73. Milala, J.; Grzelak-Błaszczyk, K.; Sójka, M.; Kosmala, M.; Dobrzyńska-Inger, A.; Rój, E. Changes of Bioactive Components in Berry Seed Oils during Supercritical CO_2 Extraction. *J. Food Process Preserv.* **2018**, *42*, e13368. [CrossRef]
74. Silva, L.d.O.; Ranquine, L.G.; Monteiro, M.; Torres, A.G. Pomegranate (*Punica granatum* L.) Seed Oil Enriched with Conjugated Linolenic Acid (CLnA), Phenolic Compounds and Tocopherols: Improved Extraction of a Specialty Oil by Supercritical CO_2. *J. Supercrit. Fluids* **2019**, *147*, 126–137. [CrossRef]
75. Borah, K.; Kumar, R.; Purohit, S.; Goud, V.V. Comparative Study on Phytochemical Extraction from Passion Fruit Wastes Using Ultrasound and Supercritical Fluid Extraction. *J. Food Meas. Charact.* **2025**, *19*, 2831–2843. [CrossRef]
76. Tobiszewski, M.; Mechlinska, A.; Namie, J. Green Analytical Chemistry—Theory and Practice. *Chem. Soc. Rev.* **2010**, *39*, 2869–2878. [CrossRef]
77. Thilakarathna, R.C.N.; Siow, L.F.; Tang, T.K.; Lee, Y.Y. A Review on Application of Ultrasound and Ultrasound Assisted Technology for Seed Oil Extraction. *J. Food Sci. Technol.* **2023**, *60*, 1222–1236. [CrossRef]

78. Bao, J.; Li, L.; Ban, L.; Pi, X.; Li, C. A Novel Efficient Method for the Simultaneous Recycle of Amygdalin and Oil from Bitter Apricot Kernels: Ultrasonication Based on Natural Deep Eutectic Solvent. *J. Mol. Liq.* **2025**, *429*, 127592. [CrossRef]
79. Marinaccio, L.; Gentile, G.; Zengin, G.; Pieretti, S.; Stefanucci, A.; Cichelli, A.; Mollica, A. Ultrasound Assisted Deep Eutectic Solvent-Based Extraction of Montepulciano d' Abruzzo Grape Seeds for the Recovery of the Grape Seed Oil and Its Biological Evaluation. *Food Chem. X* **2025**, *26*, 102273. [CrossRef]
80. Chakraborty, S.; Uppaluri, R.; Das, C. Optimization of Ultrasound-Assisted Extraction (UAE) Process for the Recovery of Bioactive Compounds from Bitter Gourd Using Response Surface Methodology (RSM). *Food Bioprod. Process.* **2020**, *120*, 114–122. [CrossRef]
81. Hu, T.; Subbiah, V.; Wu, H.; BK, A.; Rauf, A.; Alhumaydhi, F.A.; Suleria, H.A.R. Determination and Characterization of Phenolic Compounds from Australia-Grown Sweet Cherries (*Prunus avium* L.) and Their Potential Antioxidant Properties. *ACS Omega* **2021**, *6*, 34687–34699. [CrossRef]
82. Capaldi, G.; Binello, A.; Aimone, C.; Mantegna, S.; Grillo, G.; Cravotto, G. New Trends in Extraction-Process Intensification: Hybrid and Sequential Green Technologies. *Ind. Crops Prod.* **2024**, *209*, 117906. [CrossRef]
83. Yu, M.H.; Hung, T.W.; Wang, C.C.; Wu, S.W.; Yang, T.W.; Yang, C.Y.; Tseng, T.H.; Wang, C.J. Neochlorogenic Acid Attenuates Hepatic Lipid Accumulation and Inflammation via Regulating Mir-34a in Vitro. *Int. J. Mol. Sci.* **2021**, *22*, 13163. [CrossRef] [PubMed]
84. Yang, C.; Shang, K.; Lin, C.; Wang, C.; Shi, X.; Wang, H.; Li, H. Processing Technologies, Phytochemical Constituents, and Biological Activities of Grape Seed Oil (GSO): A Review. *Trends Food Sci. Technol.* **2021**, *116*, 1074–1083. [CrossRef]
85. Satriana, S.; Supardan, M.D.; Arpi, N.; Wan Mustapha, W.A. Development of Methods Used in the Extraction of Avocado Oil. *Eur. J. Lipid Sci. Technol.* **2019**, *121*, 1800210. [CrossRef]
86. Keskin Çavdar, H.; Koçak Yanık, D.; Gök, U.; Göğüş, F. Optimisation of Microwave-Assisted Extraction of Pomegranate (*Punica granatum* L.) Seed Oil and Evaluation of Its Physicochemical and Bioactive Properties. *Food Technol. Biotechnol.* **2017**, *55*, 86–94. [CrossRef]
87. da Silva, R.F.; Carneiro, C.N.; Cheila, C.B.; Gomez, J.V.F.; Espino, M.; Boiteux, J.; de los Á Fernández, M.; Silva, M.F.; Dias, F.d.S. Sustainable Extraction Bioactive Compounds Procedures in Medicinal Plants Based on the Principles of Green Analytical Chemistry: A Review. *Microchem. J.* **2022**, *175*, 107184. [CrossRef]
88. Amador-Luna, V.M.; Montero, L.; Herrero, M. Compressed Fluids for the Extraction of Bioactive Compounds from Plants, Food by-Products, Seaweeds and Microalgae—An Update from 2019 to 2023. *TrAC Trends Anal. Chem.* **2023**, *169*, 117410. [CrossRef]
89. Cheriyan, B.V.; Karunakar, K.K.; Anandakumar, R.; Murugathirumal, A.; Kumar, A.S. Eco-Friendly Extraction Technologies: A Comprehensive Review of Modern Green Analytical Methods. *Sustain. Chem. Clim. Action* **2025**, *6*, 100054. [CrossRef]
90. Višnjevec, A.M.; Barp, L.; Lucci, P.; Moret, S. Pressurized Liquid Extraction for the Determination of Bioactive Compounds in Plants with Emphasis on Phenolics. *TrAC Trends Anal. Chem.* **2024**, *173*, 117620. [CrossRef]
91. Piasecka, I.; Górska, A.; Obranović, M.; Kalisz, S.; Dobrinčić, A.; Dobroslavić, E.; Ostrowska-Ligęza, E.; Brzezińska, R.; Dragović-Uzelac, V. The Quality Assessment of Oils Obtained from Berry Fruit Seeds Using Pressurized Liquid Extraction. *Biol. Life Sci. Forum* **2023**, *26*, 84. [CrossRef]
92. Bakhshabadi, H.; Mirzaei, H.O.; Ghodsvali, A.; Jafari, S.M.; Ziaiifar, A.M. The Influence of Pulsed Electric Fields and Microwave Pretreatments on Some Selected Physicochemical Properties of Oil Extracted from Black Cumin Seed. *Food Sci. Nutr.* **2018**, *6*, 111–118. [CrossRef] [PubMed]
93. Piasecka, I.; Ostrowska-Ligęza, E.; Wiktor, A.; Górska, A. Ultrasound and Pulsed Electric Field Treatment Effect on the Thermal Properties, Oxidative Stability and Fatty Acid Profile of Oils Extracted from Berry Seeds. *J. Therm. Anal. Calorim.* **2024**, *150*, 1311–1325. [CrossRef]
94. Ferreira, I.J.B.; Alexandre, E.M.C.; Saraiva, J.A.; Pintado, M. Green Emerging Extraction Technologies to Obtain High-Quality Vegetable Oils from Nuts: A Review. *Innov. Food Sci. Emerg. Technol.* **2022**, *76*, 102931. [CrossRef]
95. Díaz-Reinoso, B.; Rivas, S.; Rivas, J.; Domínguez, H. Subcritical Water Extraction of Essential Oils and Plant Oils. *Sustain. Chem. Pharm.* **2023**, *36*, 101332. [CrossRef]
96. Quaisie, J.; Ma, H.; Golly, M.K.; Tuly, J.A.; Amaglo, N.K.; Jiaqi, Z. Effect of Ultrasound-Microwave Irradiation Hybrid Technique on Extraction, Physicochemical, Antioxidative, and Structural Properties of Stearic Acid-Rich *Allanblackia parviflora* Seed Oil. *Chem. Pap.* **2021**, *75*, 4527–4541. [CrossRef]
97. Vovk, H.; Karnpakdee, K.; Ludwig, R.; Nosenko, T. Enzymatic Pretreatment of Plant Cells for Oil Extraction. *Food Technol. Biotechnol.* **2023**, *61*, 160–178. [CrossRef]
98. Candan, A.; Arslan, D. Enzymatic Pre-Treatment in Cold Pressing: Influence on Flaxseed, Apricot Kernel and Grape Seed Oils. *Grasas Y Aceites* **2021**, *72*, 4. [CrossRef]
99. Hu, B.; Wang, H.; He, L.; Li, Y.; Li, C.; Zhang, Z.; Liu, Y.; Zhou, K.; Zhang, Q.; Liu, A.; et al. A Method for Extracting Oil from Cherry Seed by Ultrasonic-Microwave Assisted Aqueous Enzymatic Process and Evaluation of Its Quality. *J. Chromatogr. A* **2019**, *1587*, 50–60. [CrossRef]

100. Imam, M.S.; Abdelrahman, M.M. How Environmentally Friendly Is the Analytical Process? A Paradigm Overview of Ten Greenness Assessment Metric Approaches for Analytical Methods. *Trends Environ. Anal. Chem.* **2023**, *38*, e00202. [CrossRef]
101. Sajid, M.; Płotka-Wasylka, J. Green Analytical Chemistry Metrics: A Review. *Talanta* **2022**, *238*, 123046. [CrossRef]
102. Martínez, J.; Cortés, J.F.; Miranda, R. Green Chemistry Metrics, A Review. *Processes* **2022**, *10*, 1274. [CrossRef]
103. Wojnowski, W.; Tobiszewski, M.; Pena-Pereira, F.; Psillakis, E. AGREEprep—Analytical Greenness Metric for Sample Preparation. *TrAC Trends Anal. Chem.* **2022**, *149*, 116553. [CrossRef]
104. Ferrara, D.; Beccaria, M.; Cordero, C.E.; Purcaro, G. Comprehensive Comparison of Fatty Acid Methyl Ester Profile in Different Food Matrices Using Microwave-Assisted Extraction and Derivatization Methods and Comprehensive Two-Dimensional Gas Chromatography Coupled with Flame Ionization Detection. *Adv. Sample Prep.* **2024**, *11*, 100124. [CrossRef]
105. Rabiej-Kozioł, D.; Szydłowska-Czerniak, A. Antioxidant Potential Evaluation at Various Stages of Black Cumin Oil Production. *Foods* **2024**, *13*, 3518. [CrossRef]
106. Costa, F.S.; Moreira, L.S.; Ludovico, L.L.; Volpe, J.; de Oliveira, A.C.; dos Santos, M.P.; da Silva, E.G.P.; Souto, D.E.P.; Grassi, M.T.; Gonzalez, M.H.; et al. Microwave-Assisted Extraction Based on Emulsion Breaking with Natural Deep Eutectic Solvent for Vegetable Oil Sample Preparation Prior to Elemental Determination by ICP OES. *Talanta* **2024**, *266*, 125108. [CrossRef] [PubMed]
107. El Sayed Mehanni, A.-H. Impact of Enzymatic Aqueous Extraction Mango Seed Oil on Physicochemical Properties and Oxidative Stability. *J. Food Dairy Sci.* **2021**, *12*, 195–201. [CrossRef]
108. Ferreira de Mello, B.T.; Stevanato, N.; Filho, L.C.; da Silva, C. Pressurized Liquid Extraction of Radish Seed Oil Using Ethanol as Solvent: Effect of Pretreatment on Seeds and Process Variables. *J. Supercrit. Fluids* **2021**, *176*, 105307. [CrossRef]
109. Onyeaka, H.; Tamasiga, P.; Nwauzoma, U.M.; Miri, T.; Juliet, U.C.; Nwaiwu, O.; Akinsemolu, A.A. Using Artificial Intelligence to Tackle Food Waste and Enhance the Circular Economy: Maximising Resource Efficiency and Minimising Environmental Impact: A Review. *Sustainability* **2023**, *15*, 10482. [CrossRef]
110. Cheraghi, Y.; Kord, S.; Mashayekhizadeh, V. Application of Machine Learning Techniques for Selecting the Most Suitable Enhanced Oil Recovery Method; Challenges and Opportunities. *J. Pet. Sci. Eng.* **2021**, *205*, 108761. [CrossRef]
111. Iweka, S.C.; Akpomedaye, O.; Emu, A.O.; Adepoju, T.F. Impact of Optimization Parameters on Green Sustainable Oil from Ripe Palm Kernel Seeds for Energy Utilities: A Machine Learning Approach. *Sci. Afr.* **2024**, *23*, e02097. [CrossRef]
112. Shi, C.; Zhao, Z.; Jia, Z.; Hou, M.; Yang, X.; Ying, X.; Ji, Z. Artificial Neural Network-Based Shelf Life Prediction Approach in the Food Storage Process: A Review. *Crit. Rev. Food Sci. Nutr.* **2023**, *64*, 12009–12024. [CrossRef] [PubMed]
113. Agu, C.M.; Menkiti, M.C.; Ekwe, E.B.; Agulanna, A.C. Modeling and Optimization of *Terminalia catappa* L. Kernel Oil Extraction Using Response Surface Methodology and Artificial Neural Network. *Artif. Intell. Agric.* **2020**, *4*, 1–11. [CrossRef]
114. Rozina; Ahmad, M.; Zafar, M. Conversion of Waste Seed Oil of *Citrus aurantium* into Methyl Ester via Green and Recyclable Nanoparticles of Zirconium Oxide in the Context of Circular Bioeconomy Approach. *Waste Manag.* **2021**, *136*, 310–320. [CrossRef] [PubMed]
115. Nicolalde, J.F.; Martinez-Gómez, J.; Guerrero, V.H.; Chico-Proano, A. Experimental Performance of Avocado Seed Oil as a Bio-Based Phase Change Material for Refrigeration Applications. *Int. J. Refrig.* **2024**, *168*, 632–647. [CrossRef]
116. Chu, C.C.; Hasan, Z.A.B.A.; Chua, S.K.; Nyam, K.L. Formulation and Characterization of Novel Nanostructured Lipid Carriers with Photoprotective Properties Made from Carnauba Wax, Beeswax, Pumpkin Seed Oil, and UV Filters. *JAOCS J. Am. Oil Chem. Soc.* **2020**, *97*, 531–542. [CrossRef]
117. Tsiapali, O.I.; Ayfantopoulou, E.; Tzourouni, A.; Ofrydopoulou, A.; Letsiou, S.; Tsoupras, A. Unveiling the Utilization of Grape and Winery By-Products in Cosmetics with Health Promoting Properties. *Appl. Sci.* **2025**, *15*, 1007. [CrossRef]
118. Afzal, M.F.; Khalid, W.; Armghan Khalid, M.; Zubair, M.; Akram, S.; Kauser, S.; Noreen, S.; Jamal, A.; Kamran Khan, M.; Al-Farga, A. Recent Industrials Extraction of Plants Seeds Oil Used in the Development of Functional Food Products: A Review. *Int. J. Food Prop.* **2022**, *25*, 2530–2550. [CrossRef]
119. de Souza Mesquita, L.M.; Contieri, L.S.; e Silva, F.A.; Bagini, R.H.; Bragagnolo, F.S.; Strieder, M.M.; Sosa, F.H.B.; Schaeffer, N.; Freire, M.G.; Ventura, S.P.M.; et al. *Path2Green*: Introducing 12 Green Extraction Principles and a Novel Metric for Assessing Sustainability in Biomass Valorization. *Green Chem.* **2024**, *26*, 10087–10106. [CrossRef]
120. Agarwal, P.; Kumar, D.; Katiyar, R. Antecedents of Continuous Purchase Behavior for Sustainable Products: An Integrated Conceptual Framework and Review. *J. Consum. Behav.* **2025**. [CrossRef]
121. Buffi, M.; Georgakaki, A.; Hurtig, O.; Joanny, G.; Letout, S.; Motola, V.; Mountraki, A.; Scarlat, N. *Clean Energy Technology Observatory Advanced Biofuels-KJ0124063ENN*; European Commission: Bruxelles, Belgium, 2022. [CrossRef]

foods

MDPI

Article

Effects of Sodium Selenite on Accumulations of Selenium and GABA, Phenolic Profiles, and Antioxidant Activity of Foxtail Millet During Germination

Shuaiduo Sun [1], Jingjing Zhang [1], Yongji Li [1], Yunfeng Xu [1,2], Runqiang Yang [3], Lei Luo [1] and Jinle Xiang [1,2,*]

1 Faculty of Food & Bioengineering, Henan University of Science & Technology, Luoyang 471023, China; 15037912413@163.com (S.S.); z2823662022@163.com (J.Z.); 15539578706@163.com (Y.L.); xuyunfeng1573@126.com (Y.X.); 13623896431@139.com (L.L.)
2 Henan International Joint Laboratory of Food Green Processing and Safety Control, Henan University of Science & Technology, Luoyang 471023, China
3 College of Food Science & Technology, Nanjing Agricultural University, Nanjing 210000, China; yangrq@njau.edu.cn
* Correspondence: xjl5013@haust.edu.cn; Tel.: +86-152-3791-6981

Abstract: This study investigated the influence of soaking and spraying with a sodium selenite (Na_2SeO_3) solution on selenium accumulation, γ-aminobutyric acid (GABA) content, phenolic compositions, and the antioxidant activity of foxtail millet sprouts. The screening results showed that foxtail millet seeds soaked with 60 mg/L of Na_2SeO_3 solution and sprayed with 2 mg/L of Na_2SeO_3 solution were the appropriate concentrations for the germination process. Compared with the spraying method, a presoaking treatment presented far higher selenium content and significantly higher ($p < 0.05$) selenium enrichment rates in foxtail millet sprouts. The content of free and bound phenolics, as well as GABA, were significantly ($p < 0.05$) increased in foxtail millet sprouts through both soaking and spraying treatments. Correspondingly, most of the individual phenolic compounds were significantly ($p < 0.05$) increased, especially after germination for 3 days. *Trans*-ferulic acid and *trans*-*p*-coumaric acid were the predominate bound phenolic acids, feruloylquinic acid and 4-*p*-coumaroylquinic acid were the major free-form phenolic compounds, and *N*-feruloyl serotonin and *N*-(*p*-coumaroyl) serotonin were the new arising phenolic derivatives caused by germination. Both the soaking and spraying treatments induced the enrichment of these individual phenolic compositions, thus increasing the total phenolic content and in vitro antioxidant activity of foxtail millet sprouts. It was indicated that selenium-enriched germination treatment should be an effective method to produce functional selenium-enriched foxtail millet sprouts with more abundant GABA and polyphenols, thus enhancing the health benefits and added value of foxtail millet.

Keywords: selenium; foxtail millet sprouts; polyphenols; GABA; antioxidant activity

1. Introduction

Selenium (Se) is recognized as a vital trace element crucial for human health [1]. It is estimated that approximately one billion individuals globally experience inadequate selenium intake [2,3]. Se deficiency often leads to declines in cognitive ability, oxidative stress, decreased adaptive immunity, neurological disorders, and Keshan disease [4]. Edible plants are believed to be the main dietary source of Se for human, since many plants can metabolize and accumulate organic Se in their edible parts to be consumed directly, such as the leaves, flowers, fruits, seeds, and sprouts [5,6]. However, the Se content of food resources, especially organic Se, is generally very low due to the imbalance of Se in the soil of different regions [7]. Increasing the concentration of absorbable Se in edible plants or their products by Se-biofortification is a viable pathway to supplement Se for humans [3].

Se-biofortification in grain seeds through germination is believed to be a simple, rapid, and efficient method for the development of functional Se-enriched foods [8]. The germination process has also been widely applied as a low-cost and efficient method to enhance the nutritional value of grain seeds [9]. This method of Se-enriched sprouting has been proven to greatly improve the Se concentration in germinated seeds, such as mung bean sprouts [10], alfalfa sprouts [11], pea sprouts [12], and corn sprouts [13]. Additionally, Se-biofortification could also lead to accumulations of amounts of bioactive secondary metabolites during germination [5]. It has been reported that maize seeds presoaked with 50 mg/L of Na_2SeO_3 and germinated for 2 days displayed a 181% increase in bound ferulic acid when compared to germinated maize without Na_2SeO_3 treatment [14]. Se-biofortification increased the phenolic acid contents of rice sprouts, especially their soluble conjugated forms [15]. Anthocyanins were effectively enhanced by spraying radish shoots with 10 μM of Na_2SeO_3 solution during germination [16]. Compared with untreated chickpea sprouts, the total isoflavonoid content was increased by 83% through presoaking with 23.5 mg/L of Na_2SeO_3 solution [17]. The effects of Se-biofortification on other phytochemicals and nutrients have also been reported, such as chlorophyll and carotenoids [18,19], GABA [20], and vitamin C [19,20]. The accumulation of these bioactive compounds and the simultaneous enrichment of organic Se are conducive to the enhancement of health benefits for germinated grain seeds, which have the potential to be used for the development of functional foods [21].

Foxtail millet is one of the most important minor cereal crops, and it has been considered a suitable candidate for the development of functional foods [22]. Traditional and modern processing technologies, together with their resulting nutritional and functional changes to foxtail millets, have been previously reviewed [23]. In our previous study, the phenolic and GABA content and antioxidant capacity of foxtail millet were significantly increased by the germination process [24,25]. However, little research related to the Se-biofortification of cereal grain sprouts during germination has been reported, and the effects of Se-biofortification on the secondary metabolites of germinated foxtail millets remain unclear.

The main purpose of this study was to (1) screen the appropriate Na_2SeO_3 treatment method and concentration for the Se-biofortification of foxtail millet sprouts; and (2) investigate the effects of Se-biofortification on Se content, GABA concentration, phenolic profiles, and antioxidant activity during the Se-enriched germination of foxtail millet. This research could contribute to the development of Se-enriched functional foods and improve the utilization value of foxtail millet sprout as a functional ingredient.

2. Materials and Methods

2.1. Reagents and Standards

Sodium selenite (Na_2SeO_3) was purchased from Tianjin Chemical Reagent Third Factory Co., Ltd. (Tianjin, China). DAN (2,3-Diaminonaphthalene) was purchased from Shanghai Macklin Biochemical Technology Co., Ltd. (Shanghai, China). Cyclohexane was acquired from Tianjin Yongda Chemical Reagent Co., Ltd. (Tianjin, China). Folin–Ciocâlteu reagent, 6-hydroxy-2,5,7,8-tetramethylchromane-2-carboxylic acid (Trolox), 2,2′-diphenyl-1-picrylhydrazyl (DPPH), 2,2-azinobis-(3-ethylbenzothiazoline-6-sulfonate (ABTS), and TPTZ (2,4,6-tris(2-pyridyl)-s-triazine) were acquired from Sigma-Aldrich Chemical Co. (St. Louis, MO, USA). All qualitative and quantitative standard products, including γ-aminobutyric acid, L-phenylalanine, tryptophan, p-hydroxybenzaldehyde, p-hydroxybenzoic acid, protocatechuic aldehyde, protocatechuic acid, vanillic acid, syringic acid, p-coumaric acid, ferulic acid, kaempferol, caffeic acid, and rutin, were purchased from Shanghai Yuanye Biotechnology Co., Ltd., (Shanghai, China). Methanol, formic acid, and acetonitrile (chromatographic grade) were purchased from Thermo Fisher Scientific Reagent Co., Ltd., (Waltham, MA, USA). Other analytical grade chemicals and reagents, such as sodium hydroxide (NaOH), sodium carbonate (Na_2CO_3), sodium nitrite ($NaNO_2$), aluminum chloride ($AlCl_3$), ferric chloride ($FeCl_3$), ethyl acetate, n-hexane, and methanol, were purchased from Tianjin Deen Chemical Reagent Co., Ltd., (Tianjing, China).

2.2. Foxtail Millet and Germination Procedure

Foxtail millet samples were provided by Jinsu Agricultural Technology Co., Ltd. (Luoyang, China). To screen the Na_2SeO_3 presoaking solution concentration, the millet seeds were selected after cleaning and washing with pure water and soaked in 0, 20, 40, 60, 80, and 100 mg/L of Na_2SeO_3 solution at 25 °C for 12 h. The presoaked seeds were evenly placed into petri dishes and then germinated in an incubator at 25 °C for 3 days, and sprayed with pure water every 12 h to keep moisture. To screen the Na_2SeO_3 spraying solution concentration, the cleaned millet seeds were soaked in distilled water at 25 °C for 12 h, and then germinated at 25 °C in the dark. Different concentrations of Na_2SeO_3 solution (0, 2, 4, 6, 8, and 10 mg/L) were sprayed every 12 h to maintain suitable germination moisture. For investigating the effects of Na_2SeO_3 treatments on foxtail millet sprouts, three groups of germination experiments were carried out. In the control group, foxtail millet seeds were germinated with pure water; in the Se soaking group, the seeds were germinated by soaking in Na_2SeO_3 solution; and in the Se spraying group, the seeds were sprayed with Na_2SeO_3 solution during the germination process. The germination process was carried out simultaneously for 4 days. Samples were taken every day and recorded as G1, G2, G3, and G4, respectively. The ungerminated raw seed samples were recorded as G0. The morphology of foxtail millet sprouts during germination under different treatments is displayed in Figure S1.

The collected sprouts were washed with pure water to wipe off residual Na_2SeO_3, and then freeze-dried. Subsequently, the samples were crushed, sieved, and stored at −20 °C before the following analyses.

2.3. Determination of Se Content

The total Se content was determined using fluorescence spectrophotometry (Cary Eclipse; Agilent Instruments, Santa Clara, CA, USA) [26]. Briefly, foxtail millet sprout powder (0.5 g) was thermally digested overnight with 10 mL of a mixed solution of HNO_3 and $HClO_4$ (*v*/*v*, 9:1), and then 5 mL of HCl (6 M) was added to the digested solution. The thermally digested sample was reacted with 2,3-diaminonaphthalene, and the fluorescence intensity was measured after cyclohexane extraction. A selenium standard solution was used to draw the calibration curve, and the total Se content was calculated in milligrams per kilogram (mg/kg).

The extraction of inorganic Se was conducted according to the method described by C. Cheng et al. [27], with slight modifications. In short, the millet sprout powder (0.2 g) was mixed with 20 mL of ultrapure water, and then ultrasonic treatment was conducted for 15 min. The supernatant was obtained through centrifugation at 1000 rpm for 20 min to obtain the inorganic Se extract. The inorganic Se content was determined by the same procedure for the measurement of the total Se.

The organic selenium content and enrichment rate of selenium were calculated using Equations (1) and (2), respectively.

$$O_{se} = T_{se} - I_{se} \quad (1)$$

$$R_{se} = O_{se}/T_{se} \times 100\% \quad (2)$$

where T_{se} represents the total selenium content of foxtail millet sprouts, O_{se} represents the organic selenium content, I_{se} represents the inorganic selenium content, and R_{se} represents the selenium enrichment rate.

2.4. Determination of Phenolic Content

The free and bound phenolic compounds of foxtail millet sprouts were extracted, and the total phenolic content (TPC) and the total flavonoid content (TFC) were determined by our previously reported methods [28]. The freeze-dried samples were defatted twice with hexane (1:10 *w*/*v*, 15 min) at room temperature, and then mixed with 80% methanol and extracted in a water bath constant temperature oscillator (SHA-C, Jintan district Chenghui instrument factory, Changzhou, China) for 1 h. The supernatant was collected through centrifugation. This extraction process was repeated twice to obtain the free phenolic extract. After this, the residue from the free polyphenolic extraction was air-dried, and then hydrolyzed with 2 mol/L of NaOH solution for 2 h under nitrogen. The pH of the resulting slurry was adjusted to a pH of 1.5–2.0 with 6 mol/L of HCl. The hydrolysate was obtained through centrifugation and then extracted three times with ethyl acetate to obtain the bound phenolic extract. All of the liquid free or bound fractions were combined separately and evaporated to dryness by using a rotary evaporator and then re-dissolved in a 50% methanol solution for further analysis.

The TPC of the free or bound extract was determined on a 96-well microplate (PT06-96S, BIOLAND Biotechnology Co., Ltd. Hangzhou, China) by using the Folin–Ciocalteu assay. In brief, 20 μL of the appropriately diluted phenolic extract was combined with 40 μL of the Folin–Ciocalteu reagent, and then 160 μL of sodium carbonate solution (Na_2CO_3, 75 g/L, *w*/*v*) was added. The mixture was reacted for 2 h in the dark at 25 °C. A standard curve was prepared using ferulic acid, and the TPC was expressed as the milligrams of ferulic acid per 100 g of the dry weight (mg FAE/100 g DW).

The TFC was determined through the $AlCl_3$ colorimetric method. Briefly, the phenolic extract was mixed with 75 μL of $NaNO_2$ (5 g/100 mL), 150 μL of $AlCl_3$ (10 g/100 mL) was added after 5 min, and 0.5 mL of NaOH (4 g/100 mL) and pure water were then added to obtain a final volume of 2.5 mL. After reaction for 15 min, the absorbance was read at 510 nm. A standard curve was prepared using rutin, and the TFC was expressed as milligrams of rutin per 100 g of the dry weight (mg RE/100 g DW).

2.5. Determination of GABA Content

The extraction process for GABA was similar to that of the free phenolics. In brief, the defatted sample was mixed with 80% methanol solution to obtain the GABA extract. GABA quantitative analysis was performed on a UPLC (Waters UPLC H-Class) coupled to a QqQ-MS (Waters Xevo TQ-S/micro, Waters Corp., Milford, MA, USA), according to the previously described method of Li et al. [24]. The chromatography was performed on an Accucore C_{18} column (100 mm × 3 mm, Thermo Fisher Scientific, Waltham, MA, USA), with a 10 μL injection and 0.2 mL/min flow rate. The column temperature was set at 30 °C. The liquid elution procedure was as follows: 0–3 min, 95–5% B; 3–4 min, 10–90% B; 4–6 min, 10–90% B; and 6–8 min, 95–5% B, where A and B were the water containing 0.1% formic acid and acetonitrile solution, respectively. The mass spectrometry employed the Multiple Reaction Monitoring (MRM) mode under a positive ion (ESI^+) condition, with qualitative and quantitative ion pairs being 104/69.1 (*m*/*z*) and 104/87.1 (*m*/*z*), respectively. The other mass spectrometry parameters were as follows: a capillary voltage of 0.35 KV, a cone voltage of 30 V, an ion source temperature of 150 °C, and a solvent gas temperature of 500 °C. The flow rate of the solvent gas (N_2) was 800 L/h, and the flow rate of the cone gas (He) was 20 L/h. A quantitative calculation of the peak area was performed using external calibration curves to obtain the GABA content in the samples.

2.6. Identification and Quantification of Individual Polyphenols

A UPLC (Waters UPLC H-Class) coupled to a QqQ-MS (Waters Xevo TQ- S/micro) was applied for the qualitative and quantitative analysis of phenolic compounds. The equipped chromatographic column was the Accucore C_{18} column (100 mm × 3 mm, Thermo Fisher Scientific, Waltham, MA, USA). The liquid elution procedure and mass spectrometry parameters were set to be the same as the previously described method by Xiang et al. [29]. The liquid phase conditions were as follows: the binary mobile phases A and B were 0.1% formic acid water and 0.1% formic acid methanol, respectively. The column temperature was 30 °C, the injection volume was 3 μL, and the flow rate was set to 0.4 mL/min. A 25 min linear gradient was programmed as follows: 0–3.81 min, 9–14% B; 3.81–4.85 min, 14–15% B; 4.85–5.89 min, 15% B; 5.89–8.32 min, 15–17% B; 8.32–9.71 min, 17–19% B; 9.71–10.40 min, 19% B; 10.40–12.48 min, 19–26% B; 12.48–13.17 min, 26–28% B; 13.17–14.21 min, 28–35% B; 14.21–15.95 min, 35–40% B; 15.95–16.64 min, 40–48% B; 16.64–18.37 min 48–53% B; 18.37–22.53 min, 53–70% B; 22.53–22.88 min, 70–9% B; and 22.88–25.00 min, 9% B. The mass spectrum conditions were as follows: MS/MS information was collected in the negative ion mode with a scan range of 50–1000, a collision energy of 20 V, a cone voltage of 20 V, a capillary voltage of 1 kV, cone gas (He) with a flow rate of 50 L/h, desolvation gas (N_2) with a flow rate of 1000 L/h, and a temperature of 500 °C.

For phenolic compounds with the available standard, a calculation was performed using an external calibration curve. The phenolic derivatives without available standards were quantified as the equivalent concentration of the closest analog. The results were expressed in milligrams per kilogram of the dry weight (mg/kg DW) of foxtail millet sprouts.

2.7. Determination of In Vitro Antioxidant Activity

The DPPH radical scavenging capacities were determined by the method of Zheng et al. [30]. In brief, 10 μL of the polyphenol extract was added to a 96-well microplate, along with 190 μL of DPPH radical solution. After incubating at room temperature for 30 min, the absorbance at 515 nm was measured.

The $ABTS^+$ radical scavenging capacities were determined by the method of Xiang et al. [31]. In short, the polyphenol extract of 10 μL was combined with 190 μL of $ABTS^+$ working liquid on a 96-well microporous plate. After incubation at 25 °C for 30 min, the absorbance at 750 nm was measured.

The ferric-reducing antioxidant potential (FRAP) was determined according to Yuan et al. [32]. In short, 10 μL of the polyphenol extract was mixed with 300 μL of the iron-TPTZ reagent. The absorbance was measured at 600 nm after the reaction at room temperature for 2 h.

The DPPH, $ABTS^+$ radical scavenging activities, and FRAP of the polyphenol extract from foxtail millet sprouts were all expressed in micromoles of Trolox equivalents per gram (μmol TE/g). The DPPH, $ABTS^+$ radical scavenging activities, and FRAP of the foxtail millet sprouts were the sum of the corresponding DPPH, ABTS, and FRAP values of the free and bound phenolics, respectively.

2.8. Statistical Analysis

All experiments were carried out in triplicate, and the results were expressed as the mean ± standard deviation (SD). Statistical analyses were performed by SPSS 27 (v. 27, IBM Corp., Armonk, NY, USA). Significant differences were analyzed by using Tukey's test at a $p < 0.05$ level.

3. Results and Discussion

3.1. Screening of Presoaking Na_2SeO_3 Concentration for Se-Enriched Foxtail Millet Sprouts

3.1.1. Effect of Na_2SeO_3 Presoaking Concentration on Se Content

The total Se, organic Se, inorganic Se content, and Se-enrichment rate of foxtail millet sprouts at different Na_2SeO_3 soaking concentrations are shown in Figure 1. The Na_2SeO_3 soaking treatment resulted in a significant ($p < 0.05$) increase in the total Se and organic Se content of foxtail millet sprouts, which increased with the increase in Na_2SeO_3 solution concentration. The total Se content of the Se-enriched foxtail millet sprouts ranged from 7.83 to 36.46 mg/kg of DW, the organic Se content ranged from 6.27 to 26.54 mg/kg, and the Se-enrichment rate varied from 72.79% to 80.43%. Organic Se was the primary form in the foxtail millet sprouts with Na_2SeO_3 presoaking treatment, indicating that millet sprouts can effectively transform and accumulate organic Se. Similar results were observed in mung beans germinated through Na_2SeO_3 soaking treatment, where the total and organic Se content of mung bean sprouts gradually increased with the increasing Na_2SeO_3 soaking concentration [33]. However, the Se-enrichment rate decreased gradually with the increase in Na_2SeO_3 soaking solution concentration in this study. This may be attributed to the decreased ability of selenite assimilation and transformation into organic Se when the Na_2SeO_3 concentration reached a certain threshold level. Similar results have been reported for Se-enriched soybean sprouts [34].

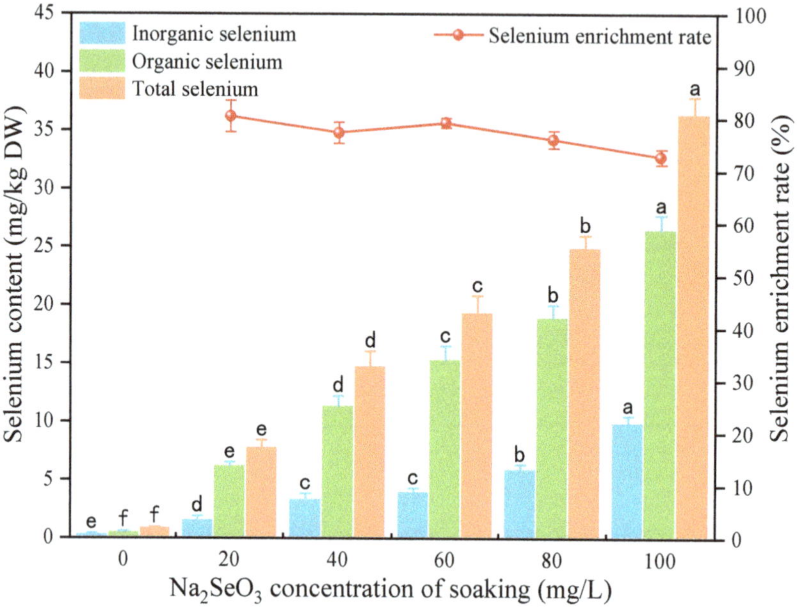

Figure 1. The effects of different concentrations of Na_2SeO_3 soaking solution on the total, organic, and inorganic Se content and the Se-enrichment rate of millet sprouts. Note: The different letters represent the difference in significance under different Na_2SeO_3 soaking concentrations, p-value < 0.05.

3.1.2. Effects of Na_2SeO_3 Presoaking Concentration on Phenolic and GABA Contents

Foxtail millet is a rich source of phenolic compounds, which exhibit wide health benefits through their antioxidant properties, thereby preventing the development of chronic diseases [22]. As shown in Figure 2, the free TPC and bound TPC of millet sprouts showed a trend of first increasing and then decreasing with the increase in the Na_2SeO_3 soaking concentration. At a Na_2SeO_3 soaking concentration of 60 mg/L, the free and bound TPC reached the maximum value, which increased by 20.70% and 28.50%, respectively, compared with the control. These results indicated that soaking with an appropriate concentration of Na_2SeO_3 had a positive effect on the production of phenolic compounds in millet sprouts and displayed greater enhancement on bound TPC. It has been reported that an appropriate concentration of Na_2SeO_3 also increased the TPC of wheat microgreen, which may be the result of an increase in the activity of phenylalanine ammonialyase (PAL) [18]. J. Huang et al. also reported that a suitable concentration of Na_2SeO_3 can promote the production of polyphenols in sprouted black beans [20]. Therefore, appropriate Na_2SeO_3 concentrations are important for promoting phenolic accumulation in plant species. Excessive Se may also produce toxic effects on plants and affect their growth and development [3].

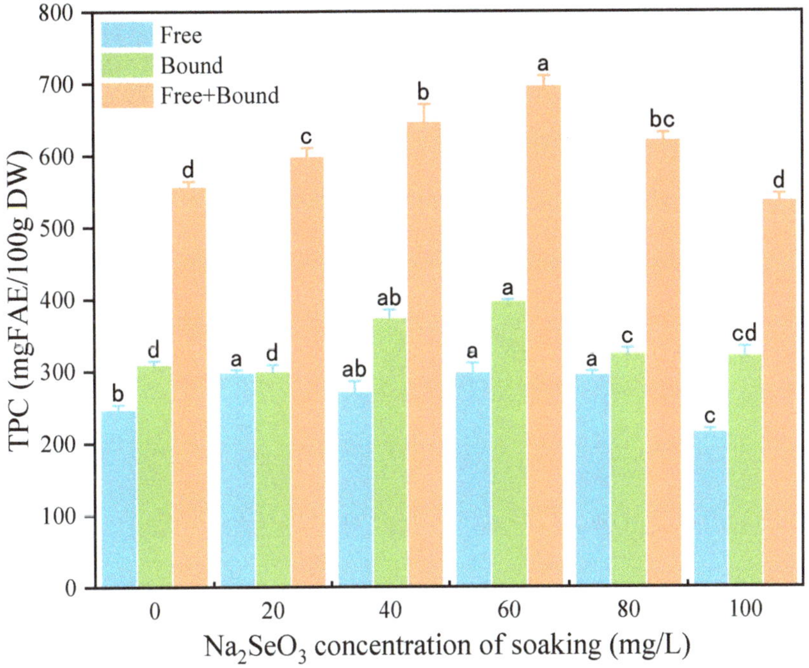

Figure 2. The effect of different concentrations of Na_2SeO_3 soaking solution on the TPC of millet sprouts. Note: The different letters represent the difference in significance under different Na_2SeO_3 soaking concentrations, *p*-value < 0.05.

Flavonoids are a unique class of phenolic compounds and are important bioactive ingredients in millet sprouts [35]. The TFC of millet sprouts germinated in the presence of different Na_2SeO_3 soaking concentrations is shown in Figure 3. The TFC significantly increased at concentrations of less than 80 mg/L Na_2SeO_3, while 80 mg/L and 100 mg/L of Na_2SeO_3 soaking solutions resulted in a significant decrease in TFC when compared with the control. Guardado-Félix et al. reported that Na_2SeO_3 significantly increased the total isoflavone concentrations in chickpea sprouts, and the accumulation of isoflavones may be related to the increase in PAL activity induced by Na_2SeO_3 [17]. Y. Huang et al. (2022) reported that high concentrations of Na_2SeO_3 inhibited the synthesis of flavonoids in soybean sprouts [34]. Therefore, the Na_2SeO_3 concentration is a key factor in promoting the synthesis of flavonoids.

GABA is an important secondary metabolite in foxtail millet and plays a vital role in human health [36]. As shown in Figure 4, compared with the control, a 60 mg/L Na_2SeO_3 soaking concentration resulted in a significant increase in GABA ($p < 0.05$), while the rest of the Na_2SeO_3 concentrations did not produce a significant difference ($p > 0.05$). The GABA content of foxtail millet sprouts presoaked with 60 mg/L of Na_2SeO_3 reached 249.02 mg/kg, which increased 1.61-fold when compared with the control. Se-biofortification has been reported to increase the GABA content of black soybean sprouts, and the increased GABA level is believed to be attributed to the increased activity of glutamate decarboxylase (GAD) [20]. Our results indicated that 60 mg/L of Na_2SeO_3 could be selected as the suitable presoaking concentration for foxtail millet sprouts.

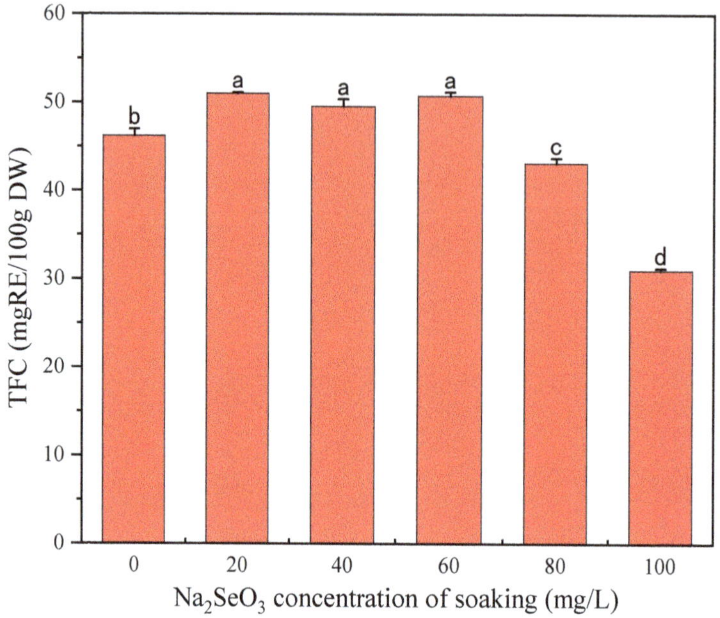

Figure 3. The effect of different concentrations of Na_2SeO_3 soaking solution on the TFC of millet sprouts. Note: The different letters represent the difference in significance under different Na_2SeO_3 soaking concentrations, *p*-value < 0.05.

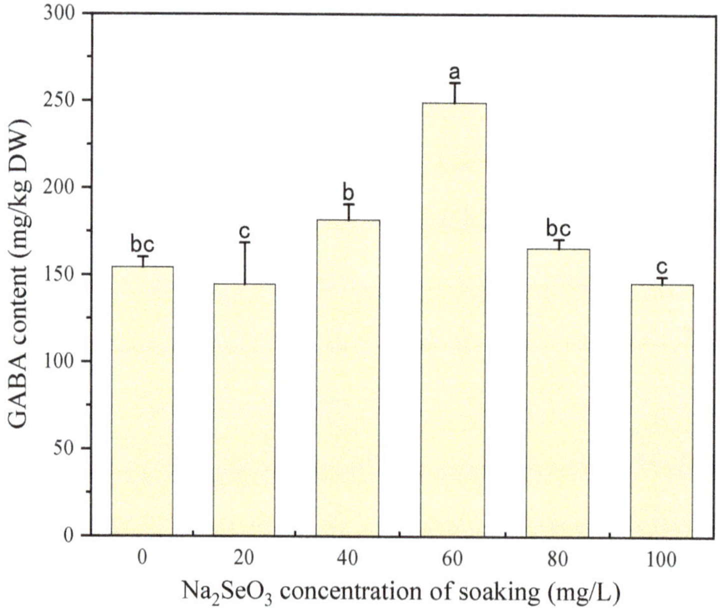

Figure 4. The effect of different concentrations of Na_2SeO_3 soaking solution on the GABA content of millet sprouts. Note: The different letters represent the difference in significance under different Na_2SeO_3 soaking concentrations, *p*-value < 0.05.

3.2. Screening of Spraying Na_2SeO_3 Concentration for Foxtail Millet Sprouts

3.2.1. Effect of Na_2SeO_3 Spraying Concentration on Se Content of Millet Sprouts

The total, organic, and inorganic Se content and the Se-enrichment rate of the foxtail millet sprouts at different Na_2SeO_3 spraying concentrations are shown in Figure 5. The total, organic, and inorganic Se content of millet sprouts increased at first and then decreased with the rise of Na_2SeO_3 spraying concentrations, and the total Se content ranged from 2.80 mg/kg to 6.79 mg/kg. It has been reported that the total Se content of lupin sprouts was in the range of 1.13–4.92 mg/kg by spraying 2–8 mg/L of Na_2SeO_3 solution during germination, but when the Na_2SeO_3 concentration surpassed 6 mg/L, the total Se content of lupin sprouts no longer significantly increased [37]. Rao et al. also reported that the total Se content of soybean shoots was reduced by a high concentration of Na_2SeO_3 solution [19]. It has been reported that with selenite application, exogenous inorganic Se is taken up by roots through passive diffusion, then accumulated and transformed into organic Se [15]. The application of high concentrations of Na_2SeO_3 during seed germination increases the density of endodermal cells in roots, thereby inhibiting the growth of both hypocotyls and roots [38]. Therefore, the decrease in Se content may be due to the damage caused by high concentration of Na_2SeO_3 to roots, thus affecting the enrichment of Se [39].

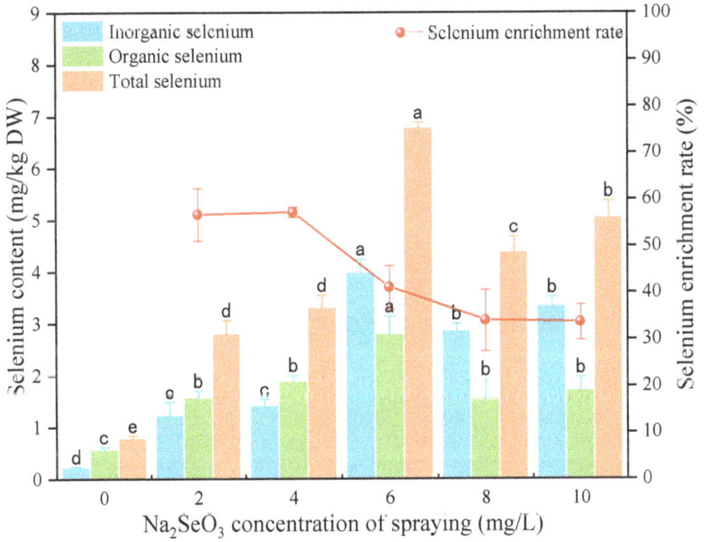

Figure 5. The effects of different concentrations of Na_2SeO_3 spraying solution on the total, organic, and inorganic Se content and the Se-enrichment rate of millet sprouts. Note: The different letters represent the difference in significance under different Na_2SeO_3 spraying concentrations, p-value < 0.05.

3.2.2. Effects of Na_2SeO_3 Spraying Concentration on Phenolic and GABA Content

Figure 6 shows the effects of different Na_2SeO_3 spraying concentrations on the free and bound TPC of millet sprouts. Similar to the results of the Na_2SeO_3 presoaking treatment, free and bound TPC showed first an increasing and then a decreasing trend with the increase in the Na_2SeO_3 spraying concentration. Compared to the control, both the free and bound TPC reached maximum values under the 2 mg/L Na_2SeO_3 spraying treatment and were increased by 25.24% and 23.45%, respectively. As shown in Figures 7 and 8, the TFC and GABA content significantly increased by 28.09% and 53.31%, respectively, only at the Na_2SeO_3 spraying concentration of 2 mg/L. It was indicated that a low concentration of Na_2SeO_3 spraying solution can promote the accumulation of polyphenols and GABA, which may also be attributed to the increased activity of PAL and GAD. In this study,

2 mg/L of Na_2SeO_3 solution was selected as the suitable spraying concentration for the germination process.

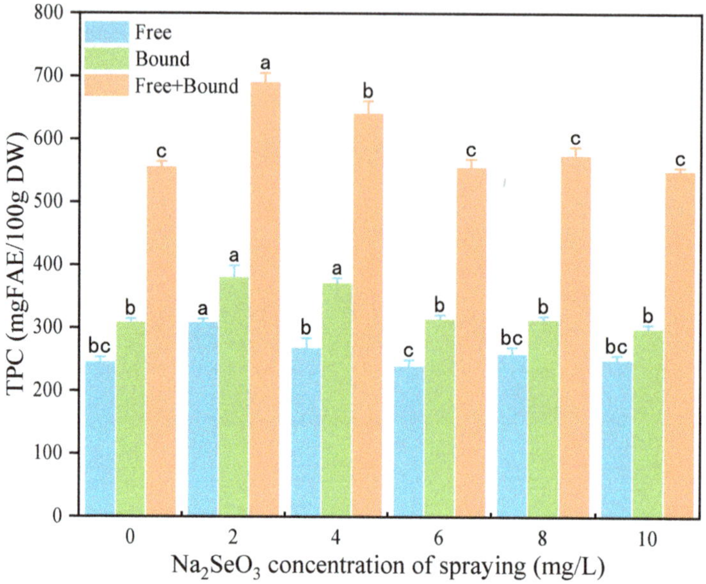

Figure 6. The effect of different concentrations of Na_2SeO_3 spraying solution on the TPC of millet sprouts. Note: The different letters represent the difference in significance under different Na_2SeO_3 soaking concentrations, p-value < 0.05.

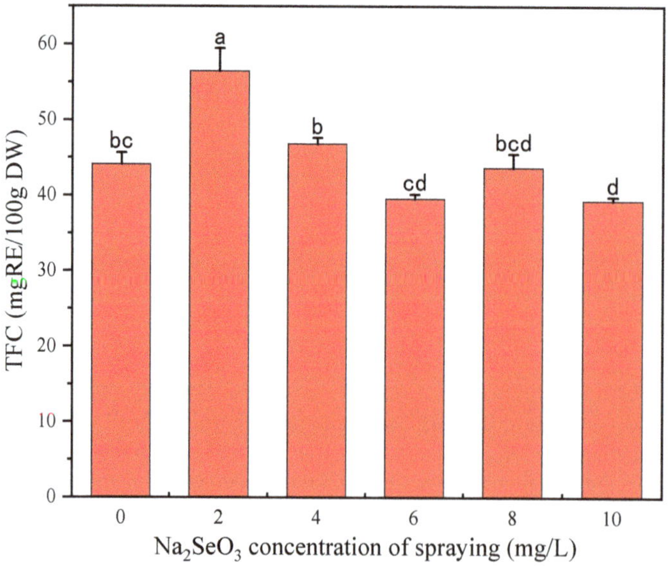

Figure 7. The effect of different concentrations of the Na_2SeO_3 spraying solution on the TFC of millet sprouts. Note: The different letters represent the difference in significance under different Na_2SeO_3 soaking concentrations, p-value < 0.05.

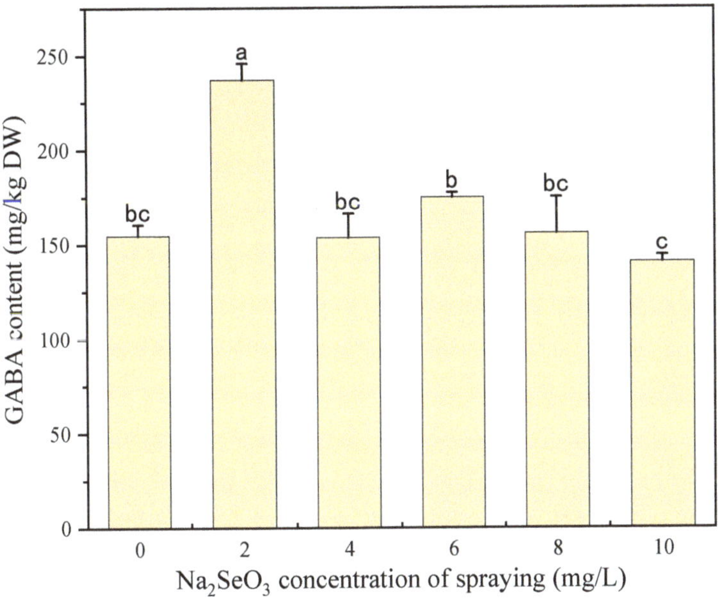

Figure 8. The effect of different concentrations of the Na_2SeO_3 spraying solution on the GABA content of millet sprouts. Note: The different letters represent the difference in significance under different Na_2SeO_3 spraying concentrations, *p*-value < 0.05.

Our results showed that the Na_2SeO_3 soaking treatment was far superior to the Na_2SeO_3 spraying treatment in promoting Se levels and the accumulation of organic Se. However, the Na_2SeO_3 spraying treatment was equally effective as the Na_2SeO_3 soaking treatment in promoting phenolic and GABA synthesis in foxtail millet sprouts.

3.3. Change in Se Content During Se-Enriched Germination of Foxtail Millet

As shown in Table 1, the application of exogenous Na_2SeO_3 significantly increased the total and organic Se content of foxtail millet sprouts. In terms of organic Se, the Na_2SeO_3 soaking treatment exhibited a significantly higher enrichment effect than the Na_2SeO_3 spraying treatment at all the germination stages. Similarly, the soaking treatment caused a significantly higher Se-enrichment rate than the spraying treatment. The Se enrichment rate exhibited a gradual rising tendency during the germination process, indicating that the germination process could effectively convert inorganic Se into organic Se. Seeds can maintain high vitality during the germination stage and remain sensitive to Se uptake and transformation. However, Se is not necessary for plants, and excessive exogenous inorganic Se could cause damage to plants. Therefore, based on self-protection, germinated seeds should rapidly convert inorganic Se to organic Se through various metabolic pathways [27].

The organic Se content of millet sprouts with the Na_2SeO_3 presoaking treatment decreased at the fourth germination day (G4), which may be due to the partial organic Se conversion to dimethyl-selenide (DMSe) or dimethyl-diselenide (DMDSe) and volatilization caused by the sulfur metabolism pathway, so as to reduce the toxic effect of excessive Se [40]. Puccinelli et al. also showed that the total Se content of the leaves, stems, and roots of basil plants decreased with a prolonged culture time [41]. Therefore, an appropriate Se treatment and germination period are key factors for the production of high-quality Se-enriched foxtail millet sprouts. Furthermore, the proper growth stage of the sprouts for harvest is equally important, as this is directly correlated to the readiness of the edible portion of sprouts [5]. According to the Se content of the edible parts of foxtail millet sprouts, it can be concluded that the Na_2SeO_3 soaking treatment was more suitable for the production of Se-enriched millet sprouts.

Table 1. Effects of Na_2SeO_3 treatments on Se content (mg/kg DW) of foxtail millet sprouts during germination process.

Se Content (mg/kg DW)/ Se-Enrichment Rate (%)	Germination Treatment	Germination Stage			
		G1	G2	G3	G4
Total Se	Control	0.81 ± 0.02^{Ac}	0.87 ± 0.01^{Ac}	0.83 ± 0.05^{Ac}	1.02 ± 0.06^{Ac}
	Na_2SeO_3 (soaking)	12.58 ± 0.18^{Aa}	12.60 ± 0.25^{Aa}	12.89 ± 0.44^{Aa}	10.32 ± 0.39^{Ba}
	Na_2SeO_3 (spraying)	4.16 ± 0.03^{Bb}	4.33 ± 0.02^{Bb}	5.35 ± 0.04^{Ab}	5.01 ± 0.06^{Ab}
Organic Se	Control	0.56 ± 0.03^{Ac}	0.61 ± 0.14^{Ac}	0.54 ± 0.06^{Ac}	0.64 ± 0.08^{Ad}
	Na_2SeO_3 (soaking)	8.81 ± 0.26^{Ca}	10.11 ± 0.20^{Ba}	10.97 ± 0.44^{Aa}	8.37 ± 0.45^{Cc}
	Na_2SeO_3 (spraying)	2.21 ± 0.15^{Bb}	2.71 ± 0.09^{Bb}	3.23 ± 0.18^{Ab}	3.43 ± 0.11^{Ab}
Inorganic Se	Control	0.25 ± 0.05^{Ac}	0.26 ± 0.12^{Ac}	0.28 ± 0.06^{Ac}	0.37 ± 0.04^{Ac}
	Na_2SeO_3 (soaking)	3.77 ± 0.10^{Aa}	2.48 ± 0.07^{Ba}	1.92 ± 0.08^{Cb}	1.95 ± 0.06^{Ca}
	Na_2SeO_3 (spraying)	1.94 ± 0.15^{Bb}	1.62 ± 0.07^{Bb}	2.12 ± 0.14^{Aa}	1.58 ± 0.08^{Bb}
Se-enrichment rate	Control	-	-	-	-
	Na_2SeO_3 (soaking)	70.02	80.31	85.10	81.01
	Na_2SeO_3 (spraying)	53.21	62.53	60.35	68.44

Different capital letters indicate significant differences in different germination days ($p < 0.05$), while different lowercase letters indicate significant differences in different germination treatments ($p < 0.05$).

3.4. Changes in TPC During Se-Enriched Germination of Foxtail Millet

The changes in the free and bound TPC of millet sprouts during the Se-enriched germination process are shown in Figure 9. The bound TPC of the millet sprouts decreased slightly during the first germination day (G1) and the second day (G2) but increased rapidly after three germination days. The free TPC slightly reduced at the G1 stage without a statistical difference, compared with ungerminated seeds (G0) ($p > 0.05$), and then increased significantly ($p < 0.05$) with the extension of the germination process, which is in agreement with our previous study [24]. Germination contributed to the accumulation of phenolic compounds, which was ascribed to the PAL activated during the germination process [29]. The phenolic acids bound to non-starch polysaccharides in the grain cell wall could be released under the action of the cell wall degrading enzyme activated during germination, which would cause a rapid increase in free TPC [42].

During early germination stages (G1 and G2), both the bound and free TPC of the millet sprouts displayed no significant difference ($p > 0.05$) between the Na_2SeO_3 soaking or spraying treatments and the control. In comparison, during the later stages of germination (G3 and G4), the bound and free TPC of the two Na_2SeO_3 treatments were significantly higher ($p < 0.05$) than those of the control. For the bound TPC, the highest level of 462.87 mg FAE/100 g was observed in foxtail millet sprouts with the Na_2SeO_3 spraying treatment at the G4 stage, followed by the Na_2SeO_3 soaking treatment, and both of them were significantly higher ($p < 0.001$) than the control. As for the free phenolics, compared with the control, the Na_2SeO_3 soaking treatment showed a significant increase ($p < 0.05$) at G4, reaching 437.65 mg FAE/100 g, which was superior to the Na_2SeO_3 spraying treatment.

A similar phenomenon was observed in soybean sprouts; after being presoaked with Na_2SeO_3, the TPC of soybean sprouts exhibited a slight increase during the initial germination stages, but increased dramatically after three days of germination, compared to the control group [34]. It was indicated that the effect of Na_2SeO_3 on the TPC of millet sprouts is not only related to the method of Na_2SeO_3 treatment but also to the germination stage.

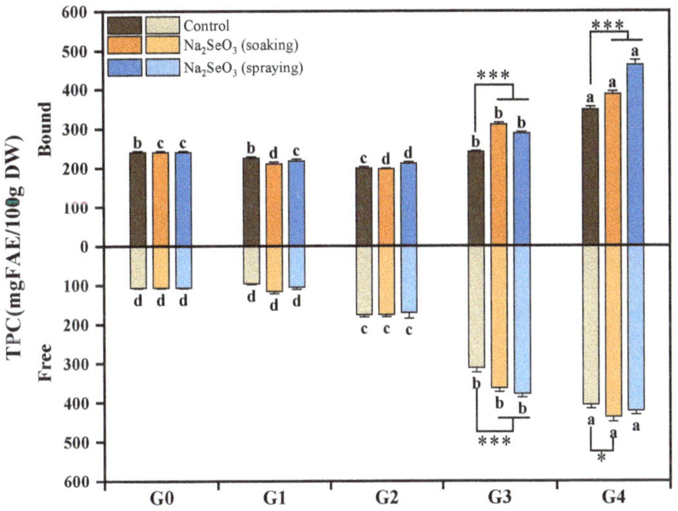

Figure 9. Dynamic changes in TPC of foxtail millet during Se-enriched germination. Different lower cases indicate significant differences at different germination days ($p < 0.05$), and asterisks indicate statistically significant differences between Na_2SeO_3 treatment and control (* $p < 0.05$, *** $p < 0.001$).

3.5. *Changes in TFC During Se-Enriched Germination of Foxtail Millet*

The changes in the TFC of millet sprouts during Se-enriched germination are shown in Figure 10. The TFC of foxtail millet sprouts with all the different treatments showed first a decreasing and then an increasing trend during the germination process. The early decrease in the TFC was probably due to the leaching of water-soluble flavonoid glycosides and the activation of polyphenol oxidase and peroxidase, while the increased TFC after two germination days was attributed to the activation of flavonoid synthetase [29].

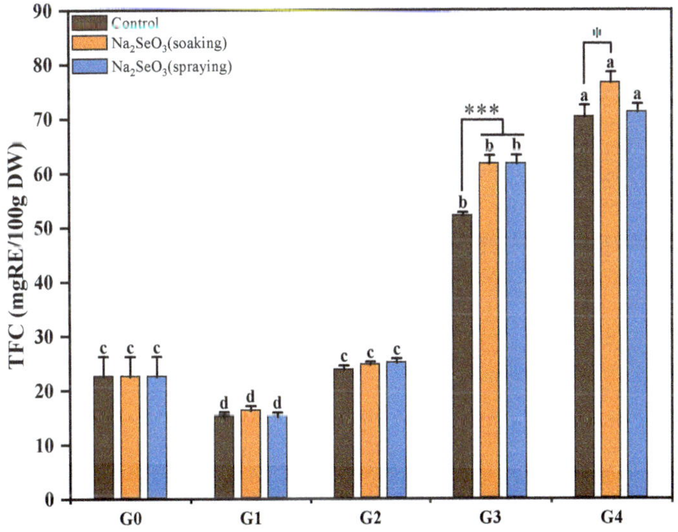

Figure 10. Dynamic changes in TFC of foxtail millet during Se-enriched germination. Different lower cases indicate significant differences at different germination days ($p < 0.05$), and asterisks indicate statistically significant differences between Na_2SeO_3 treatment and control (* $p < 0.05$, *** $p < 0.001$).

Consistent with the performance of the free and bound TPC, the TFC of millet sprouts did not show a significant difference between the Na_2SeO_3 treatment groups and the control during the early germination stages. After three days of germination, the TFC of the millet sprouts with Na_2SeO_3 soaking and spraying treatments were significantly higher ($p < 0.001$) than the control. However, after four days of germination, only the Na_2SeO_3 soaking treatment group showed a significantly higher ($p < 0.05$) TFC than the control, reaching the highest value of 76.56 mg RE/100 g of DW.

Guardado-Félix et al. observed that the total isoflavonoid concentration of Se-enriched chickpea sprouts was far higher than that of the control group during the later stages of germination, which indicates that the enrichment effect of Na_2SeO_3 treatment on flavonoids was also related to the seed germination cycle [17]. Similar effects regarding the enhancement of flavonoid biosynthesis in germinated seeds by Na_2SeO_3, such as radish sprouts [16], black soybean sprouts [20], and buckwheat sprouts [43], have been detected. It has been reported that Na_2SeO_3 enhances flavonoid accumulation by increasing the expression levels of flavonoid biosynthesis structural genes [10].

3.6. Changes in GABA Content During Se-Enriched Germination of Foxtail Millet

The changes in the GABA content of the foxtail millet with three different treatments during germination are shown in Figure 11. The GABA levels of millet sprouts with both Na_2SeO_3 treatments were significantly higher ($p < 0.001$) than that of the control at the same germination stage, separately. The Na_2SeO_3 spraying treatment displayed a superior accumulative effect on GABA content compared to the Na_2SeO_3 soaking treatment at the G2 and G3 stages. However, after four days of germination, the GABA of the Na_2SeO_3 soaking sprouts showed the highest level of 254.35 mg/kg of DW, followed by the Na_2SeO_3 spraying treatment.

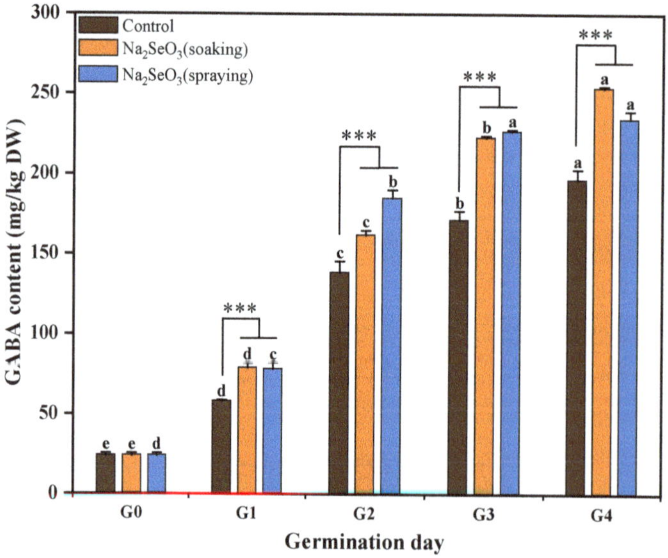

Figure 11. Dynamic changes in GABA content of foxtail millet during Se-enriched germination. Different lower cases indicate significant differences at different germination days ($p < 0.05$), and asterisks indicate statistically significant differences between Na_2SeO_3 treatment and control (*** $p < 0.001$).

It has been reported that excessive concentrations of selenite increases reactive oxygen species (ROS) levels in peanut sprouts [38]; thus, a high concentration of exogenous Na_2SeO_3 causes a similar effect as an abiotic stress [15], while increased GABA in germinated seeds helps to enhance the antioxidant defense system and reduce ROS levels under stress conditions, thereby protecting plants from oxidative damage [44]. Given the limited research on the impact of Na_2SeO_3 on GABA of germinated seeds, our study clearly indicates that the appropriate concentration of Na_2SeO_3 soaking or spraying can effectively induce the accumulation of GABA in germinating seeds.

3.7. Changes in Individual Phenolic Compounds of Foxtail Millet Sprouts During Se-Enriched Germination

Typical phenolic profiles of the free and bound extracts from the Se-enriched foxtail millet sprouts are provided in Figure S2. The phenolic compounds identified in Se-enriched foxtail millet sprouts were consistent with our previous research [24], and their characteristic data of retention time (RT), MS ([M-H]$^-$), and MS/MS data, and the identification results of free and bound phenolic compounds are listed in Table S1 and Table S2, respectively. Foxtail millet sprouts with Na_2SeO_3 soaking or spraying treatment exhibited similar phenolic profiles with the control at the same germination period, with a typical chromatogram shown in Figure 12A,B.

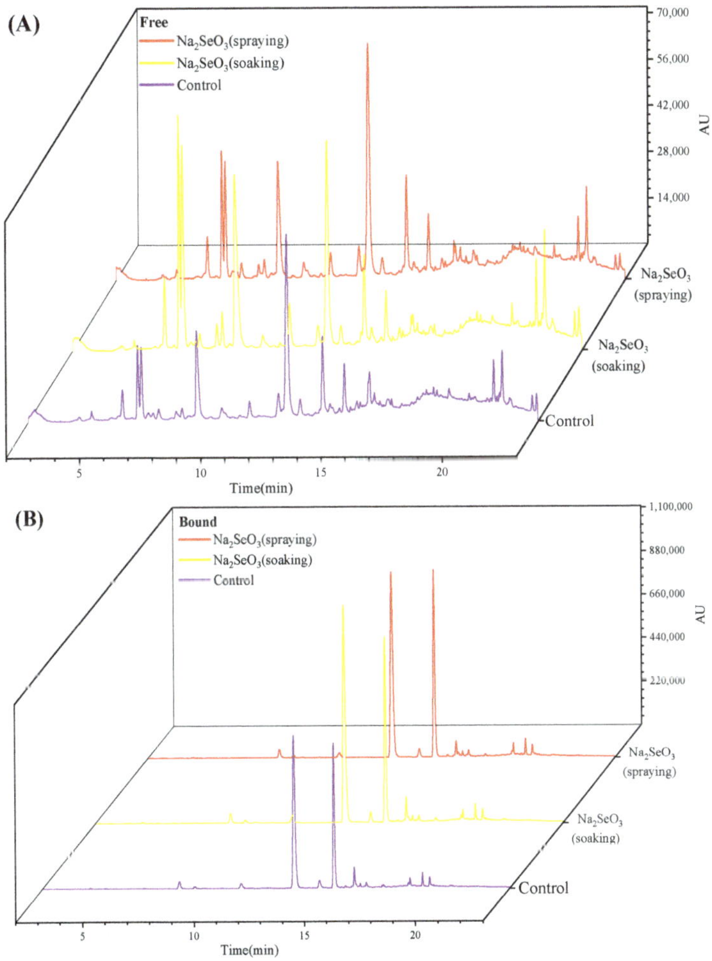

Figure 12. UPLC chromatograms of free (**A**) and bound (**B**) phenolics from foxtail millet sprouts after germination for three days. The detection wavelengths of free and bound phenolics were 320 nm and 280 nm, respectively.

3.7.1. Free Phenolic Compounds

The content changes in the individual free phenolic compounds during Se-enriched germination are presented in Table S3. The predominant phenolic compounds of free phenolic fractions were feruloylquinic acid, 4-p-coumarinoylquinic acid, and 3,7-dimethylquercetin, which were rapidly accumulated during the germination process. N-feruloyl serotonin and N-(p-coumaroyl) serotonin emerged during the second day and accumulated gradually with the prolongation of the germination process. These results were consistent with our previous study [24]. In this study, feruloylquinic acid and 4-p-coumaroylquinic acid reached their highest levels at the G3 stage, while 3,7-dimethylquercetin reached its highest level at the G4 stage. These three major free phenolic compounds were significantly increased ($p < 0.05$) by the two Se-enriched treatments. Feruloylquinic acid and 4-p-coumaroylquinic acid displayed the highest levels of 18.59 mg/kg of DW and 14.02 mg/kg of DW, respectively, through the Na_2SeO_3 spraying treatment. In comparison, the Na_2SeO_3 soaking treatment resulted in the highest level for 3,7-dimethylquercetin, reaching 25.70 mg/kg of DW.

In addition, both Na_2SeO_3 treatments also contributed to the accumulation of N-feruloyl serotonin and N-(p-coumaroyl) serotonin. The content of p-hydroxybenzoic acid was increased significantly through the Na_2SeO_3 soaking treatment, with the highest increase observed during the later stages of germination, and the highest concentration reached was approximately 2.86-fold higher than that of the control. The content of 1-O-p-coumaroyl-3-O-feruloylglycerol was significantly increased by Na_2SeO_3 soaking and spraying during the later stages of germination. Hydroxycinnamic acid amides (HCAAs) are a unique class of phenolic compounds found in foxtail millet sprouts, including N′-caffeoylspermidine, N′-p-coumaroyl-N″-feruloylspermidine, N′, N″-di-p-coumaroylspermidine, and N′, N″-di feruloylspermine. Most of the HACC levels of Se-enriched foxtail millet sprouts were not elevated or decreased compared to the control, with apigenin-C-pentosyl-C-hexoside, trans-p-coumaric acid, and trans-ferulic acid exhibiting similar performance.

3.7.2. Bound Phenolic Compounds

The content changes in the individual bound phenolic compounds during Se-enriched germination are presented in Table S4. The concentrations of the bound phenolic compounds were mostly enhanced by Na_2SeO_3 soaking and spraying treatments, particularly during the later stages of germination. Among them, trans-p-coumaric acid and trans-ferulic acid were the most abundant bound phenolics in foxtail millet sprouts, and their levels increased with the prolongation of the germination process. Trans-p-coumaric acid and trans-ferulic acid in foxtail millet sprouts reached the highest levels at the G4 stage and exhibited the highest increases through the Na_2SeO_3 spraying treatment, which increased by 97.40% and 26.24%, respectively. Similar results of Na_2SeO_3 treatment increasing levels of ferulic acid and *p*-coumaric acid have been observed in rice sprouts [15] and black bean sprouts [20]. In addition, during the later stages of germination, protocatechuic acid, *p*-hydroxybenzaldehyde, vanillic acid, syringic acid, 3,7-dimethylquercetin, ferulic acid dimers (DFAs), and ferulic acid trimers (TFAs) also increased to varying degrees by Na_2SeO_3 soaking or spraying treatments. However, *cis*-ferulic acid significantly ($p < 0.05$) reduced with the Na_2SeO_3 spraying treatment during germination, which showed the opposite result compared with the Na_2SeO_3 soaking treatment.

Overall, both Na_2SeO_3 treatments promoted the enrichment of most phenolic acids in millet sprouts during the later stages of germination. Phenolic acids play an important role in plant resistance to biological and abiotic stresses due to their powerful antioxidant capability [45]. The higher enrichment of these phenolic acids may be attributed to defending the stress of a high-concentration Na_2SeO_3 solution. Although the two Na_2SeO_3 treatments differed in the regulation of some phenolic components in foxtail millet sprouts, the effects on the enrichment of the predominate free and bound phenolic compounds were positive. Phenolic acids are mainly produced via the phenylpropanoid pathway, which could be mediated by proper Na_2SeO_3 treatment through upregulating PAL, trans-cinnamate-4-hydroxylase, and cinnamyl-alcohol dehydrogenase [38].

3.8. Changes in Antioxidant Activity of Foxtail Millet Sprouts During Se-Enriched Germination

The antioxidant activity of phenolic extracts was evaluated by DPPH, ABTS, and FRAP assays. As shown in Table 2, germination with or without Na_2SeO_3 treatment caused the enhancement of the antioxidant capacity of foxtail millet sprouts. Compared with the control, Se-enriched millet sprouts exhibited significantly higher ($p < 0.05$) antioxidative capacity.

Table 2. Effects of Na_2SeO_3 treatments on antioxidant activity (μmol TE/g) of foxtail millet sprouts during germination process.

Antioxidant Assay	Germination Treatment	Germination Stage			
		G1	G2	G3	G4
DPPH	Control	5.97 ± 0.60 ^{Ca}	5.28 ± 0.49 ^{Ca}	8.95 ± 0.55 ^{Bb}	12.86 ± 0.40 ^{Ab}
	Na_2SeO_3 (soaking)	5.50 ± 0.29 ^{Da}	6.64 ± 0.08 ^{Ca}	11.76 ± 0.23 ^{Ba}	14.90 ± 0.23 ^{Aa}
	Na_2SeO_3 (spraying)	6.16 ± 0.23 ^{Ca}	6.07 ± 0.42 ^{Ca}	11.64 ± 0.50 ^{Ba}	15.30 ± 0.40 ^{Aa}
ABTS	Control	6.45 ± 0.25 ^{Db}	8.92 ± 0.17 ^{Cb}	12.22 ± 0.26 ^{Bb}	13.78 ± 0.39 ^{Aa}
	Na_2SeO_3 (soaking)	8.03 ± 0.29 ^{Da}	9.69 ± 0.24 ^{Cab}	13.42 ± 0.18 ^{Ba}	14.25 ± 0.36 ^{Aa}
	Na_2SeO_3 (spraying)	7.62 ± 0.18 ^{Da}	9.72 ± 0.29 ^{Ca}	12.85 ± 0.45 ^{Bab}	14.42 ± 0.32 ^{Aa}
FRAP	Control	7.86 ± 0.22 ^{Da}	8.40 ± 0.09 ^{Ca}	13.95 ± 0.10 ^{Bc}	19.31 ± 0.16 ^{Ab}
	Na_2SeO_3 (soaking)	7.85 ± 0.12 ^{Da}	8.76 ± 0.06 ^{Ca}	17.03 ± 0.06 ^{Ba}	21.49 ± 0.29 ^{Aa}
	Na_2SeO_3 (spraying)	7.69 ± 0.14 ^{Ca}	7.77 ± 0.19 ^{Cb}	15.96 ± 0.11 ^{Bb}	21.76 ± 0.14 ^{Aa}

Different capital letters indicate significant differences in different germination days ($p < 0.05$), while different lowercase letters indicate significant differences in different germination treatments ($p < 0.05$). Results are expressed as mean ± SD.

Regarding the DPPH radical scavenging activity, DPPH values of the phenolics from Se-enriched foxtail millet sprouts showed no significant difference ($p > 0.05$) from the control at the early germination stages. However, during the later stages (G3 and G4), these values were significantly higher ($p < 0.05$) than those of the control. Among them, the phenolics of millet sprouts with Na_2SeO_3 spraying treatment displayed the highest DPPH value at the G4 stage, reaching 15.30 μmol TE/g.

Unlike the DPPH assay, the ABTS values of phenolics from Se-enriched millet sprouts were significantly higher ($p < 0.05$) than those of the control during the initial three days of germination, but there was no significant difference ($p > 0.05$) at the G4 stage when comparing the control. The results of various antioxidant assays showed some variability, which may be attributed to the distinct mechanisms of action inherent to each assay.

For ferric-reducing antioxidant power (FRAP), similar to the DPPH assay, the FRAP values of phenolics from the foxtail millet sprouts with the two Na_2SeO_3 treatments were observed to be significantly higher ($p < 0.05$) than those of the control during the later germination stages. The millet sprouts with Na_2SeO_3 spraying treatment exhibited the highest FRAP value at the G4 stage, reaching 21.76 μmol TE/g.

Overall, the Na_2SeO_3 soaking or spraying treatments enhanced the antioxidant activity of foxtail millet sprouts, which was mainly attributed to the accumulation and synthesis of diverse polyphenols [17]. It has been reported that phenolic extracts from Se-enriched black soybean sprouts have higher antioxidant activities than those of germinated black soybean sprouts without Se treatment [20].

4. Conclusions

This study demonstrated that soaking and spraying with Na_2SeO_3 significantly influenced the Se content of foxtail millet sprouts, with the Na_2SeO_3 soaking treatment exhibiting superior Se-enrichment effectiveness. The appropriate concentration of Na_2SeO_3 can increase the TPC, TFC, and GABA content of foxtail millet sprouts. The main phenolic compounds (such as trans-ferulic acid, trans-p-coumaric acid, and feruloylquinic acid) and serotonin derivatives were further enriched by two Na_2SeO_3 treatments. The antioxidant activity of Se-enriched millet sprouts was further enhanced compared to the general foxtail millet sprouts. Foxtail millet is one of the world's most important grains. The application of Se-biofortification can increase the levels of polyphenols, GABA, and antioxidant activity during the germination process, which provides certain ideas for the development of high-quality and high-added value Se-enriched functional food. Nevertheless, the form in which organic Se exists following Se-enriched treatments of foxtail millet sprouts, as well as its bioavailability, remains unclear. Further research is required to develop foxtail millet sprouts as a functional Se-enriched food.

Supplementary Materials: The following supporting information can be downloaded at https://www.mdpi.com/article/10.3390/foods13233916/s1, Figure S1: Changes in millet growth morphology under different germination treatments. Figure S2: UPLC chromatograms of free and bound phenolics extracts from Se-enriched foxtail millet sprout samples. Table S1: Individual compounds identified using UPLC-ESI-MS/MS in the free fraction of Se-enriched foxtail millet sprout samples. Table S2: Individual compounds identified using UPLC-ESI-MS/MS in the bound fraction of Se-enriched foxtail millet sprout samples. Table S3: Effect of Na_2SeO_3 on individual free phenolic compounds contents (mg/kg DW) of foxtail millet sprouts during germination process. Table S4: Effect of Na_2SeO_3 on individual bound phenolic compounds contents (mg/kg DW) of foxtail millet sprouts during germination process.

Author Contributions: S.S.: Methodology, Investigation, Formal analysis, Writing—original draft. J.Z.: Investigation, Formal analysis, Writing—review and editing. Y.L.: Investigation, Formal analysis, Writing—review and editing. Y.X.: Investigation, Data curation, Writing—review and editing. R.Y.: Conceptualization, Investigation, Writing—review and editing, Supervision. L.L.: Investigation, Writing—review and editing. J.X.: Conceptualization, Project administration, Funding acquisition, Resources, Writing—review and editing, Supervision. All authors have read and agreed to the published version of the manuscript.

Funding: This research was supported by the Henan Provincial Science and Technology Research Project (No. 242102111059).

Institutional Review Board Statement: Not applicable.

Informed Consent Statement: Not applicable.

Data Availability Statement: The original contributions presented in the study are included in the article; further inquiries can be directed to the corresponding author.

Acknowledgments: We acknowledge Jinsu Agricultural Technology Co., Ltd. (Henan Province, China) for donating the foxtail millet materials used for experiments.

Conflicts of Interest: The authors declare no conflicts of interest.

Abbreviations

TPC	total phenolic content
TFC	total flavonoid content
GABA	γ-aminobutyric acid
FAE	ferulic acid equivalents
RE	rutin equivalents
TE	Trolox equivalents
DW	dry weight of sample

References

1. Genchi, G.; Lauria, G.; Catalano, A.; Sinicropi, M.S.; Carocci, A. Biological activity of selenium and its impact on human health. *Int. J. Mol. Sci.* **2023**, *24*, 2633. [CrossRef] [PubMed]
2. Ekumah, J.-N.; Ma, Y.; Akpabli-Tsigbe, N.D.K.; Kwaw, E.; Ma, S.; Hu, J. Global soil distribution, dietary access routes, bioconversion mechanisms and the human health significance of selenium: A review. *Food Biosci.* **2021**, *41*, 100960. [CrossRef]

3. Hossain, A.; Skalicky, M.; Brestic, M.; Maitra, S.; Sarkar, S.; Ahmad, Z.; Vemuri, H.; Garai, S.; Mondal, M.; Bhatt, R.; et al. Selenium biofortification: Roles, mechanisms, responses and prospects. *Molecules* **2021**, *26*, 881. [CrossRef]
4. Alshammari, M.K.; Fatima, W.; Alraya, R.A.; Khuzaim Alzahrani, A.; Kamal, M.; Alshammari, R.S.; Alshammari, S.A.; Alharbi, L.M.; Alsubaie, N.S.; Alosaimi, R.B.; et al. Selenium and covid-19: A spotlight on the clinical trials, inventive compositions, and patent literature. *J. Infect. Public Health* **2022**, *15*, 1225–1233. [CrossRef] [PubMed]
5. D'Amato, R.; Regni, L.; Falcinelli, B.; Mattioli, S.; Benincasa, P.; Dal Bosco, A.; Pacheco, P.; Proietti, P.; Troni, E.; Santi, C.; et al. Current knowledge on selenium biofortification to improve the nutraceutical profile of food: A comprehensive review. *J. Agric. Food Chem.* **2020**, *68*, 4075–4097. [CrossRef]
6. Duborská, E.; Šebesta, M.; Matulová, M.; Zvěřina, O.; Urík, M. Current strategies for selenium and iodine biofortification in crop plants. *Nutrients* **2022**, *14*, 4717. [CrossRef] [PubMed]
7. De Vita, P.; Platani, C.; Fragasso, M.; Ficco, D.B.M.; Colecchia, S.A.; Del Nobile, M.A.; Padalino, L.; Di Gennaro, S.; Petrozza, A. Selenium-enriched durum wheat improves the nutritional profile of pasta without altering its organoleptic properties. *Food Chem.* **2017**, *214*, 374–382. [CrossRef]
8. Deng, B.; Tian, S.; Li, S.; Guo, M.; Liu, H.; Li, Y.; Wang, Q.; Zhao, X. A simple, rapid and efficient method for essential element supplementation based on seed germination. *Food Chem.* **2020**, *325*, 126827. [CrossRef]
9. Lintschinger, J.; Fuchs, N.; Moser, J.; Kuehnelt, D.; Goessler, W. Selenium-enriched sprouts. A raw material for fortified cereal-based diets. *J. Agric. Food Chem.* **2000**, *48*, 5362–5368. [CrossRef]
10. Zhao, F.; Jin, J.; Yang, M.; Santiago, F.E.M.; Xue, J.; Xu, L.; Duan, Y. Selenium differentially regulates flavonoid accumulation and antioxidant capacities in sprouts of twenty diverse mungbean (*Vigna radiata* (l.) Wilczek)) genotypes. *Phyton* **2024**, *93*, 611–625. [CrossRef]
11. Ren, Y.; Zhang, Q.; Li, X.; Zhang, T.; Tian, D.; Liu, L.; Dong, X.; Wang, Z.-Y.; Chai, M. Effects of selenium content on growth, antioxidant activity, and key selenium-enriched gene expression in alfalfa sprouts. *Foods* **2024**, *13*, 2261. [CrossRef]
12. Tan, C.; Zhang, L.; Duan, X.; Chai, X.; Huang, R.; Kang, Y.; Yang, X. Effects of exogenous sucrose and selenium on plant growth, quality, and sugar metabolism of pea sprouts. *J. Sci. Food Agric.* **2021**, *102*, 2855–2863. [CrossRef]
13. Guardado-Félix, D.; Pérez-Carrillo, E.; Heredia-Olea, E.; Serna-Saldivar, S.O. Comparison of regular and selenium-enriched tortillas produced from sprouted corn kernels. *Plant Foods Hum. Nutr.* **2022**, *77*, 226–232. [CrossRef]
14. Gastelum-Hernández, A.C.; Serrano-Sandoval, S.N.; Dávila-Vega, J.P.; Antunes-Ricardo, M.; Serna-Saldivar, S.O.; Gutiérrez-Uribe, J.A.; Milán-Carrillo, J.; Díaz-Páez, J.A.; Guardado-Félix, D. Comparison of minerals, ferulic acid, antioxidant, and anti-inflammatory properties of products from raw maize and germinated maize with selenium. *ACS Food Sci. Technol.* **2023**, *3*, 318–325. [CrossRef]
15. D'Amato, R.; Fontanella, M.C.; Falcinelli, B.; Beone, G.M.; Bravi, E.; Marconi, O.; Benincasa, P.; Businelli, D. Selenium biofortification in rice (*Oryza sativa* L.) sprouting: Effects on se yield and nutritional traits with focus on phenolic acid profile. *J. Agric. Food Chem.* **2018**, *66*, 4082–4090. [CrossRef]
16. Chen, J.; Chen, H.; Wang, H.; Zhan, J.; Yuan, X.; Cui, J.; Su, N. Selenium treatment promotes anthocyanin accumulation in radish sprouts (*Raphanus sativus* L.) by its regulation of photosynthesis and sucrose transport. *Food Res. Int.* **2023**, *165*, 112551. [CrossRef]
17. Guardado-Felix, D.; Serna-Saldivar, S.O.; Cuevas-Rodríguez, E.O.; Jacobo-Velázquez, D.A.; Gutiérrez-Uribe, J.A. Effect of sodium selenite on isoflavonoid contents and antioxidant capacity of chickpea (*Cicer arietinum* L.) sprouts. *Food Chem.* **2017**, *226*, 69–74. [CrossRef] [PubMed]
18. Islam, M.Z.; Park, B.-J.; Kang, H.-M.; Lee, Y.-T. Influence of selenium biofortification on the bioactive compounds and antioxidant activity of wheat microgreen extract. *Food Chem.* **2020**, *309*, 125763. [CrossRef] [PubMed]
19. Rao, S.; Xiao, X.; Wang, Y.; Xiong, Y.; Cheng, H.; Li, L.; Cheng, S. Comparative study of the effects of selenium nanoparticles and selenite on selenium content and nutrient quality in soybean sprouts. *Folia Hortic.* **2022**, *34*, 223–234. [CrossRef]
20. Huang, J.; Qian, J.; Wang, S.; Li, Y.; Zhai, X.; Olajide, T.M.; Shen, G.X.; Liao, X. Effect of selenium biofortification on bioactive compounds and antioxidant activity in germinated black soybean. *J. Food Sci.* **2022**, *87*, 1009–1019. [CrossRef] [PubMed]
21. Galieni, A.; Falcinelli, B.; Stagnari, F.; Datti, A.; Benincasa, P. Sprouts and microgreens: Trends, opportunities, and horizons for novel research. *Agronomy* **2020**, *10*, 1424. [CrossRef]
22. Sharma, R.; Sharma, S.; Dar, B.N.; Singh, B. Millets as potential nutri-cereals: A review of nutrient composition, phytochemical profile and techno-functionality. *Int. J. Food Sci. Technol.* **2021**, *56*, 3703–3718. [CrossRef]
23. Yousaf, L.; Hou, D.; Liaqat, H.; Shen, Q. Millet: A review of its nutritional and functional changes during processing. *Food Res. Int.* **2021**, *142*, 110197. [CrossRef] [PubMed]
24. Li, C.; Zhang, Y.; Xiang, J.; Zhang, J.; Johnson, J.B.; Xu, Y.; Beta, T. Accumulation of γ-aminobutyric acid and modifications of phenolic profiles and antioxidant capacity of foxtail millets during germination. *J. Cereal Sci.* **2023**, *114*, 103815. [CrossRef]
25. Yu, S.; Li, C.; Wang, X.; Herrera-Balandrano, D.D.; Johnson, J.B.; Xiang, J. Optimization of germination conditions for enriched γ-aminobutyric acid and phenolic compounds of foxtail millet sprouts by response surface methodology. *Foods* **2024**, *13*, 3340. [CrossRef]
26. Diowksz, A.; Kordialik-Bogacka, E.; Ambroziak, W. Se-enriched sprouted seeds as functional additives in sourdough fermentation. *LWT-Food Sci. Technol.* **2014**, *56*, 524–528. [CrossRef]
27. Cheng, C.; Coldea, T.E.; Yang, H.; Zhao, H. Selenium uptake, translocation, and metabolization pattern during barley malting: A comparison of selenate, selenite, and selenomethionine. *J. Agric. Food Chem.* **2023**, *71*, 5240–5249. [CrossRef]

28. Zhang, J.; Liu, J.; Han, Z.; He, X.; Herrera-Balandrano, D.D.; Xiang, J. Comprehensive evaluation on phenolic derivatives and antioxidant activities of diverse yellow maize varieties. *Food Chem.* **2025**, *464*, 141602. [CrossRef]
29. Xiang, J.; Yuan, Y.; Du, L.; Zhang, Y.; Li, C.; Beta, T. Modification on phenolic profiles and enhancement of antioxidant activity of proso millets during germination. *Food Chem. X* **2023**, *18*, 100628. [CrossRef]
30. Zheng, B.; Yuan, Y.; Xiang, J.; Jin, W.; Johnson, J.B.; Li, Z.; Wang, C.; Luo, D. Green extraction of phenolic compounds from foxtail millet bran by ultrasonic-assisted deep eutectic solvent extraction: Optimization, comparison and bioactivities. *LWT* **2022**, *154*, 112740. [CrossRef]
31. Xiang, J.; Zhang, M.; Apea-Bah, F.B.; Beta, T. Hydroxycinnamic acid amide (HCAA) derivatives, flavonoid c-glycosides, phenolic acids and antioxidant properties of foxtail millet. *Food Chem.* **2019**, *295*, 214–223. [CrossRef] [PubMed]
32. Yuan, Y.; Xiang, J.; Zheng, B.; Sun, J.; Luo, D.; Li, P.; Fan, J. Diversity of phenolics including hydroxycinnamic acid amide derivatives, phenolic acids contribute to antioxidant properties of proso millet. *LWT* **2022**, *154*, 112611. [CrossRef]
33. Cheng, H.; Li, L.; Dong, J.; Wang, S.; Wu, S.; Rao, S.; Li, L.; Cheng, S.; Li, L. Transcriptome and physiological determination reveal the effects of selenite on the growth and selenium metabolism in mung bean sprouts. *Food Res. Int.* **2023**, *169*, 112880. [CrossRef] [PubMed]
34. Huang, Y.; Lei, N.; Xiong, Y.; Liu, Y.; Tong, L.; Wang, F.; Fan, B.; Maesen, P.; Blecker, C. Influence of selenium biofortification of soybeans on speciation and transformation during seed germination and sprouts quality. *Foods* **2022**, *11*, 1200. [CrossRef]
35. Triki, T.; Tlahig, S.; Benabderrahim, M.A.; Elfalleh, W.; Mabrouk, M.; Bagues, M.; Yahia, H.; Belhouchette, K.; Guasmi, F.; Loumerem, M.; et al. Variation in phenolic, mineral, dietary fiber, and antioxidant activity across southern tunisian pearl millet germplasm. *J. Food Qual.* **2022**, *2022*, 1437306. [CrossRef]
36. Bai, Q.; Yang, R.; Zhang, L.; Gu, Z. Salt stress induces accumulation of γ-aminobutyric acid in germinated foxtail millet (*Setaria italica* L.). *Cereal Chem.* **2023**, *90*, 145–149. [CrossRef]
37. Juana, F.; Piotr, G.; Cristina, M.-V.; Radosław, P.; Enrique, B.; BegonA, J.; Krzysztof, G.; Concepcion, V.-V. Influence of germination with different selenium solutions on nutritional value and cytotoxicity of lupin seeds. *J. Agric. Food Chem.* **2009**, *57*, 1319–1325.
38. Wang, G.; Wu, L.; Zhang, H.; Wu, W.; Zhang, M.; Li, X.; Wu, H. Regulation of the phenylpropanoid pathway: A mechanism of selenium tolerance in peanut (*Arachis hypogaea* L.) seedlings. *J. Agric. Food Chem.* **2016**, *64*, 3626–3635. [CrossRef]
39. Benincasa, P.; Falcinelli, B.; Lutts, S.; Stagnari, F.; Galieni, A. Sprouted grains: A comprehensive review. *Nutrients* **2019**, *11*, 421. [CrossRef]
40. Rattanamanee, C.; Sunisa, S.; Panupong, P. Selenium bio-fortification: An alternative to improve phytochemicals and bioactivities of plant foods. *Funct. Foods Health Dis.* **2017**, *7*, 263–279.
41. Puccinelli, M.; Malorgio, F.; Rosellini, I.; Pezzarossa, B. Uptake and partitioning of selenium in basil (*Ocimum basilicum* L.) plants grown in hydroponics. *Sci. Hortic.* **2017**, *225*, 271–276. [CrossRef]
42. Sharma, S.; Saxena, D.C.; Riar, C.S.; Yildiz, F. Antioxidant activity, total phenolics, flavonoids and antinutritional characteristics of germinated foxtail millet (*Setaria italica*). *Cogent Food Agric.* **2015**, *1*, 1081728. [CrossRef]
43. Nguyen, M.H.; Xiao, H.; Tan, X.; Chen, F.; Shi, X. Comparative growth, biochemical and sensory analyses reveal the feasibility of largescale development of selenium-enriched tartary buckwheat sprouts. *Food Sci. Technol.* **2022**, *42*, e81722. [CrossRef]
44. Ahmad, S.; Fariduddin, Q. Deciphering the enigmatic role of gamma-aminobutyric acid (GABA) in plants: Synthesis, transport, regulation, signaling, and biological roles in interaction with growth regulators and abiotic stresses. *Plant Physiol. Biochem.* **2024**, *208*, 108502. [CrossRef] [PubMed]
45. Jacobo-Velázquez, D.A.; Martínez-Hernández, G.B.; del C. Rodríguez, S.; Cao, C.-M.; Cisneros-Zevallos, L. Plants as biofactories: Physiological role of reactive oxygen species on the accumulation of phenolic antioxidants in carrot tissue under wounding and hyperoxia stress. *J. Agric. Food Chem.* **2011**, *59*, 6583–6593.

MDPI AG
Grosspeteranlage 5
4052 Basel
Switzerland
Tel.: +41 61 683 77 34

Foods Editorial Office
E-mail: foods@mdpi.com
www.mdpi.com/journal/foods

www.ingramcontent.com/pod-product-compliance
Lightning Source LLC
LaVergne TN
LVHW070454061126
851858LV00011B/237